METRIC IN MEASURE SPACES

METRIC IN MEASURE SPACES

J Yeh
University of California, Irvine, USA

World Scientific

NEW JERSEY • LONDON • SINGAPORE • BEIJING • SHANGHAI • HONG KONG • TAIPEI • CHENNAI • TOKYO

Published by

World Scientific Publishing Co. Pte. Ltd.
5 Toh Tuck Link, Singapore 596224
USA office: 27 Warren Street, Suite 401-402, Hackensack, NJ 07601
UK office: 57 Shelton Street, Covent Garden, London WC2H 9HE

Library of Congress Control Number: 2019049780

British Library Cataloguing-in-Publication Data
A catalogue record for this book is available from the British Library.

METRIC IN MEASURE SPACES

Copyright © 2020 by World Scientific Publishing Co. Pte. Ltd.

All rights reserved. This book, or parts thereof, may not be reproduced in any form or by any means, electronic or mechanical, including photocopying, recording or any information storage and retrieval system now known or to be invented, without written permission from the publisher.

For photocopying of material in this volume, please pay a copying fee through the Copyright Clearance Center, Inc., 222 Rosewood Drive, Danvers, MA 01923, USA. In this case permission to photocopy is not required from the publisher.

ISBN 978-981-3200-39-5
ISBN 978-981-3200-40-1 (pbk)

For any available supplementary material, please visit
https://www.worldscientific.com/worldscibooks/10.1142/10289#t=suppl

Printed in Singapore

To my wife
Betty Liu Yeh

Preface

Measure and metric are two fundamental concepts in measuring the size of a mathematical object. Yet there has been no systematic investigation of this relation. This book closes that gap by dealing with real analysis aspects of measure theory, integration, and differentiation, all in the context of metric spaces. Topics include metric, Borel, Radon and Lebesgue outer measures (§3), Hausdorff measures (§§4-5), covering theorems (§6), differentiation of measures and integrals (§§7-9), maximal functions (§10), Lipschitz mappings (§13), and theorems of Lusin and Egorov (§14).

The material is of interest to mathematicians and accessible to graduate students in mathematics. For the first half of the book (§§1-7), which is concerned with measure theory, no knowledge of basic measure theory is assumed since what is needed is presented, with complete proofs, in §§1-2. As with my previous book *Real Analysis. Theory of Measure and Integration. Third Edition* (periodically referred to in the text by [LRA]) §§1-6 has as prerequisite only advanced calculus, as found for example in R. C. Buck's book *Advanced Calculus.* On the other hand, for the second half of the book (§§7-14), which is concerned primarily with differentiation and integration, knowledge of the Lebesgue integral on a general measure space (§§4,7-9 in [LRA]) is assumed.

J. Yeh

Contents

§1 Measure and Outer Measure

[I] Measure on a σ-algebra of Subsets of a Set

Notations. We write $\mathbb{N}$ for both the sequence $(1, 2, 3, \ldots)$ and the set $\{1, 2, 3, \ldots\}$. Whether a sequence or a set is meant by $\mathbb{N}$ will be clear from the context. We write $\mathbb{Z}$ for the set $\{0, 1, -1, 2, -2, \cdots\}$. We write $\mathbb{Z}_+$ for both the set $\{0, 1, 2, \ldots\}$ and the sequence $(0, 1, 2, \cdots)$.

Definition 1.1. (Algebra and σ-algebra) *Let X be an arbitrary set. A collection $\mathfrak{A}$ of subsets of X is called an algebra of subsets of X if it satisfies the following conditions:*

$1°$ $X \in \mathfrak{A}$,

$2°$ $A \in \mathfrak{A} \Rightarrow A^c \in \mathfrak{A}$,

$3°$ $A, B \in \mathfrak{A} \Rightarrow A \cup B \in \mathfrak{A}$.

An algebra $\mathfrak{A}$ of subsets of a set X is called a σ-algebra if it satisfies the additional condition:

$4°$ $(A_n : n \in \mathbb{N}) \subset \mathfrak{A} \Rightarrow \bigcup_{n \in \mathbb{N}} A_n \in \mathfrak{A}$.

Definition 1.2. *Let $\mathfrak{E}$ be a collection of subsets of a set X. Let γ be a nonnegative extended real-valued set function on $\mathfrak{E}$. We say that*

(a) *γ is monotone on $\mathfrak{E}$ if $\gamma(E_1) \leq \gamma(E_2)$ for $E_1, E_2 \in \mathfrak{E}$ such that $E_1 \subset E_2$,*

(b) *γ is additive on $\mathfrak{E}$ if $\gamma(E_1 \cup E_2) = \gamma(E_1) + \gamma(E_2)$ for $E_1, E_2 \in \mathfrak{E}$ such that $E_1 \cap E_2 = \emptyset$ and $E_1 \cup E_2 \in \mathfrak{E}$,*

(c) *γ is finitely additive on $\mathfrak{E}$ if $\gamma(\bigcup_{k=1}^n E_k) = \sum_{k=1}^n \gamma(E_k)$ for every disjoint finite sequence $(E_k : k = 1, \ldots, n)$ in $\mathfrak{E}$ such that $\bigcup_{k=1}^n E_k \in \mathfrak{E}$,*

(d) *γ is countably additive on $\mathfrak{E}$ if $\gamma(\bigcup_{n \in \mathbb{N}} E_n) = \sum_{n \in \mathbb{N}} \gamma(E_n)$ for every disjoint sequence $(E_n : n \in \mathbb{N})$ in $\mathfrak{E}$ such that $\bigcup_{n \in \mathbb{N}} E_n \in \mathfrak{E}$,*

(e) *γ is subadditive on $\mathfrak{E}$ if $\gamma(E_1 \cup E_2) \leq \gamma(E_1) + \gamma(E_2)$ for $E_1, E_2 \in \mathfrak{E}$ such that $E_1 \cup E_2 \in \mathfrak{E}$,*

(f) *γ is finitely subadditive on $\mathfrak{E}$ if $\gamma(\bigcup_{k=1}^n E_k) \leq \sum_{k=1}^n \gamma(E_k)$ for every finite sequence $(E_k : k = 1, \ldots, n)$ in $\mathfrak{E}$ such that $\bigcup_{k=1}^n E_k \in \mathfrak{E}$,*

(g) *γ is countably subadditive on $\mathfrak{E}$ if $\gamma(\bigcup_{n \in \mathbb{N}} E_n) \leq \sum_{n \in \mathbb{N}} \gamma(E_n)$ for every sequence $(E_n : n \in \mathbb{N})$ in $\mathfrak{E}$ such that $\bigcup_{n \in \mathbb{N}} E_n \in \mathfrak{E}$.*

Observe that if (i). $\emptyset \in \mathfrak{E}$, (ii). γ is additive on $\mathfrak{E}$ and (iii). there exists $E \in \mathfrak{E}$ with $\gamma(E) < \infty$, then by (b) we have $\gamma(E) = \gamma(E \cup \emptyset) = \gamma(E) + \gamma(\emptyset)$ so that $\gamma(\emptyset) = 0$.

Note that in (c) while $\bigcup_{k=1}^n E_k \in \mathfrak{E}$ is required, it is not required that any of the partial unions $\bigcup_{k=1}^2 E_k, \bigcup_{k=1}^3 E_k, \ldots, \bigcup_{k=1}^{n-1} E_k$ be in $\mathfrak{E}$. If we assume the collection $\mathfrak{E}$ is an algebra then the partial unions are all in $\mathfrak{E}$. Note also that (c) implies (b) and (f) implies (e).

Observation 1.3. Let γ be a nonnegative extended real-valued set function on a collection $\mathfrak{E}$ of subsets of a set X. Assume that $\emptyset \in \mathfrak{E}$ and $\gamma(\emptyset) = 0$.

(a) If γ is countably additive on $\mathfrak{E}$, then it is finitely additive on $\mathfrak{E}$.

(b) If γ is countably subadditive on $\mathfrak{E}$, then it is finitely subadditive on $\mathfrak{E}$.

Proof. Suppose γ is countably additive on $\mathfrak{E}$. To show that it is finitely additive on $\mathfrak{E}$, let $(E_k : k = 1, \ldots, n)$ be a disjoint finite sequence in $\mathfrak{E}$ such that $\bigcup_{k=1}^n E_k \in \mathfrak{E}$. Consider the infinite sequence $(F_k : k \in \mathbb{N})$ in $\mathfrak{E}$ defined by $F_k = E_k$ for $k = 1, \ldots, n$ and $F_k = \emptyset$ for $k \geq n+1$. Since $\emptyset \in \mathfrak{E}$, $(F_k : k \in \mathbb{N})$ is a disjoint sequence in $\mathfrak{E}$ with $\bigcup_{k \in \mathbb{N}} F_k = \bigcup_{k=1}^n E_k \in \mathfrak{E}$. Thus by the countable additivity of γ on $\mathfrak{E}$ and by the fact that $\gamma(\emptyset) = 0$, we have $\gamma(\bigcup_{k=1}^n E_k) = \gamma(\bigcup_{k \in \mathbb{N}} F_k) = \sum_{k \in \mathbb{N}} \gamma(F_k) = \sum_{k=1}^n \gamma(E_k)$. This proves the finite additivity of γ on $\mathfrak{E}$. We show similarly that if γ is countably subadditive on $\mathfrak{E}$, then it is finitely subadditive on $\mathfrak{E}$. ■

Lemma 1.4. *Let $(E_n : n \in \mathbb{N})$ be an arbitrary sequence in an algebra $\mathfrak{A}$ of subsets of a set X. Then there exists a disjoint sequence $(F_n : n \in \mathbb{N})$ in $\mathfrak{A}$ such that*

(1) $\displaystyle\bigcup_{n=1}^N E_n = \bigcup_{n=1}^N F_n$ *for every $N \in \mathbb{N}$,*

and

(2) $\displaystyle\bigcup_{n \in \mathbb{N}} E_n = \bigcup_{n \in \mathbb{N}} F_n.$

In particular, if $\mathfrak{A}$ is a σ-algebra, then $\bigcup_{n \in \mathbb{N}} F_n = \bigcup_{n \in \mathbb{N}} E_n \in \mathfrak{A}$.

Proof. Let $F_1 = E_1$ and $F_n = E_n \setminus (E_1 \cup \ldots \cup E_{n-1})$ for $n \geq 2$. Since $\mathfrak{A}$ is an algebra, $F_n \in \mathfrak{A}$ for $n \in \mathbb{N}$. Let us prove (1) and (2) and then the disjointness of $(F_n : n \in \mathbb{N})$.

Let us prove (1) by induction. To start with, (1) is valid when $N = 1$ since $F_1 = E_1$. Next, assume that (1) is valid for some $N \in \mathbb{N}$, that is, $\bigcup_{n=1}^N E_n = \bigcup_{n=1}^N F_n$. Then we have

$$\bigcup_{n=1}^{N+1} F_n = \left(\bigcup_{n=1}^N F_n\right) \cup F_{N+1} = \left(\bigcup_{n=1}^N E_n\right) \cup \left(E_{N+1} \setminus \bigcup_{n=1}^N E_n\right) = \bigcup_{n=1}^{N+1} E_n,$$

that is, (1) holds for $N+1$. Thus by induction, (1) holds for every $N \in \mathbb{N}$.

To prove (2), let $x \in \bigcup_{n \in \mathbb{N}} E_n$. Then $x \in E_n$ for some $n \in \mathbb{N}$ and thus we have $x \in \bigcup_{k=1}^n E_k = \bigcup_{k=1}^n F_k \subset \bigcup_{n \in \mathbb{N}} F_n$ by (1). We show similarly that if $x \in \bigcup_{n \in \mathbb{N}} F_n$ then $x \in \bigcup_{n \in \mathbb{N}} E_n$. Thus we have $\bigcup_{n \in \mathbb{N}} E_n = \bigcup_{n \in \mathbb{N}} F_n$. This proves (2).

Finally let us show that $(F_n : n \in \mathbb{N})$ is a disjoint sequence. Consider F_n and F_m where $n \neq m$, say $n < m$. We have $F_m = E_m \setminus (E_1 \cup \cdots \cup E_{m-1})$. By (1) and by the fact that $n < m$, we have $E_1 \cup \cdots \cup E_{m-1} = F_1 \cup \cdots \cup F_{m-1} \supset F_n$. Thus we have $F_n \bigcap F_m = \emptyset$. This proves the disjointness of $(F_n : n \in \mathbb{N})$. ■

Lemma 1.5. *Let γ be a nonnegative extended real-valued set function on an algebra $\mathfrak{A}$ of subsets of a set X.*

(a) *If γ is additive on $\mathfrak{A}$, it is finitely additive, monotone, and finitely subadditive on $\mathfrak{A}$.*

(b) *If γ is countably additive on $\mathfrak{A}$, then it is countably subadditive on $\mathfrak{A}$.*

Proof. 1. Suppose γ is additive on $\mathfrak{A}$. Let $(E_k : k = 1, \ldots, n)$ be a disjoint finite sequence in $\mathfrak{A}$. Since $\mathfrak{A}$ is an algebra, we have $\bigcup_{i=1}^{k} E_i \in \mathfrak{A}$ for $k = 1, \ldots, n$. By the disjointness of $\bigcup_{k=1}^{n-1} E_k$ and E_n and by the additivity of γ on $\mathfrak{A}$, we have

$$\gamma\left(\bigcup_{k=1}^{n} E_k\right) = \gamma\left(\bigcup_{k=1}^{n-1} E_k\right) + \gamma(E_n).$$

Repeating the argument, we have $\gamma(\bigcup_{k=1}^{n} E_k) = \sum_{k=1}^{n} \gamma(E_k)$. This proves the finite additivity of γ on $\mathfrak{A}$. To prove the monotonicity of γ on $\mathfrak{A}$, let $E_1, E_2 \in \mathfrak{A}$ and $E_1 \subset E_2$. Then $E_1, E_2 \setminus E_1 \in \mathfrak{A}$, $E_1 \cap (E_2 \setminus E_1) = \emptyset$, and $E_1 \cup (E_2 \setminus E_1) = E_2 \in \mathfrak{A}$ so that by the additivity of γ on $\mathfrak{A}$, we have $\gamma(E_1) + \gamma(E_2 \setminus E_1) = \gamma(E_2)$. Then since $\gamma(E_2 \setminus E_1) \geq 0$, we have $\gamma(E_1) \leq \gamma(E_2)$. This proves the monotonicity of γ on $\mathfrak{A}$.

To show the finite subadditivity of γ on $\mathfrak{A}$, let $(E_k : k = 1, \ldots, n)$ be a finite sequence in $\mathfrak{A}$. If we let $F_1 = E_1$ and $F_k = E_k \setminus (E_1 \cup \cdots \cup E_{k-1})$ for $k = 2, \ldots, n$, then as we showed in the Proof of Lemma 1.4, $(F_k : k = 1, \ldots, n)$ is a disjoint finite sequence in $\mathfrak{A}$ with $\bigcup_{k=1}^{n} F_k = \bigcup_{k=1}^{n} E_k$ so that by the finite additivity and the monotonicity of γ on $\mathfrak{A}$, we have $\gamma(\bigcup_{k=1}^{n} E_k) = \gamma(\bigcup_{k=1}^{n} F_k) = \sum_{k=1}^{n} \gamma(F_k) \leq \sum_{k=1}^{n} \gamma(E_k)$. This proves the finite subadditivity of γ on $\mathfrak{A}$.

2. Suppose γ is countably additive on $\mathfrak{A}$. To show that it is countable subadditive on $\mathfrak{A}$, let $(E_n : n \in \mathbb{N})$ be a sequence in $\mathfrak{A}$ such that $\bigcup_{n\in\mathbb{N}} E_n \in \mathfrak{A}$. Let $F_1 = E_1$ and $F_n = E_n \setminus (E_1 \cup \ldots E_{n-1})$ for $n \geq 2$. Then by Lemma 1.4, $(F_n : n \in \mathbb{N})$ is a disjoint sequence in $\mathfrak{A}$ and $\bigcup_{n\in\mathbb{N}} F_n = \bigcup_{n\in\mathbb{N}} E_n$. Thus by the countable additivity and the monotonicity of γ on $\mathfrak{A}$ by (a), we have $\gamma(\bigcup_{n\in\mathbb{N}} E_n) = \gamma(\bigcup_{n\in\mathbb{N}} F_n) = \sum_{n\in\mathbb{N}} \gamma(F_n) \leq \sum_{n\in\mathbb{N}} \gamma(E_n)$. This proves the countable subadditivity of γ on $\mathfrak{A}$. ∎

Theorem 1.6. *Let γ be a nonnegative extended real-valued set function on an algebra $\mathfrak{A}$ of subsets of a set X. If γ is additive and countably subadditive on $\mathfrak{A}$ then γ is countably additive on $\mathfrak{A}$.*

Proof. Suppose γ is additive and countably subadditive on $\mathfrak{A}$. To show that γ is countably additive on $\mathfrak{A}$, let $(E_n : n \in \mathbb{N})$ be a disjoint sequence in $\mathfrak{A}$ such that $\bigcup_{n\in\mathbb{N}} E_n \in \mathfrak{A}$. The additivity of γ on $\mathfrak{A}$ implies its monotonicity and finite additivity on $\mathfrak{A}$ by (a) of Lemma 1.5. Thus for every $N \in \mathbb{N}$, we have $\gamma(\bigcup_{n\in\mathbb{N}} E_n) \geq \gamma(\bigcup_{n=1}^{N} E_n) = \sum_{n=1}^{N} \gamma(E_n)$. Since this holds for every $N \in \mathbb{N}$, we have $\gamma(\bigcup_{n\in\mathbb{N}} E_n) \geq \sum_{n\in\mathbb{N}} \gamma(E_n)$. On the other hand, by the countable subadditivity of γ on $\mathfrak{A}$, we have $\gamma(\bigcup_{n\in\mathbb{N}} E_n) \leq \sum_{n\in\mathbb{N}} \gamma(E_n)$. Thus $\gamma(\bigcup_{n\in\mathbb{N}} E_n) = \sum_{n\in\mathbb{N}} \gamma(E_n)$. This proves the countable additivity of γ on $\mathfrak{A}$. ∎

Definition 1.7. (Measure) *Let $\mathfrak{A}$ be a σ-algebra of subsets of a set X. A set function μ defined on $\mathfrak{A}$ is called a measure if it satisfies the following conditions:*

1° *nonnegative extended real-valued:* $\mu(E) \in [0, \infty]$ *for every* $E \in \mathfrak{A}$,

2° $\mu(\emptyset) = 0$,

3° *countable additivity:* $(E_n : n \in \mathbb{N}) \subset \mathfrak{A}$, *disjoint* $\Rightarrow \mu(\bigcup_{n\in\mathbb{N}} E_n) = \sum_{n\in\mathbb{N}} \mu(E_n)$.

Observe that countable additivity implies finite additivity and in particular additivity. Indeed, given a finite disjoint sequence $(E_1, \ldots, E_k) \subset \mathfrak{A}$, consider a disjoint sequence $(E_n : n \in \mathbb{N}) \subset \mathfrak{A}$ defined by setting $E_n = \emptyset$ for $n \geq k+1$. Then we have $\bigcup_{n=1}^k E_n = \bigcup_{n \in \mathbb{N}} E_n$ and thus by the countable additivity and by $\mu(\emptyset) = 0$, we have

$$\mu\left(\bigcup_{n=1}^{k} E_n\right) = \mu\left(\bigcup_{n \in \mathbb{N}} E_n\right) = \sum_{n \in \mathbb{N}} \mu(E_n) = \sum_{n=1}^{k} \mu(E_n).$$

Lemma 1.8. *A measure μ on a σ-algebra $\mathfrak{A}$ of subsets of a set X has the following properties:*

(1) *finite additivity: $(E_1, \ldots, E_n) \subset \mathfrak{A}$, disjoint $\Rightarrow \mu(\bigcup_{k=1}^n E_k) = \sum_{k=1}^n \mu(E_k)$,*

(2) *monotonicity: $E_1, E_2 \in \mathfrak{A}, E_1 \subset E_2 \Rightarrow \mu(E_1) \leq \mu(E_2)$,*

(3) *$E_1, E_2 \in \mathfrak{A}, E_1 \subset E_2, \mu(E_1) < \infty \Rightarrow \mu(E_2 \setminus E_1) = \mu(E_2) - \mu(E_1)$,*

(4) *countable subadditivity: $(E_n : n \in \mathbb{N}) \subset \mathfrak{A} \Rightarrow \mu(\bigcup_{n \in \mathbb{N}} E_n) \leq \sum_{n \in \mathbb{N}} \mu(E_n)$,*

and in particular

(5) *finite subadditivity: $(E_1, \ldots, E_n) \subset \mathfrak{A} \Rightarrow \mu(\bigcup_{k=1}^n E_k) \leq \sum_{k=1}^n \mu(E_k)$.*

Proof. The countable additivity of μ on $\mathfrak{A}$ implies its finite additivity on $\mathfrak{A}$ by (a) of Observation 1.3. The finite additivity of μ on $\mathfrak{A}$ implies its additivity on $\mathfrak{A}$ and then its monotonicity on $\mathfrak{A}$ by (a) of Lemma 1.5.

To prove (3), let $E_1, E_2 \in \mathfrak{A}$ and $E_1 \subset E_2$. Then E_1 and $E_2 \setminus E_1$ are two disjoint members of $\mathfrak{A}$ whose union is equal to E_2. Thus by the additivity of μ on $\mathfrak{A}$, we have $\mu(E_2) = \mu(E_1) + \mu(E_2 \setminus E_1)$. If $\mu(E_1) < \infty$, then subtracting $\mu(E_1)$ for both sides of the last equality, we have $\mu(E_2) - \mu(E_1) = \mu(E_2 \setminus E_1)$. This proves (3).

The countable additivity of μ on $\mathfrak{A}$ implies its countable subadditivity on $\mathfrak{A}$ by (b) of Lemma 1.5. This then implies the finite subadditivity of μ on $\mathfrak{A}$ by (b) of Observation 1.3. ∎

Regarding (3) of Lemma 1.8, let us note that if $\mu(E_1) = \infty$ then by the monotonicity of μ we also have $\mu(E_2) = \infty$ so that $\mu(E_2) - \mu(E_1)$ is not defined.

Theorem 1.9. (Monotone Convergence Theorem for Sequences of Measurable Sets) *Let μ be a measure on a σ-algebra $\mathfrak{A}$ of subsets of a set X and let $(E_n : n \in \mathbb{N})$ be a monotone sequence in $\mathfrak{A}$.*

(a) *If $E_n \uparrow$, then $\lim_{n \to \infty} \mu(E_n) = \mu(\lim_{n \to \infty} E_n)$.*

(b) *If $E_n \downarrow$, then $\lim_{n \to \infty} \mu(E_n) = \mu(\lim_{n \to \infty} E_n)$, provided that there exists a set $A \in \mathfrak{A}$ with $\mu(A) < \infty$ such that $E_1 \subset A$.*

Proof. If $E_n \uparrow$, then $\lim_{n \to \infty} E_n = \bigcup_{n \in \mathbb{N}} E_n \in \mathfrak{A}$. If $E_n \downarrow$, then $\lim_{n \to \infty} E_n = \bigcap_{n \in \mathbb{N}} E_n \in \mathfrak{A}$.

Note also that if $(E_n : n \in \mathbb{N})$ is a monotone sequence in $\mathfrak{A}$, then $(\mu(E_n) : n \in \mathbb{N})$ is a monotone sequence in $[0, \infty]$ by the monotonicity of μ so that $\lim_{n\to\infty} \mu(E_n)$ exists in $[0, \infty]$.

1. Suppose $E_n \uparrow$. Then we have $\mu(E_n) \uparrow$. Consider first the case where $\mu(E_{n_0}) = \infty$ for some $n_0 \in \mathbb{N}$. In this case we have $\lim_{n\to\infty} \mu(E_n) = \infty$. Since $E_{n_0} \subset \bigcup_{n\in\mathbb{N}} E_n = \lim_{n\to\infty} E_n$, we have $\mu(\lim_{n\to\infty} E_n) \geq \mu(E_{n_0}) = \infty$. Thus $\mu(\lim_{n\to\infty} E_n) = \infty = \lim_{n\to\infty} \mu(E_n)$.

Consider next the case where $\mu(E_n) < \infty$ for every $n \in \mathbb{N}$. Let $E_0 = \emptyset$ and consider a disjoint sequence $(F_n : n \in \mathbb{N})$ in $\mathfrak{A}$ defined by $F_n = E_n \setminus E_{n-1}$ for $n \in \mathbb{N}$. We have $\bigcup_{n=1}^{N} E_n = \bigcup_{n=1}^{N} F_n$ for every $N \in \mathbb{N}$ and hence $\bigcup_{n\in\mathbb{N}} E_n = \bigcup_{n\in\mathbb{N}} F_n$. Then we have

$$\mu\left(\lim_{n\to\infty} E_n\right) = \mu\left(\bigcup_{n\in\mathbb{N}} E_n\right) = \mu\left(\bigcup_{n\in\mathbb{N}} F_n\right) = \sum_{n\in\mathbb{N}} \mu(F_n)$$
$$= \sum_{n\in\mathbb{N}} \mu(E_n \setminus E_{n-1}) = \sum_{n\in\mathbb{N}} \{\mu(E_n) - \mu(E_{n-1})\},$$

where the third equality is by the countable additivity of μ and the fifth equality is by (3) of Lemma 1.8. Since the sum of a series is the limit of the sequence of partial sums we have

$$\sum_{n\in\mathbb{N}} \{\mu(E_n) - \mu(E_{n-1})\} = \lim_{n\to\infty} \sum_{k=1}^{n} \{\mu(E_k) - \mu(E_{k-1})\}$$
$$= \lim_{n\to\infty} \{\mu(E_n) - \mu(E_0)\} = \lim_{n\to\infty} \mu(E_n).$$

Thus we have $\mu(\lim_{n\to\infty} E_n) = \lim_{n\to\infty} \mu(E_n)$.

2. Suppose $E_n \downarrow$ and assume the existence of a containing set A with finite measure. Define a disjoint sequence $(F_n : n \in \mathbb{N})$ in $\mathfrak{A}$ by setting $F_n = E_n \setminus E_{n+1}$ for $n \in \mathbb{N}$. Then

$$(1) \qquad E_1 \setminus \bigcap_{n\in\mathbb{N}} E_n = \bigcup_{n\in\mathbb{N}} F_n.$$

To show this, let $x \in E_1 \setminus \bigcap_{n\in\mathbb{N}} E_n$. Then $x \in E_1$ and x is not in every E_n. Since $E_n \downarrow$, there exists the first set E_{n_0+1} in the sequence not containing x. Then $x \in E_{n_0} \setminus E_{n_0+1} = F_{n_0} \subset \bigcup_{n\in\mathbb{N}} F_n$. This shows that $E_1 \setminus \bigcap_{n\in\mathbb{N}} E_n \subset \bigcup_{n\in\mathbb{N}} F_n$. Conversely if $x \in \bigcup_{n\in\mathbb{N}} F_n$, then $x \in F_{n_0} = E_{n_0} \setminus E_{n_0+1}$ for some $n_0 \in \mathbb{N}$. Now $x \in E_{n_0} \subset E_1$. Since $x \notin E_{n_0+1}$, we have $x \notin \bigcap_{n\in\mathbb{N}} E_n$. Thus $x \in E_1 \setminus \bigcap_{n\in\mathbb{N}} E_n$. This shows that $\bigcup_{n\in\mathbb{N}} F_n \subset E_1 \setminus \bigcap_{n\in\mathbb{N}} E_n$. Therefore (1) holds. Now by (1), we have

$$(2) \qquad \mu\left(E_1 \setminus \bigcap_{n\in\mathbb{N}} E_n\right) = \mu\left(\bigcup_{n\in\mathbb{N}} F_n\right).$$

Since $\mu(\bigcap_{n\in\mathbb{N}} E_n) \leq \mu(E_1) \leq \mu(A) < \infty$, we have by (3) of Lemma 1.8

$$(3) \qquad \mu\left(E_1 \setminus \bigcap_{n\in\mathbb{N}} E_n\right) = \mu(E_1) - \mu\left(\bigcap_{n\in\mathbb{N}} E_n\right) = \mu(E_1) - \mu\left(\lim_{n\to\infty} E_n\right).$$

By the countable additivity of μ, we have

$$(4)\qquad \mu\left(\bigcup_{n\in\mathbb{N}} F_n\right) = \sum_{n\in\mathbb{N}} \mu(F_n) = \sum_{n\in\mathbb{N}} \mu(E_n \setminus E_{n+1})$$

$$= \sum_{n\in\mathbb{N}} \{\mu(E_n) - \mu(E_{n+1})\} = \lim_{n\to\infty} \sum_{k=1}^{n} \{\mu(E_k) - \mu(E_{k+1})\}$$

$$= \lim_{n\to\infty} \{\mu(E_1) - \mu(E_{n+1})\} = \mu(E_1) - \lim_{n\to\infty} \mu(E_{n+1}).$$

Substituting (3) and (4) in (2), we have

$$\mu(E_1) - \mu\left(\lim_{n\to\infty} E_n\right) = \mu(E_1) - \lim_{n\to\infty} \mu(E_{n+1}) = \mu(E_1) - \lim_{n\to\infty} \mu(E_n).$$

Subtracting $\mu(E_1) \in \mathbb{R}$ from both sides we have $\mu(\lim_{n\to\infty} E_n) = \lim_{n\to\infty} \mu(E_n)$. ∎

Remark 1.10. (b) of Theorem 1.9 has the following particular cases. Let $(E_n : n \in \mathbb{N})$ be a decreasing sequence in $\mathfrak{A}$. Then $\lim_{n\to\infty} \mu(E_n) = \mu(\lim_{n\to\infty} E_n)$ if any one of the following conditions is satisfied:
(a) $\mu(X) < \infty$,
(b) $\mu(E_1) < \infty$,
(c) $\mu(E_{n_0}) < \infty$ for some $n_0 \in \mathbb{N}$.

Proof. (a) and (b) are particular cases of (b) of Theorem 1.9 in which X and E_1 respectively are the containing set $A \in \mathfrak{A}$ with $\mu(A) < \infty$.

To prove (c), suppose $\mu(E_{n_0}) < \infty$ for some $n_0 \in \mathbb{N}$. Let $(F_n : n \in \mathbb{N})$ be a decreasing sequence in $\mathfrak{A}$ obtained by dropping the first n_0 terms from $(E_n : n \in \mathbb{N})$, that is, we set $F_n = E_{n_0+n}$ for $n \in \mathbb{N}$. Then we have $\liminf_{n\to\infty} F_n = \liminf_{n\to\infty} E_n$ and $\limsup_{n\to\infty} F_n = \limsup_{n\to\infty} E_n$ and thus $\lim_{n\to\infty} F_n = \lim_{n\to\infty} E_n$. Now since $(F_n : n \in \mathbb{N})$ is a decreasing sequence and $F_n \subset E_{n_0}$ for $n \in \mathbb{N}$ and since $\mu(E_{n_0}) < \infty$, (b) of Theorem 1.9 applies so that $\lim_{n\to\infty} \mu(F_n) = \mu(\lim_{n\to\infty} F_n) = \mu(\lim_{n\to\infty} E_n)$. Since $(\mu(F_n) : n \in \mathbb{N})$ is a sequence obtained by dropping the first n_0 terms of $(\mu(E_n) : n \in \mathbb{N})$, we have $\lim_{n\to\infty} \mu(F_n) = \lim_{n\to\infty} \mu(E_n)$. Therefore we have $\lim_{n\to\infty} \mu(E_n) = \mu(\lim_{n\to\infty} E_n)$. ∎

Theorem 1.11. *Let μ be a measure on a σ-algebra $\mathfrak{A}$ of subsets of a set X.*
(a) *For an arbitrary sequence $(E_n : n \in \mathbb{N})$ in $\mathfrak{A}$, we have*

$$(1)\qquad \mu\left(\liminf_{n\to\infty} E_n\right) \le \liminf_{n\to\infty} \mu(E_n).$$

(b) *If there exists $A \in \mathfrak{A}$ with $\mu(A) < \infty$ such that $E_n \subset A$ for $n \in \mathbb{N}$, then*

$$(2)\qquad \mu\left(\limsup_{n\to\infty} E_n\right) \ge \limsup_{n\to\infty} \mu(E_n).$$

(c) *If both* $\lim_{n\to\infty} E_n$ *and* $\lim_{n\to\infty} \mu(E_n)$ *exist, then*

$$\mu\left(\lim_{n\to\infty} E_n\right) \le \lim_{n\to\infty} \mu(E_n). \tag{3}$$

(d) *If* $\lim_{n\to\infty} E_n$ *exist and if there exists* $A \in \mathfrak{A}$ *with* $\mu(A) < \infty$ *such that* $E_n \subset A$ *for* $n \in \mathbb{N}$, *then* $\lim_{n\to\infty} \mu(E_n)$ *exists and*

$$\mu\left(\lim_{n\to\infty} E_n\right) = \lim_{n\to\infty} \mu(E_n). \tag{4}$$

Proof. 1. Recall that $\liminf_{n\to\infty} E_n = \bigcup_{n\in\mathbb{N}} \bigcap_{k\ge n} E_k = \lim_{n\to\infty} \bigcap_{k\ge n} E_k$ by the fact that $(\bigcap_{k\ge n} E_k : n \in \mathbb{N})$ is an increasing sequence in $\mathfrak{A}$. Then by (a) of Theorem 1.9, we have $\mu(\liminf_{n\to\infty} E_n) = \lim_{n\to\infty} \mu(\bigcap_{k\ge n} E_k) = \liminf_{n\to\infty} \mu(\bigcap_{k\ge n} E_k)$ since the limit of a sequence, if it exists, is equal to the limit inferior of the sequence. Since $\bigcap_{k\ge n} E_k \subset E_n$, we have $\mu(\bigcap_{k\ge n} E_k) \le \mu(E_n)$ for $n \in \mathbb{N}$ by the monotonicity of μ. This then implies $\liminf_{n\to\infty} \mu(\bigcap_{k\ge n} E_k) \le \liminf_{n\to\infty} \mu(E_n)$. Continuing the chain of equalities above with this inequality, we have (1).

2. Assume that there exists $A \in \mathfrak{A}$ with $\mu(A) < \infty$ such that $E_n \subset A$ for $n \in \mathbb{N}$. Now $\limsup_{n\to\infty} E_n = \bigcap_{n\in\mathbb{N}} \bigcup_{k\ge n} E_k = \lim_{n\to\infty} \bigcup_{k\ge n} E_k$ by the fact that $(\bigcup_{k\ge n} E_k : n \in \mathbb{N})$ is a decreasing sequence in $\mathfrak{A}$. Since $E_n \subset A$ for all $n \in \mathbb{N}$, we have $\bigcup_{k\ge n} E_k \subset A$ for all $n \in \mathbb{N}$. Thus we have $\mu(\limsup_{n\to\infty} E_n) = \mu(\lim_{n\to\infty} \bigcup_{k\ge n} E_k) = \lim_{n\to\infty} \mu(\bigcup_{k\ge n} E_k)$ by (b) of Theorem 1.9. Now $\lim_{n\to\infty} \mu(\bigcup_{k\ge n} E_k) = \limsup_{n\to\infty} \mu(\bigcup_{k\ge n} E_k)$ since the limit of a sequence, if it exists, is equal to the limit superior of the sequence. Then by $\bigcup_{k\ge n} E_k \supset E_n$, we have $\mu(\bigcup_{k\ge n} E_k) \ge \mu(E_n)$. Thus $\limsup_{n\to\infty} \mu(\bigcup_{k\ge n} E_k) \ge \limsup_{n\to\infty} \mu(E_n)$. Continuing the chain of equalities above with this inequality, we have (2).

3. If $\lim_{n\to\infty} E_n$ and $\lim_{n\to\infty} \mu(E_n)$ exist, then $\lim_{n\to\infty} E_n = \liminf_{n\to\infty} E_n$ and $\lim_{n\to\infty} \mu(E_n) = \liminf_{n\to\infty} \mu(E_n)$ so that (1) reduces to (3).

4. If $\lim_{n\to\infty} E_n$ exists, then $\limsup_{n\to\infty} E_n = \lim_{n\to\infty} E_n = \liminf_{n\to\infty} E_n$. If there exists $A \in \mathfrak{A}$ with $\mu(A) < \infty$ such that $E_n \subset A$ for $n \in \mathbb{N}$, then by (2) and (1) we have

$$\begin{aligned} \limsup_{n\to\infty} \mu(E_n) &\le \mu\left(\limsup_{n\to\infty} E_n\right) = \mu\left(\lim_{n\to\infty} E_n\right) \\ &= \mu\left(\liminf_{n\to\infty} E_n\right) \le \liminf_{n\to\infty} \mu(E_n). \end{aligned} \tag{5}$$

Since $\liminf_{n\to\infty} \mu(E_n) \le \limsup_{n\to\infty} \mu(E_n)$ the inequalities (5) imply

$$\liminf_{n\to\infty} \mu(E_n) = \mu\left(\lim_{n\to\infty} E_n\right) = \limsup_{n\to\infty} \mu(E_n). \tag{6}$$

Thus $\lim_{n\to\infty} \mu(E_n)$ exists and then by (6) we have $\mu(\lim_{n\to\infty} E_n) = \lim_{n\to\infty} \mu(E_n)$. This proves (4). ∎

Definition 1.12. (a) (Measurable Space) *If $\mathfrak{A}$ is a σ-algebra of subsets of a set X then we call the pair $(X, \mathfrak{A})$ a measurable space.*
(b) (Measure Space) *If μ is a measure on a σ-algebra $\mathfrak{A}$ of subsets of a set X then we call the triple $(X, \mathfrak{A}, \mu)$ a measure space.*

Definition 1.13. (Complete σ-algebra) *Let $(X, \mathfrak{A}, \mu)$ be a measure space. We say that the σ-algebra $\mathfrak{A}$ is complete with respect to the measure μ if $E \in \mathfrak{A}$, $\mu(E) = 0$, and $E_0 \subset E$ imply $E_0 \in \mathfrak{A}$. If $\mathfrak{A}$ is complete with respect to μ then we call the measure space $(X, \mathfrak{A}, \mu)$ a complete measure space.*

[II] Outer Measure on a Set

Definition 1.14. (Outer Measure) *Let X be an arbitrary set. A set function μ defined on the collection $\mathfrak{P}(X)$ of all subsets of X is called an outer measure on X if it satisfies the following conditions:*

1° *nonnegative extended real-valued:* $\mu(E) \in [0, \infty]$ *for every* $E \in \mathfrak{P}(X)$,

2° $\mu(\emptyset) = 0$,

3° *monotonicity:* $E_1, E_2 \in \mathfrak{P}(X), E_1 \subset E_2 \Rightarrow \mu(E_1) \leq \mu(E_2)$,

4° *countable subadditivity:* $(E_n : n \in \mathbb{N}) \subset \mathfrak{P}(X) \Rightarrow \mu(\bigcup_{n\in\mathbb{N}} E_n) \leq \sum_{n\in\mathbb{N}} \mu(E_n)$.

Observation 1.15. Let μ be an outer measure on a set X. Let $A, B \in \mathfrak{P}(X)$. If $\mu(A \setminus B) = 0$, then $\mu(A) \leq \mu(B)$.

Proof. For any $A, B \in \mathfrak{P}(X)$, we have $A = (A \cap B) \cup (A \setminus B)$. Then we have

$$\mu(A) \leq \mu(A \cap B) + \mu(A \setminus B) = \mu(A \cap B) \leq \mu(B)$$

by the subadditivity of μ, the assumption that $\mu(A \setminus B) = 0$, and the monotonicity of μ. ∎

Definition 1.16. (Carathéodory Measurability Condition) *Let μ be an outer measure on a set X. We say a set that $E \in \mathfrak{P}(X)$ is μ-measurable (or measurable with respect to μ) if it satisfies the following Carathéodory condition:*

$$\mu(A) = \mu(A \cap E) + \mu(A \cap E^c) \quad \textit{for every } A \in \mathfrak{P}(X).$$

The set A is called a testing set in the Carathéodory measurability condition. We write $\mathfrak{M}(\mu)$ for the collection of all μ-measurable $E \in \mathfrak{P}(X)$.

Observation 1.17. The countable subadditivity of μ implies its finite subadditivity on $\mathfrak{P}(X)$. Thus $\mu(A) \leq \mu(A \cap E) + \mu(A \cap E^c)$ for any $E, A \in \mathfrak{P}(X)$. Therefore to verify the Carathéodory condition for $E \in \mathfrak{P}(X)$, it suffices to verify $\mu(A) \geq \mu(A \cap E) + \mu(A \cap E^c)$ for every $A \in \mathfrak{P}(X)$.

Our next goal is to show that given an outer measure μ on X the collection $\mathfrak{M}(\mu)$ of all μ-measurable $E \in \mathfrak{P}(X)$ constitutes a σ-algebra of subsets of X.

Lemma 1.18. *Let μ be an outer measure on a set X. Consider the collection $\mathfrak{M}(\mu)$ of all μ-measurable $E \in \mathfrak{P}(X)$.*
(a) *If $E_1, E_2 \in \mathfrak{M}(\mu)$, then $E_1 \cup E_2 \in \mathfrak{M}(\mu)$.*
(b) *The set function μ is additive on $\mathfrak{M}(\mu)$, that is, $\mu(E_1 \cup E_2) = \mu(E_1) + \mu(E_2)$ for $E_1, E_2 \in \mathfrak{M}(\mu)$ such that $E_1 \cap E_2 = \emptyset$.*

Proof. 1. Suppose $E_1, E_2 \in \mathfrak{M}(\mu)$. Let $A \in \mathfrak{P}(X)$. Since $E_1 \in \mathfrak{M}(\mu)$ we have

$$\mu(A) = \mu(A \cap E_1) + \mu(A \cap E_1^c). \tag{1}$$

With $A \cap E_1^c$ as a testing set for $E_2 \in \mathfrak{M}(\mu)$, we have

$$\mu(A \cap E_1^c) = \mu(A \cap E_1^c \cap E_2) + \mu(A \cap E_1^c \cap E_2^c). \tag{2}$$

Substituting (2) in (1) we have

$$\mu(A) = \mu(A \cap E_1) + \mu(A \cap E_1^c \cap E_2) + \mu(A \cap E_1^c \cap E_2^c). \tag{3}$$

Regarding the first two terms on the right hand side of (3) note that

$$\begin{aligned}(A \cap E_1) \cup (A \cap E_1^c \cap E_2) &= A \cap (E_1 \cup (E_1^c \cap E_2)) \\ &= A \cap (E_1 \cup (E_2 \setminus E_1)) = A \cap (E_1 \cup E_2).\end{aligned}$$

Then by the subadditivity of μ we have

$$\begin{aligned}\mu(A \cap E_1) + \mu(A \cap E_1^c \cap E_2) &\geq \mu\left((A \cap E_1) \cup (A \cap E_1^c \cap E_2)\right) \\ &= \mu\left(A \cap (E_1 \cup E_2)\right).\end{aligned} \tag{4}$$

No $\mu(A \cap E_1^c \cap E_2^c) = \mu(A \cap (E_1 \cup E_2)^c)$. Substituting this and (4) in (3), we have

$$\mu(A) \geq \mu\left(A \cap (E_1 \cup E_2)\right) + \mu\left(A \cap (E_1 \cup E_2)^c\right).$$

According to Observation 1.17 this shows that $E_1 \cup E_2$ satisfies the Carathéodory condition. Hence $E_1 \cup E_2 \in \mathfrak{M}(\mu)$.

2. To prove the additivity of μ on $\mathfrak{M}(\mu)$, let $E_1, E_2 \in \mathfrak{M}(\mu)$ and $E_1 \cap E_2 = \emptyset$. Since $E_1 \in \mathfrak{M}(\mu)$, we have $\mu(A) = \mu(A \cap E_1) + \mu(A \cap E_1^c)$ for every $A \in \mathfrak{P}(X)$. In particular, with $A = E_1 \cup E_2$ we have

$$\mu(E_1 \cup E_2) = \mu\left((E_1 \cup E_2) \cap E_1\right) + \mu\left((E_1 \cup E_2) \cap E_1^c\right).$$

Now the disjointness of E_1 and E_2 implies that $(E_1 \cup E_2) \cap E_1 = E_1$ and $(E_1 \cup E_2) \cap E_1^c = E_2$. Thus $\mu(E_1 \cup E_2) = \mu(E_1) + \mu(E_2)$. ■

Lemma 1.19. *Let μ be an outer measure on a set X. If $E, F \in \mathfrak{P}(X)$ and $\mu(F) = 0$, then $\mu(E \cup F) = \mu(E)$.*

Proof. By the subadditivity of μ, we have $\mu(E \cup F) \le \mu(E) + \mu(F) = \mu(E)$. On the other hand by the monotonicity of μ, we have $\mu(E) \le \mu(E \cup F)$. Therefore we have $\mu(E \cup F) = \mu(E)$. ■

Theorem 1.20. *Let μ be an outer measure on a set X. Then the collection $\mathfrak{M}(\mu)$ of all μ-measurable subsets of X is a σ-algebra of subsets of X.*

Proof. 1. For any $A \in \mathfrak{P}(X)$, we have $\mu(A \cap X) + \mu(A \cap X^c) = \mu(A) + \mu(\emptyset) = \mu(A)$. This shows that X satisfies the Carathéodory condition so that $X \in \mathfrak{M}(\mu)$.

2. Let $E \in \mathfrak{M}(\mu)$. Then for every $A \in \mathfrak{P}(X)$ we have

$$\mu(A) = \mu(A \cap E) + \mu(A \cap E^c) = \mu\left(A \cap (E^c)^c\right) + \mu(A \cap E^c).$$

This shows that E^c satisfies the Carathéodory condition so that $E^c \in \mathfrak{M}(\mu)$.

3. Let $(E_n : n \in \mathbb{N}) \subset \mathfrak{M}(\mu)$. By Observation 1.17, to show that $\bigcup_{n\in\mathbb{N}} E_n \in \mathfrak{M}(\mu)$, it suffices to show that for every $A \in \mathfrak{P}(X)$, we have

$$\mu(A) \ge \mu\left(A \cap \left[\bigcup_{n\in\mathbb{N}} E_n\right]\right) + \mu\left(A \cap \left[\bigcup_{n\in\mathbb{N}} E_n\right]^c\right).$$

Let us show first that for every $k \in \mathbb{N}$, we have

$$\mu(A) = \sum_{j=1}^{k} \mu\left(A \cap \left[\bigcup_{i=1}^{j-1} E_i\right]^c \cap E_j\right) + \mu\left(A \cap \left[\bigcup_{j=1}^{k} E_j\right]^c\right) \tag{1}$$

with the understanding that $\bigcup_{i=1}^{0} E_i = \emptyset$. Let us prove (1) by induction on $k \in \mathbb{N}$. Now for $k = 1$, since $E_1 \in \mathfrak{M}(\mu)$, we have $\mu(A) = \mu(A \cap E_1) + \mu(A \cap E_1^c)$. Thus (1) is valid for $k = 1$. Next assume that (1) is valid for some $k \in \mathbb{N}$. Let us show that (1) is valid for $k + 1$.

Now with $A \cap [\bigcup_{j=1}^{k} E_j]^c$ as a testing set for $E_{k+1} \in \mathfrak{M}(\mu)$, we have

$$\begin{aligned}\mu\left(A \cap \left[\bigcup_{j=1}^{k} E_j\right]^c\right) &= \mu\left(A \cap \left[\bigcup_{j=1}^{k} E_j\right]^c \cap E_{k+1}\right) + \mu\left(A \cap \left[\bigcup_{j=1}^{k} E_j\right]^c \cap E_{k+1}^c\right) \\ &= \mu\left(A \cap \left[\bigcup_{j=1}^{k} E_j\right]^c \cap E_{k+1}\right) + \mu\left(A \cap \left[\bigcup_{j=1}^{k+1} E_j\right]^c\right).\end{aligned}$$

Substituting this equality in (1) which is valid for k by our assumption, we have

$$\begin{aligned}\mu(A) &= \sum_{j=1}^{k} \mu\left(A \cap \left[\bigcup_{i=1}^{j-1} E_i\right]^c \cap E_j\right) + \mu\left(A \cap \left[\bigcup_{j=1}^{k} E_j\right]^c \cap E_{k+1}\right) + \mu\left(A \cap \left[\bigcup_{j=1}^{k+1} E_j\right]^c\right) \\ &= \sum_{j=1}^{k+1} \mu\left(A \cap \left[\bigcup_{i=1}^{j-1} E_i\right]^c \cap E_j\right) + \mu\left(A \cap \left[\bigcup_{j=1}^{k+1} E_j\right]^c\right).\end{aligned}$$

This shows that (1) is valid for $k+1$ under the assumption that it is valid for k. Thus by induction, (1) is valid for every $k \in \mathbb{N}$.

Since $\bigcup_{j=1}^{k} E_i \subset \bigcup_{j\in\mathbb{N}} E_j$, we have $[\bigcup_{j=1}^{k} E_i]^c \supset [\bigcup_{j\in\mathbb{N}} E_j]^c$, and then by the monotonicity of μ, we have $\mu(A \cap [\bigcup_{j=1}^{k} E_i]^c) \geq \mu(A \cap [\bigcup_{j\in\mathbb{N}} E_j]^c)$. Using this in (1), we have

$$\mu(A) \geq \sum_{j=1}^{k} \mu\left(A \cap \left[\bigcup_{i=1}^{j-1} E_i\right]^c \cap E_j\right) + \mu\left(A \cap \left[\bigcup_{j\in\mathbb{N}} E_j\right]^c\right).$$

Since this holds for every $k \in \mathbb{N}$, we have

$$\begin{aligned}\mu(A) &\geq \sum_{j\in\mathbb{N}} \mu\left(A \cap \left[\bigcup_{i=1}^{j-1} E_i\right]^c \cap E_j\right) + \mu\left(A \cap \left[\bigcup_{j\in\mathbb{N}} E_j\right]^c\right) \\ &\geq \mu\left(\bigcup_{j\in\mathbb{N}}\left(A \cap \left[\bigcup_{i=1}^{j-1} E_i\right]^c \cap E_j\right)\right) + \mu\left(A \cap \left[\bigcup_{j\in\mathbb{N}} E_j\right]^c\right) \\ &= \mu\left(A \cap \bigcup_{j\in\mathbb{N}}\left(\left[\bigcup_{i=1}^{j-1} E_i\right]^c \cap E_j\right)\right) + \mu\left(A \cap \left[\bigcup_{j\in\mathbb{N}} E_j\right]^c\right),\end{aligned}$$

where the second inequality is by the countable subadditivity of μ. Now by Lemma 1.4

$$\bigcup_{j\in\mathbb{N}}\left(\left[\bigcup_{i=1}^{j-1} E_i\right]^c \cap E_j\right) = \bigcup_{j\in\mathbb{N}}\left(E_j \setminus \left[\bigcup_{i=1}^{j-1} E_i\right]\right) = \bigcup_{j\in\mathbb{N}} E_j.$$

Therefore we have

$$\mu(A) \geq \mu\left(A \cap \left[\bigcup_{j\in\mathbb{N}} E_j\right]\right) + \mu\left(A \cap \left[\bigcup_{j\in\mathbb{N}} E_j\right]^c\right).$$

This shows that $\bigcup_{j \in \mathbb{N}} E_j$ satisfies the Carathéodory condition so that $\bigcup_{j \in \mathbb{N}} E_j \in \mathfrak{M}(\mu)$. ∎

We have just shown that if μ is an outer measure on a set X, then the collection $\mathfrak{M}(\mu)$ of all μ-measurable subsets of X is a σ-algebra of subsets of X. For an arbitrary measurable space $(X, \mathfrak{A})$, we say that a subset of X is $\mathfrak{A}$-measurable if it is a member of $\mathfrak{A}$. Since $\mathfrak{M}(\mu)$ is the collection of all μ-measurable subsets of X by Definition 1.16, μ-measurability and $\mathfrak{M}(\mu)$-measurability of a subset of X are equivalent.

Given an outer measure on a set X, the collection $\mathfrak{M}(\mu)$ of all μ-measurable subsets of X is a σ-algebra of subsets of X according to Theorem 1.20. We show next that the restriction of μ to $\mathfrak{M}(\mu)$ is a measure on $\mathfrak{M}(\mu)$ and moreover the measure space $(X, \mathfrak{M}(\mu), \mu)$ is a complete measure space.

Lemma 1.21. *Let μ be an outer measure on a set X. If $E \in \mathfrak{P}(X)$ and $\mu(E) = 0$, then every subset E_0 of E, and in particular E itself, is a member of $\mathfrak{M}(\mu)$.*

Proof. If $\mu(E) = 0$, then for any subset E_0 of E, we have $\mu(E_0) = 0$ by the monotonicity of μ. Then for every $A \in \mathfrak{P}(X)$, we have $\mu(A \cap E_0) + \mu(A \cap E_0^c) \leq \mu(E_0) + \mu(A) = \mu(A)$ by the monotonicity of μ. This shows that E_0 satisfies the Carathéodory condition according to Observation 1.17. Hence $E_0 \in \mathfrak{M}(\mu)$. ∎

Theorem 1.22. *Let μ be an outer measure on a set X. The restriction of μ to the σ-algebra $\mathfrak{M}(\mu)$ is a measure on $\mathfrak{M}(\mu)$ and furthermore $(X, \mathfrak{M}(\mu), \mu)$ is a complete measure space.*

Proof. Since μ is countably subadditive on $\mathfrak{P}(X)$, its restriction on $\mathfrak{M}(\mu)$ is countably subadditive on $\mathfrak{M}(\mu)$. By Lemma 1.18, μ is additive on $\mathfrak{M}(\mu)$. Thus μ is additive and countably subadditive on $\mathfrak{M}(\mu)$ and this implies that μ is countably additive on a σ-algebra $\mathfrak{M}(\mu)$ according to Theorem 1.6. Thus μ is a measure on $\mathfrak{M}(\mu)$. If we write μ for the restriction of μ to $\mathfrak{M}(\mu)$, then we have a measure space $(X, \mathfrak{M}(\mu), \mu)$. According to Lemma 1.21, if $E \in \mathfrak{M}(\mu)$ and $\mu(E) = 0$, then every subset of E is a member of $\mathfrak{M}(\mu)$. Thus $(X, \mathfrak{M}(\mu), \mu)$ is a complete measure space. ∎

Given an outer measure on a set X, the collection $\mathfrak{M}(\mu)$ of all μ-measurable subsets of X is a σ-algebra of subsets of X and if we restrict μ to the σ-algebra $\mathfrak{M}(\mu)$ then we have a measure space $(X, \mathfrak{M}(\mu), \mu)$ according to Theorem 1.22. Now we have always $\mathfrak{M}(\mu) \subset \mathfrak{P}(X)$. We inquire next the possibility of $\mathfrak{M}(\mu) = \mathfrak{P}(X)$.

Theorem 1.23. *Let μ be an outer measure on a set X. Consider the following two conditions:*

(i) *$\mathfrak{M}(\mu) = \mathfrak{P}(X)$, that is, every subset of X is μ-measurable.*

(ii) *μ is additive on $\mathfrak{P}(X)$.*

Then (i) and (ii) are equivalent.
Thus there exist non μ-measurable sets in X if and only if μ is not additive on $\mathfrak{P}(X)$.

Proof. 1. Suppose μ is additive on $\mathfrak{P}(X)$. Let $E \in \mathfrak{P}(X)$. Then for an arbitrary $A \in \mathfrak{P}(X)$, the two sets $A \cap E$ and $A \cap E^c$ are disjoint members of $\mathfrak{P}(X)$ whose union is equal to A so that by the additivity of μ on $\mathfrak{P}(X)$ we have $\mu(A) = \mu(A \cap E) + \mu(A \cap E^c)$. This shows that every $E \in \mathfrak{P}(X)$ satisfies the Carathéodory condition. Thus $E \in \mathfrak{M}(\mu)$ and then $\mathfrak{P}(X) \subset \mathfrak{M}(\mu)$. On the other hand, since $\mathfrak{P}(X)$ is the collection of all subsets of X we have $\mathfrak{M}(\mu) \subset \mathfrak{P}(X)$. Therefore we have $\mathfrak{M}(\mu) = \mathfrak{P}(X)$.

2. Conversely suppose $\mathfrak{P}(X) = \mathfrak{M}(\mu)$. Then since μ is additive on $\mathfrak{M}(\mu)$ by Lemma 1.18, μ is additive on $\mathfrak{P}(X)$. ∎

Below is an example of a measure space in which every subset of the space is measurable.

Theorem 1.24. (Counting-measure Space) *Let X be an arbitrary set. Let ν be a set function on $\mathfrak{P}(X)$ defined by setting $\nu(E)$ to be equal to the number of elements in E for every $E \in \mathfrak{P}(X)$. Then $(X, \mathfrak{P}(X), \nu)$ is a measure space.*

Proof. The set function νsatisfies conditions 1° - 4° in Definition 1.14 and is thus an outer measure on X. Moreover ν is additive on $\mathfrak{P}(X)$. Thus $(X, \mathfrak{P}(X), \nu)$ is a measure space by Theorem 1.23. ∎

[III] σ-finite Measure

Definition 1.25. (σ-finite Measure) *Given a measure space $(X, \mathfrak{A}, \mu)$. We say that the measure μ is σ-finite if there exists a disjoint sequence $(A_n : n \in \mathbb{N}) \subset \mathfrak{A}$ such that $\bigcup_{n \in \mathbb{N}} A_n = X$ and $\mu(A_n) < \infty$ for every $n \in \mathbb{N}$.*
We call $(X, \mathfrak{A}, \mu)$ a σ-finite measure space when μ is a σ-finite measure.

Remark 1.26. (Existence of Non σ-finite Measure) There exist non σ-finite measures. For example, if X is an uncountable set then the counting-measure space $(X, \mathfrak{P}(X), \nu)$ as defined in Theorem 1.24 is not a σ-finite measure space.

Proof. Let X be an uncountable set and consider the counting-measure space $(X, \mathfrak{P}(X), \nu)$. Let us show that $(X, \mathfrak{P}(X), \nu)$ is not a σ-finite measure space. Suppose it were. Then there exists a disjoint sequence $(A_n : n \in \mathbb{N}) \subset \mathfrak{P}(X)$ such that $\bigcup_{n \in \mathbb{N}} A_n = X$ and $\nu(A_n) < \infty$ for every $n \in \mathbb{N}$. Then $\nu(X) = \nu(\bigcup_{n \in \mathbb{N}} A_n) = \sum_{n \in \mathbb{N}} \nu(A_n)$. Since $\nu(A_n)$ is a nonnegative integer for every $n \in \mathbb{N}$, $\nu(X)$ is a countable number. This contradicts the fact that $\nu(X)$ is an uncountable number. ∎

Theorem 1.27. (Sufficient Condition for σ-finite Measure) *Let $(X, \mathfrak{A}, \mu)$ be a measure space. If there exists a collection $\{E_n : n \in \mathbb{N}\} \subset \mathfrak{A}$ such that $\bigcup_{n \in \mathbb{N}} E_n = X$ and $\mu(E_n) < \infty$ for every $n \in \mathbb{N}$, then μ is σ-finite.*

Proof. In order to show that μ is σ-finite, we show that there exists a disjoint sequence $(F_n : n \in \mathbb{N}) \subset \mathfrak{A}$ such that $\bigcup_{n\in\mathbb{N}} F_n = X$ and $\mu(F_n) < \infty$ for every $n \in \mathbb{N}$.

Now by Lemma 1.4 there exists a disjoint sequence $(F_n : n \in \mathbb{N}) \subset \mathfrak{A}$ such that

$$\text{(1)} \qquad \bigcup_{k=1}^{n} E_k = \bigcup_{k=1}^{n} F_k \quad \text{for every } n \in \mathbb{N}$$

and

$$\text{(2)} \qquad \bigcup_{k\in\mathbb{N}} E_k = \bigcup_{k\in\mathbb{N}} F_k.$$

Then we have

$$\text{(3)} \qquad \bigcup_{n\in\mathbb{N}} F_n = \bigcup_{n\in\mathbb{N}} E_n = X.$$

Since $\mu(E_n) < \infty$ for every $n \in \mathbb{N}$, we also have

$$\text{(4)} \qquad \mu\left(\bigcup_{k=1}^{n} F_k\right) = \mu\left(\bigcup_{k=1}^{n} E_k\right) \leq \sum_{k=1}^{n} \mu(E_k) < \infty.$$

This implies that for every $n \in \mathbb{N}$ we have

$$\text{(5)} \qquad \mu(F_n) \leq \mu\left(\bigcup_{k=1}^{n} F_k\right) < \infty.$$

This completes the proof. ∎

Here is an important property of a σ-finite measure space.

Theorem 1.28. *Let $(X, \mathfrak{A}, \mu)$ be a measure space. Let $\mathfrak{E} = \{E_\lambda : \lambda \in \Lambda\}$ be a disjoint collection in $\mathfrak{A}$ such that $\bigcup_{\lambda\in\Lambda} E_\lambda = X$ and $\mu(E_\lambda) > 0$ for every $\lambda \in \Lambda$. If the measure space $(X, \mathfrak{A}, \mu)$ is σ-finite then $\mathfrak{E}$ is a countable collection.*

Proof. Let us show that if $\mathfrak{E}$ is an uncountable collection then μ is not σ-finite. Assume that $\mathfrak{E}$ is an uncountable collection. To show that μ is not σ-finite, we show that for any countable disjoint collection $\{A_k : k \in \mathbb{N}\}$ in $\mathfrak{A}$ such that $\bigcup_{k\in\mathbb{N}} A_k = X$ there exists some $k_0 \in \mathbb{N}$ such that $\mu(A_{k_0}) = \infty$.

Let $F_{\lambda,k} = E_\lambda \cap A_k$ and consider the collection $\mathfrak{F} = \{F_{\lambda,k} : \lambda \in \Lambda, k \in \mathbb{N}\}$. Now for every $\lambda \in \Lambda$, we have $\mu(E_\lambda) > 0$ and $E_\lambda = \bigcup_{k\in\mathbb{N}}(E_\lambda \cap A_k) = \bigcup_{k\in\mathbb{N}} F_{\lambda,k}$. Thus for each $\lambda \in \Lambda$ there exists $k \in \mathbb{N}$ depending on λ such that $\mu(F_{\lambda,k}) > 0$. Since Λ is an uncountable set, there exist uncountably many members $F_{\lambda,k}$ of $\mathfrak{F}$ such that $\mu(F_{\lambda,k}) > 0$. Now since $\{A_k : k \in \mathbb{N}\}$ is a disjoint collection with $\bigcup_{k\in\mathbb{N}} A_k = X$ and $\mathfrak{F} = \{F_{\lambda,k} : \lambda \in \Lambda, k \in \mathbb{N}\}$ is a disjoint collection with $\bigcup_{\lambda\in\Lambda}\bigcup_{k\in\mathbb{N}} F_{\lambda,k} = X$ and for every $k \in \mathbb{N}$ the set A_k is a union of members of $\mathfrak{F}$, there exists $k_0 \in \mathbb{N}$ such that A_{k_0} contains uncountably many members $F_{\lambda,k}$ of $\mathfrak{F}$ with $\mu(F_{\lambda,k}) > 0$. Now $(0, \infty] = \bigcup_{\ell \in \mathbb{N}}[\frac{1}{\ell}, \infty]$. Thus there exists $\ell_0 \in \mathbb{N}$ such that A_{k_0} contains infinitely many members $F_{\lambda,k}$ of $\mathfrak{F}$ with $\mu(F_{\lambda,k}) \geq \frac{1}{\ell_0}$. Then $\mu(A_{k_0}) \geq \frac{1}{\ell_0} + \frac{1}{\ell_0} + \frac{1}{\ell_0} + \cdots = \infty$. ∎

Definition 1.29. **(σ-finite Outer Measure)** *Let μ be an outer measure on a set X and let $\mathfrak{M}(\mu)$ be the σ-algebra of μ-measurable subsets of X. We say that the outer measure μ is σ-finite on $\mathfrak{M}(\mu)$ if there exists a disjoint sequence $(A_n : n \in \mathbb{N}) \subset \mathfrak{M}(\mu)$ such that $\bigcup_{n \in \mathbb{N}} A_n = X$ and $\mu(A_n) < \infty$ for every $n \in \mathbb{N}$.*

Theorem 1.30. **(Sufficient Condition for σ-finite Outer Measure)** *Let μ be an outer measure on a set X. If there exists a collection $\{E_n : n \in \mathbb{N}\} \subset \mathfrak{M}(\mu)$ such that $\bigcup_{n \in \mathbb{N}} E_n = X$ and $\mu(E_n) < \infty$ for every $n \in \mathbb{N}$, then μ is σ-finite on $\mathfrak{M}(\mu)$.*

Proof. This theorem is proved in the same way as Theorem 1.27. ∎

Observation 1.31. Let μ be an outer measure on a set X and consider the measure space $(X, \mathfrak{M}(\mu), \mu)$. If the outer measure μ is σ-finite on $\mathfrak{M}(\mu)$ then the measure μ is σ-finite. This is an immediate consequence of Definition 1.29.

[IV] Truncated Outer Measure

Proposition 1.32. *Let μ be an outer measure on a set X. Let $A \subset X$ and define a set function $\mu|_A$ on $\mathfrak{P}(X)$ by setting*

$$\mu|_A(E) = \mu(E \cap A) \quad \text{for } E \in \mathfrak{P}(X).$$

Then $\mu|_A$ is an outer measure on X.

Proof. Note that for $A, E \in \mathfrak{P}(X)$, we have

$$\mu|_A(E) = \mu(E \cap A) \in [0, \infty].$$

$$\mu|_A(\emptyset) = \mu(\emptyset \cap A) = \mu(\emptyset) = 0.$$

The monotonicity of $\mu|_A$ follows immediately from the definition of $\mu|_A$ and the monotonicity of μ. Indeed if $E_1, E_2 \in \mathfrak{P}(X)$ and $E_1 \subset E_2$, then

$$\mu|_A(E_1) = \mu(E_1 \cap A) \leq \mu(E_2 \cap A) = \mu|_A(E_2).$$

To verify the countable subadditivity of $\mu|_A$, let $(E_n : n \in \mathbb{N})$ be a sequence in $\mathfrak{P}(X)$.

Then we have

$$\mu|_A\left(\bigcup_{n\in\mathbb{N}} E_n\right) = \mu\left(\bigcup_{n\in\mathbb{N}} E_n \cap A\right) = \mu\left(\bigcup_{n\in\mathbb{N}} (E_n \cap A)\right)$$
$$\leq \sum_{n\in\mathbb{N}} \mu(E_n \cap A) = \sum_{n\in\mathbb{N}} \mu|_A(E_n).$$

This completes the proof that $\mu|_A$ is an outer measure on X. ∎

Definition 1.33. (Truncation of Outer Measure) *Let μ be an outer measure on a set X and let A be an arbitrary subset of X. We call the outer measure on X defined by*

$$\mu|_A(E) = \mu(E \cap A) \quad \textit{for } E \in \mathfrak{P}(X)$$

a truncated outer measure. (Note that $\mu|_A$ is an outer measure on X, not just on A.)

Theorem 1.34. *Let μ be an outer measure on a set X. With an arbitrary subset A of X, consider the outer measure $\mu|_A$ on X. Then we have $\mathfrak{M}(\mu) \subset \mathfrak{M}(\mu|_A)$.*

Proof. Recall that $\mathfrak{M}(\mu)$ is the σ-algebra of all subsets E of X satisfying the condition

$$(1) \qquad \mu(B) = \mu(B \cap E) + \mu(B \cap E^c) \quad \text{for every } B \subset X$$

and similarly $\mathfrak{M}(\mu|_A)$ is the σ-algebra of all subsets E of X satisfying the condition

$$\mu|_A(B) = \mu|_A(B \cap E) + \mu|_A(B \cap E^c) \quad \text{for every } B \subset X,$$

that is,

$$(2) \qquad \mu(B \cap A) = \mu(B \cap A \cap E) + \mu(B \cap A \cap E^c) \quad \text{for every } B \subset X.$$

Now if E satisfies (1) then E satisfies (2) since $B \cap A \subset X$. This shows that $\mathfrak{M}(\mu) \subset \mathfrak{M}(\mu|_A)$. ∎

[V] Regular Outer Measures

In a measure space $(X, \mathfrak{A}, \mu)$, if $(E_n : n \in \mathbb{N})$ is an increasing sequence in $\mathfrak{A}$, then $\lim_{n\to\infty} \mu(E_n) = \mu(\lim_{n\to\infty} E_n)$ according to (a) of Theorem 1.9. This is a consequence of the countable additivity of μ on $\mathfrak{A}$. If μ is an outer measure on a set X, then for an increasing sequence $(E_n : n \in \mathbb{N})$ in $\mathfrak{P}(X)$ we may not have $\lim_{n\to\infty} \mu(E_n) = \mu(\lim_{n\to\infty} E_n)$.

Definition 1.35. (Regular Outer Measure) *An outer measure μ on a set X is called a regular outer measure if every subset of X is contained in a μ-measurable set with equal* *outer measure, that is, for every $E \in \mathfrak{P}(X)$, there exists $F \in \mathfrak{M}(\mu)$ such that $F \supset E$ and $\mu(F) = \mu(E)$.*

Theorem 1.36. *Let μ be an outer measure on a set X and let $(E_n : n \in \mathbb{N})$ be an increasing sequence of subsets of X. Then*

$$\lim_{n\to\infty} \mu(E_n) \le \mu\left(\lim_{n\to\infty} E_n\right). \tag{1}$$

If μ is a regular outer measure then

$$\lim_{n\to\infty} \mu(E_n) = \mu\left(\lim_{n\to\infty} E_n\right). \tag{2}$$

Proof. 1. Let $(E_n : n \in \mathbb{N})$ be an increasing sequence of subsets of X. Then we have $\lim_{n\to\infty} E_n = \bigcup_{n\in\mathbb{N}} E_n$ and the existence of $\lim_{n\to\infty} E_n$ implies that we have

$$\lim_{n\to\infty} E_n = \liminf_{n\to\infty} E_n = \limsup_{n\to\infty} E_n. \tag{3}$$

Now by the monotonicity of μ, $(\mu(E_n) : n \in \mathbb{N})$ is an increasing sequence of nonnegative extended real numbers so that $\lim_{n\to\infty} \mu(E_n)$ exists. Also by the monotonicity of μ we have $\mu(E_n) \le \mu(\bigcup_{n\in\mathbb{N}} E_n)$ for every $n \in \mathbb{N}$ so that $\lim_{n\to\infty} \mu(E_n) \le \mu(\bigcup_{n\in\mathbb{N}} E_n) = \mu(\lim_{n\to\infty} E_n)$. This proves (1).

2. Assume that μ is a regular outer measure. Then for every $n \in \mathbb{N}$, there exists $F_n \in \mathfrak{M}(\mu)$ such that $E_n \subset F_n$ and $\mu(E_n) = \mu(F_n)$. Then we have

$$\begin{aligned}
\mu\left(\lim_{n\to\infty} E_n\right) &= \mu\left(\liminf_{n\to\infty} E_n\right) && \text{by (3)} \\
&\le \mu\left(\liminf_{n\to\infty} F_n\right) && \text{by } E_n \subset F_n \text{ and monotonicity of } \mu \\
&\le \liminf_{n\to\infty} \mu(F_n) && \text{by (a) of Theorem 1.11} \\
&= \liminf_{n\to\infty} \mu(E_n) && \text{since } \mu(E_n) = \mu(F_n) \\
&= \lim_{n\to\infty} \mu(E_n) && \text{since } \lim_{n\to\infty} \mu(E_n) \text{ exists.}
\end{aligned} \tag{4}$$

Then (1) and (4) imply (2). ∎

In a measure space $(X, \mathfrak{A}, \mu)$, if $E, F \in \mathfrak{A}$, $E \subset F$, and $\mu(E) = \mu(F) < \infty$, then $\mu(F \setminus E) = \mu(F) - \mu(E) = 0$. This follows from the additivity of μ on $\mathfrak{A}$. The assumption $\mu(E) < \infty$ is to ensure that the difference $\mu(F) - \mu(E)$ is defined. If μ is an outer measure on a set X, and if $E \in \mathfrak{P}(X)$, $F \in \mathfrak{M}(\mu)$, $E \subset F$, and $\mu(E) = \mu(F) < \infty$, do we have $\mu(F \setminus E) = 0$?

Proposition 1.37. *Let μ be a regular and σ-finite outer measure on a set X. Then the following two conditions are equivalent:*

(i) $\mathfrak{M}(\mu) = \mathfrak{P}(X)$.

(ii) $E \in \mathfrak{P}(X), F \in \mathfrak{M}(\mu), E \subset F, \mu(E) = \mu(F) < \infty \Rightarrow \mu(F \setminus E) = 0$.

Proof. 1. To show that (i) implies (ii), assume $\mathfrak{M}(\mu) = \mathfrak{P}(X)$. Suppose E and F satisfy the hypothesis of (ii). Then E and F are members of $\mathfrak{M}(\mu)$. Since μ is a measure on the σ-algebra $\mathfrak{M}(\mu)$, it is additive on $\mathfrak{M}(\mu)$. This implies that $\mu(F) = \mu(F \setminus E) + \mu(E)$. Subtracting $\mu(E) \in \mathbb{R}$ from both sides, we have $\mu(F \setminus E) = \mu(F) - \mu(E) = 0$. This shows that (i) implies (ii).

2. To show that (ii) implies (i), let us assume (ii). Let E be an arbitrary member of $\mathfrak{P}(X)$. Since μ is σ-finite on $\mathfrak{P}(X)$, there exists a sequence $(A_n : n \in \mathbb{N})$ in $\mathfrak{P}(X)$ such that $\bigcup_{n\in\mathbb{N}} A_n = X$ and $\mu(A_n) < \infty$ for every $n \in \mathbb{N}$. Let $E_n = E \cap A_n$ for $n \in \mathbb{N}$. Then $(E_n : n \in \mathbb{N})$ is a sequence in $\mathfrak{P}(X)$ with $\bigcup_{n\in\mathbb{N}} E_n = E$ and $\mu(E_n) \le \mu(A_n) < \infty$ for every $n \in \mathbb{N}$ by the monotonicity of μ on $\mathfrak{P}(X)$. Now since μ is a regular outer measure, there exists $F_n \in \mathfrak{M}(\mu)$ such that $F_n \supset E_n$ and $\mu(F_n) = \mu(E_n)$ for every $n \in \mathbb{N}$. By (ii), we have $\mu(F_n \setminus E_n) = 0$. This implies that $F_n \setminus E_n \in \mathfrak{M}(\mu)$ by Lemma 1.21. Since $E_n = F_n \setminus (F_n \setminus E_n)$ and F_n and $F_n \setminus E_n$ are members of the σ-algebra $\mathfrak{M}(\mu)$, E_n is a member of $\mathfrak{M}(\mu)$ for every $n \in \mathbb{N}$. Then $E = \bigcup_{n\in\mathbb{N}} E_n \in \mathfrak{M}(\mu)$. This shows that every member of $\mathfrak{P}(X)$ is a member of $\mathfrak{M}(\mu)$. Thus (ii) implies (i). ∎

[VI] Extension of a Measure to an Outer Measure

We show here that the measure μ of an arbitrary measure space $(X, \mathfrak{A}, \mu)$ can be extended to an outer measure on X. For this purpose let us observe first that for every $A \in \mathfrak{P}(X)$ there exists $E \in \mathfrak{A}$ such that $A \subset E$. Indeed $X \in \mathfrak{A}$ and $A \subset X$.

Theorem 1.38. *Let $(X, \mathfrak{A}, \mu)$ be a measure space. Define a set function $\widetilde{\mu}$ on $\mathfrak{P}(X)$ by*

(1) $$\widetilde{\mu}(A) := \inf\{\mu(E) : A \subset E \in \mathfrak{A}\} \quad \text{for every } A \in \mathfrak{P}(X).$$

Then we have:

(a) *$\widetilde{\mu} = \mu$ on $\mathfrak{A}$.*

(b) *$\widetilde{\mu}$ is an outer measure on X.*

(c) *For every $A \in \mathfrak{P}(X)$ there exists $E \in \mathfrak{A}$ such that $A \subset E$ and $\widetilde{\mu}(A) = \mu(E)$.*

(d) *$\mathfrak{A} \subset \mathfrak{M}(\widetilde{\mu})$.*

(e) *$\widetilde{\mu}$ is a regular outer measure.*

Proof. 1. Let us prove (a). Let $E \in \mathfrak{A}$. Then by (1) we have

$$\widetilde{\mu}(E) = \inf\{\mu(F) : E \subset F \in \mathfrak{A}\} = \mu(E).$$

This shows that $\widetilde{\mu} = \mu$ on $\mathfrak{A}$.

2. Let us prove (b). It is immediate from (1) that $\widetilde{\mu}(A) \in [0, \infty]$ for every $A \in \mathfrak{P}(X)$. Next, since $\emptyset \subset \emptyset \in \mathfrak{A}$ and $\mu(\emptyset) = 0$, (1) implies that $\widetilde{\mu}(\emptyset) = 0$.

Monotonicity of $\widetilde{\mu}$ is verified as follows. Let $A_1, A_2 \in \mathfrak{P}(X)$ and $A_1 \subset A_2$. Then $A_2 \subset E \in \mathfrak{A}$ implies $A_1 \subset E$ and thus

$$\{E \in \mathfrak{A} : A_1 \subset E\} \supset \{E \in \mathfrak{A} : A_2 \subset E\},$$

and then

$$\inf\{\mu(E) : A_1 \subset E \in \mathfrak{A}\} \leq \inf\{\mu(E) : A_2 \subset E \in \mathfrak{A}\},$$

that is, $\widetilde{\mu}(A_1) \leq \widetilde{\mu}(A_2)$.

It remains to verify the countable subadditivity of $\widetilde{\mu}$. Thus we want to show that for an arbitrary sequence $(A_n : n \in \mathbb{N}) \subset \mathfrak{P}(X)$ we have $\widetilde{\mu}(\bigcup_{n\in\mathbb{N}} A_n) \leq \sum_{n\in\mathbb{N}} \widetilde{\mu}(A_n)$.

Now for every $n \in \mathbb{N}$, (1) implies

$$\widetilde{\mu}(A_n) = \inf\{\mu(E); A_n \subset E \in \mathfrak{A}\}. \tag{2}$$

Thus for an arbitrary $\varepsilon > 0$ there exists $E_n \in \mathfrak{A}$ such that $A_n \subset E_n$ and

$$\widetilde{\mu}(A_n) \geq \mu(E_n) - \frac{\varepsilon}{2^n}. \tag{3}$$

Then we have $\bigcup_{n\in\mathbb{N}} A_n \subset \bigcup_{n\in\mathbb{N}} E_n$ and moreover

$$\begin{aligned} \widetilde{\mu}\left(\bigcup_{n\in\mathbb{N}} A_n\right) &\leq \widetilde{\mu}\left(\bigcup_{n\in\mathbb{N}} E_n\right) && \text{by the monotonicity of } \widetilde{\mu} \\ &= \mu\left(\bigcup_{n\in\mathbb{N}} E_n\right) && \text{since } \widetilde{\mu} = \mu \text{ on } \mathfrak{A} \\ &\leq \sum_{n\in\mathbb{N}} \mu(E_n) && \text{by the countable subadditivity of the measure } \mu \\ &\leq \sum_{n\in\mathbb{N}} \left\{\widetilde{\mu}(A_n) + \frac{\varepsilon}{2^n}\right\} && \text{by (3)} \\ &= \sum_{n\in\mathbb{N}} \widetilde{\mu}(A_n) + \varepsilon. \end{aligned} \tag{4}$$

Since (4) holds for an arbitrary $\varepsilon > 0$, we have

$$\widetilde{\mu}\left(\bigcup_{n\in\mathbb{N}} A_n\right) \leq \sum_{n\in\mathbb{N}} \widetilde{\mu}(A_n).$$

This proves the countable subadditivity of $\widetilde{\mu}$ and completes the proof that $\widetilde{\mu}$ is an outer measure on X.

3. Let us prove (c). Let $A \in \mathfrak{P}(X)$. By (1) for every $k \in \mathbb{N}$ there exists $E_k \in \mathfrak{A}$ such that $A \subset E_k$ and $\mu(E_k) \leq \widetilde{\mu}(A) + \frac{1}{k}$. Let $E = \bigcap_{k \in \mathbb{N}} E_k$. Then we have $E \in \mathfrak{A}$ and $A \subset E$. By (1) we have

$$\widetilde{\mu}(A) \leq \mu(E) \leq \mu(E_k) \leq \widetilde{\mu}(A) + \frac{1}{k}.$$

Letting $k \to \infty$ in the last inequality, we have $\widetilde{\mu}(A) \leq \mu(E) \leq \widetilde{\mu}(A)$ and thus $\widetilde{\mu}(A) \leq \mu(E)$.

4. Let us prove (d), that is, $\mathfrak{A} \subset \mathfrak{M}(\widetilde{\mu})$, that is, every $E \in \mathfrak{A}$ is $\widetilde{\mu}$-measurable, that is,

$$\widetilde{\mu}(B) = \widetilde{\mu}(B \cap E) + \widetilde{\mu}(B \cap E^c) \quad \text{for every } B \in \mathfrak{P}(X). \tag{5}$$

Let us prove (5). Let $B \in \mathfrak{P}(X)$. By the subadditivity of the outer measure $\widetilde{\mu}$ we have

$$\widetilde{\mu}(B) \leq \widetilde{\mu}(B \cap E) + \widetilde{\mu}(B \cap E^c).$$

It remains to prove the reverse inequality. Now by the definition of $\widetilde{\mu}$ by (1), for every $\varepsilon > 0$ there exists $F \in \mathfrak{A}$ such that $B \subset F$ and $\mu(F) \leq \widetilde{\mu}(B) + \varepsilon$. Now $E, F \in \mathfrak{A}$ and μ is a measure on the σ-algebra $\mathfrak{A}$ so that by the additivity of μ on $\mathfrak{A}$ we have

$$\mu(F) = \mu(F \cap E) + \mu(F \cap E^c).$$

Thus we have

$$\mu(F \cap E) + \mu(F \cap E^c) = \mu(F) \leq \widetilde{\mu}(B) + \varepsilon. \tag{6}$$

Since $B \subset F$, by the monotonicity of the outer measure $\widetilde{\mu}$ and then by the fact that $\widetilde{\mu} = \mu$ on $\mathfrak{A}$ we have

$$\widetilde{\mu}(B \cap E) \leq \widetilde{\mu}(F \cap E) = \mu(F \cap E),$$

$$\widetilde{\mu}(B \cap E^c) \leq \widetilde{\mu}(F \cap E^c) = \mu(F \cap E^c).$$

Adding these two inequalities side by side and then applying (6), we have

$$\widetilde{\mu}(B \cap E) + \widetilde{\mu}(B \cap E^c) \leq \mu(F \cap E) + \mu(F \cap E^c) \leq \widetilde{\mu}(B) + \varepsilon.$$

Since this holds for every $\varepsilon > 0$, we have

$$\widetilde{\mu}(B \cap E) + \widetilde{\mu}(B \cap E^c) \leq \widetilde{\mu}(B).$$

This completes the proof of (5).

5. Let us show that $\widetilde{\mu}$ is a regular outer measure, that is, for every $A \in \mathfrak{P}(X)$ there exists $E \in \mathfrak{M}(\widetilde{\mu})$ such that $A \subset E$ and $\widetilde{\mu}(A) = \widetilde{\mu}(E)$. But this follows immediately from (a), (c) and (d). ∎

Theorem 1.39. *Given a measure space $(X, \mathfrak{A}, \mu)$. The measure μ on the σ-algebra $\mathfrak{A}$ of subsets of X is always equal to the restriction to $\mathfrak{A}$ of some outer measure on X. (Thus every measure is a restriction of some outer measure.)*

Proof. According to Theorem 1.38, the set function $\widetilde{\mu}$ on $\mathfrak{P}(X)$ defined by setting $\widetilde{\mu}(A) := \inf\{\mu(E) : A \subset E \in \mathfrak{A}\}$ for every $A \in \mathfrak{P}(X)$ is an outer measure on X and furthermore $\mathfrak{A} \subset \mathfrak{M}(\widetilde{\mu})$ and $\widetilde{\mu} = \mu$ on $\mathfrak{A}$. ∎

Remark 1.40. Given a measure space $(X, \mathfrak{A}, \mu)$. In Theorem 1.38, we obtained an outer measure $\widetilde{\mu}$ on X. By setting

$$\widetilde{\mu}(A) := \inf\{\mu(E) : A \subset E \in \mathfrak{A}\} \quad \text{for every } A \in \mathfrak{P}(X).$$

Let us define a set function ν on $\mathfrak{P}(X)$ by setting

$$\nu(A) := \inf\left\{\sum_{n\in\mathbb{N}} \mu(E_n) : (E_n : n \in \mathbb{N}) \subset \mathfrak{A}, A \subset \bigcup_{n\in\mathbb{N}} E_n\right\}.$$

Then we have $\widetilde{\mu} = \nu$. (This is relevant in construction of outer measures as we show below.)

Proof. Let us show that $\widetilde{\mu} = \nu$.

1. Since ν is the infimum of a greater collection than the collection for $\widetilde{\mu}$, we have $\nu \leq \widetilde{\mu}$.

2. Let us show $\widetilde{\mu} \leq \nu$. Let $(E_n : n \in \mathbb{N})$ be a sequence in $\mathfrak{A}$ such that $A \subset \bigcup_{n\in\mathbb{N}} E_n$. Let $E = \bigcup_{n\in\mathbb{N}} E_n$. Then $E \in \mathfrak{A}$, $A \subset E$ and moreover by the countable subadditivity of the measure μ we have $\mu(E) \leq \sum_{n\in\mathbb{N}} \mu(E_n)$. Form this it follows that $\widetilde{\mu} \leq \nu$. ∎

Theorem 1.41. (Regularization of Outer Measure) *Let μ be an outer measure on a set X. There exists a regular outer measure $\widetilde{\mu}$ on X such that $\widetilde{\mu} = \mu$ on $\mathfrak{M}(\mu)$.*

Proof. Let μ be an outer measure on a set X. By Theorem 1.22, $(X, \mathfrak{M}(\mu), \mu)$ is a measure space. Define a set function $\widetilde{\mu}$ on $\mathfrak{P}(X)$ by setting

$$\widetilde{\mu}(A) := \inf\{\mu(E) : A \subset E \in \mathfrak{M}(\mu)\} \quad \text{for every } A \in \mathfrak{P}(X).$$

Then according to Theorem 1.38, $\widetilde{\mu}$ is a regular outer measure on X and furthermore $\widetilde{\mu} = \mu$ on $\mathfrak{M}(\mu)$. ∎

[VII] Completion of Measure Space

Theorem 1.42. (Completion of Measure Space) *Let $(X, \mathfrak{A}, \mu)$ be a measure space. There exists a complete measure space $(X, \mathfrak{F}, \widetilde{\mu})$ such that $\mathfrak{F} \supset \mathfrak{A}$ and $\widetilde{\mu} = \mu$ on $\mathfrak{A}$.*

Proof. Let us define a set function $\widetilde{\mu)}$ on $\mathfrak{P}(X)$ by setting

$$\widetilde{\mu}(A) := \inf\{\mu(E) : A \subset E \in \mathfrak{A}\} \quad \text{for every } A \in \mathfrak{P}(X).$$

Then according to Theorem 1.38, we have:

(a) $\widetilde{\mu} = \mu$ on $\mathfrak{A}$.

(b) $\widetilde{\mu}$ is an outer measure on X.

(c) $\mathfrak{A} \subset \mathfrak{M}(\widetilde{\mu})$.
Then by Theorem 1.22, $(X, \mathfrak{M}(\widetilde{\mu}), \widetilde{\mu})$ is a complete measure space. Let $\mathfrak{F} = \mathfrak{M}(\widetilde{\mu})$. ∎

[VIII] Support of Outer Measures

Definition 1.43. *Let μ be an outer measure on a topological space X. Let $\mathfrak{V}$ be the collection of all non-empty open sets V in X such that $\mu(V) = 0$. The support of μ, denoted by* $\operatorname{supp} \mu$, *is defined by*

$$\operatorname{supp} \mu = \Big(\bigcup_{V \in \mathfrak{V}} V \Big)^c .$$

Proposition 1.44. *Let μ be an outer measure on a topological space X. Then* $\operatorname{supp} \mu = X$ *if and only if for every non-empty open set V in X we have $\mu(V) > 0$.*

Proof. 1. Suppose for every non-empty open set V in X we have $\mu(V) > 0$. Then the collection $\mathfrak{V}$ of all non-empty open sets V in X such that $\mu(V) = 0$ is an empty collection and thus $\bigcup_{V \in \mathfrak{V}} V = \emptyset$ and then $\operatorname{supp} \mu = (\bigcup_{V \in \mathfrak{V}} V)^c = \emptyset^c = X$.

2. Suppose there exists a non-empty open set V_0 in X such that $\mu(V_0) = 0$. Let $\mathfrak{V}$ be the collection of all non-empty open sets V in X such that $\mu(V) = 0$. Then $V_0 \in \mathfrak{V}$ and hence $\bigcup_{V \in \mathfrak{V}} V \supset V_0$ and $\operatorname{supp} \mu = (\bigcup_{V \in \mathfrak{V}} V)^c \subset V_0^c$. Since $V_0 \neq \emptyset$ we have $V_0^c \neq X$ and hence $\operatorname{supp} \mu$ is contained in a proper subset V_0^c of X so that $\operatorname{supp} \mu \neq X$. ∎

Theorem 1.45. *Let μ be an outer measure on a metric space (X, ρ). For $x \in X$ and $r \in (0, r)$, let $B_r(x) = \{y \in X : \rho(x, y) < r\}$ and $C_r(x) = \{y \in X : \rho(x, y) \leq r\}$.*
(a) $\operatorname{supp} \mu = X$ *if and only if $\mu(B_r(x)) > 0$ for every $x \in X$ and $r \in (0, \infty)$.*
(b) $\operatorname{supp} \mu = X$ *if and only if $\mu(C_r(x)) > 0$ for every $x \in X$ and $r \in (0, \infty)$.*

Proof. 1. Let us prove (a). If $\operatorname{supp} \mu = X$, then by Proposition 1.44 we have $\mu(V) > 0$ for every non-empty open set V and hence $\mu(B_r(x)) > 0$ for every $x \in X$ and $r \in (0, \infty)$. Conversely suppose $\mu(B_r(x)) > 0$ for every $x \in X$ and $r \in (0, \infty)$. Then since every non-empty open set V contains some $B_r(x)$ we have $\mu(V) \geq \mu(B_r(x)) > 0$ by the monotonicity of the outer measure μ. Thus $\operatorname{supp} \mu = X$ by Proposition 1.44.

2. To probe (b), let us observe that $\mu(B_r(x)) > 0$ for every $x \in X$ and $r \in (0, \infty)$ if and only if $\mu(C_r(x)) > 0$ for every $x \in X$ and $r \in (0, \infty)$. (Indeed if $\mu(B_r(x)) > 0$ for every $x \in X$ and $r \in (0, \infty)$ then $\mu(C_r(x)) \geq \mu(B_r(x)) > 0$, and if $\mu(C_r(x)) > 0$ for every $x \in X$ and $r \in (0, \infty)$ then $\mu(B_r(x)) \geq \mu(C_{r/2}(x)) > 0$.) Then (b) follows from (a). ∎

§2 Construction of Outer Measures

[I] Construction of Outer Measures on an Arbitrary Set

Definition 2.1. (Covering Class) *A collection $\mathfrak{V}$ of subsets of a set X is called a covering class for X if it satisfies the following conditions:*

1° *there exists $(V_n : n \in \mathbb{N}) \subset \mathfrak{V}$ such that $\bigcup_{n\in\mathbb{N}} V_n = X$,*

2° $\emptyset \in \mathfrak{V}$.

For $E \in \mathfrak{P}(X)$, a sequence $(V_n : n \in \mathbb{N}) \subset \mathfrak{V}$ such that $\bigcup_{n\in\mathbb{N}} V_n \supset E$ is called a covering sequence for E.

Observation 2.2. (Existence of Covering Class) For an arbitrary set X a covering class $\mathfrak{V}$ always exists. Indeed if we decompose X into countably many subsets, say $\{V_n : n \in \mathbb{N}\}$, then since $\bigcup_{n\in\mathbb{N}} V_n = X$, $\mathfrak{V} = \{\emptyset, V_n : n \in \mathbb{N}\}$ is a covering class for X. Moreover, since $\mathfrak{V} \subset \mathfrak{P}(X)$, $\mathfrak{P}(X)$ is a covering class for X.

Definition 2.3. (Premeasure) *Let X be an arbitrary set and let $\mathfrak{V}$ be a covering class for X. A set function γ on $\mathfrak{V}$ is called a premeasure on $\mathfrak{V}$ if it satisfies the following conditions:*

1° *nonnegative extended real-valued: $\gamma(V) \in [0, \infty]$ for every $V \in \mathfrak{V}$.*

2° $\gamma(\emptyset) = 0$.

The empty set $\emptyset \in \mathfrak{P}(X)$ does not cover any point $x \in X$. Then why do we insist that a covering class $\mathfrak{V}$ for X must include $\emptyset$ as a member? The reason is as follows. For every measure μ we have $\mu(\emptyset) = 0$. Thus for every premeasure γ we must have $\gamma(\emptyset) = 0$. Then since a premeasure is to be defined on a covering class $\mathfrak{V}$ for X, $\mathfrak{V}$ must include $\emptyset$ as a member.

With a premeasure γ on a covering class $\mathfrak{V}$ for a set X, we can construct an outer measure μ on X as follows.

Theorem 2.4. *Let γ be a premeasure on a covering class $\mathfrak{V}$ for a set X. Let us define a set function $\mu_{[\gamma,\mathfrak{V}]}$ on $\mathfrak{P}(X)$ by setting for every $E \in \mathfrak{P}(X)$*

$$\mu_{[\gamma,\mathfrak{V}]}(E) := \inf\left\{\sum_{n\in\mathbb{N}} \gamma(V_n) : (V_n : n \in \mathbb{N}) \subset \mathfrak{V}, \bigcup_{n\in\mathbb{N}} V_n \supset E\right\}.$$

Then $\mu_{[\gamma,\mathfrak{V}]}$ is an outer measure on X.

Proof. Let us show that $\mu_{[\gamma,\mathfrak{V}]}$ satisfies conditions 1°, 2°, 3°, and 4° of Definition 1.14. For brevity in notations, let us abbreviate $\mu_{[\gamma,\mathfrak{V}]}$ as μ for the rest of this proof.

Clearly $\mu(E) \in [0, \infty]$ for every $E \in \mathfrak{P}(X)$. Since $\emptyset \in \mathfrak{V}$, $(\emptyset)$ is a one term covering sequence in $\mathfrak{V}$ for $\emptyset$ and thus $\mu(\emptyset) \leq \gamma(\emptyset) = 0$.

To show the monotonicity of μ on $\mathfrak{P}(X)$, let $E_1, E_2 \in \mathfrak{P}(X)$ and $E_1 \subset E_2$. Then every covering sequence in $\mathfrak{V}$ for E_2 is also a covering sequence for E_1. This implies that the collection of extended nonnegative numbers on which we take the infimum to obtain $\mu(E_1)$ is greater than the collection on which we take infimum to obtain $\mu(E_2)$. Now if A and B are two collections of extended real numbers and $A \supset B$ then $\inf A \leq \inf B$. Thus $\mu(E_1) \leq \mu(E_2)$. This proves the monotonicity of μ.

To show the countable subadditivity of μ on $\mathfrak{P}(X)$, let $(E_n : n \in \mathbb{N}) \subset \mathfrak{P}(X)$. Let $\varepsilon > 0$. For each $n \in \mathbb{N}$, by the definition of $\mu(E_n)$ as an infimum there exists a sequence $(V_{n,k} : k \in \mathbb{N}) \subset \mathfrak{V}$ such that $\bigcup_{k \in \mathbb{N}} V_{n,k} \supset E_n$ and $\sum_{k \in \mathbb{N}} \gamma(V_{n,k}) \leq \mu(E_n) + \frac{\varepsilon}{2^n}$. Then $\bigcup_{n \in \mathbb{N}} \left(\bigcup_{k \in \mathbb{N}} V_{n,k}\right) \supset \bigcup_{n \in \mathbb{N}} E_n$. This implies

$$\mu\left(\bigcup_{n \in \mathbb{N}} E_n\right) \leq \sum_{n \in \mathbb{N}} \left[\sum_{k \in \mathbb{N}} \gamma(V_{n,k})\right] \leq \sum_{n \in \mathbb{N}} \left\{\mu(E_n) + \frac{\varepsilon}{2^n}\right\} = \sum_{n \in \mathbb{N}} \mu(E_n) + \varepsilon.$$

By the arbitrariness of $\varepsilon > 0$, we have $\mu\left(\bigcup_{n \in \mathbb{N}} E_n\right) \leq \sum_{n \in \mathbb{N}} \mu(E_n)$. This proves the countable subadditivity of μ. This completes the proof that $\mu_{[\gamma,\mathfrak{V}]}$ is an outer measure on X. ∎

Definition 2.5. (Outer Measure Generated by a Premeasure on a Covering Class) *Let X be an arbitrary set and let $\mathfrak{V}$ be a covering class for X. Let γ be a premeasure on the covering class $\mathfrak{V}$. We call the outer measure $\mu_{[\gamma,\mathfrak{V}]}$ on X defined by setting for every $E \in \mathfrak{P}(X)$*

$$\mu_{[\gamma,\mathfrak{V}]}(E) := \inf\left\{\sum_{n \in \mathbb{N}} \gamma(V_n) : (V_n : n \in \mathbb{N}) \subset \mathfrak{V}, \bigcup_{n \in \mathbb{N}} V_n \supset E\right\}$$

the outer measure on X generated by the premeasure γ on the covering class $\mathfrak{V}$.

For an outer measure $\mu_{[\gamma,\mathfrak{V}]}$, the scope of testing set in the Carathéodory condition for $\mu_{[\gamma,\mathfrak{V}]}$-measurability is considerably reduced. Indeed we need only use the members of the covering class $\mathfrak{V}$ as testing sets. We show this below.

Proposition 2.6. *Consider an outer measure $\mu_{[\gamma,\mathfrak{V}]}$ on a set X generated by a premeasure γ on a covering class $\mathfrak{V}$ for X. We have*

$$\mu_{[\gamma,\mathfrak{V}]}(V) \leq \gamma(V) \quad \textit{for every } V \in \mathfrak{V}.$$

Proof. Let $V^* \in \mathfrak{V}$. Then by Definition 2.5 we have

$$\mu_{[\gamma,\mathfrak{V}]}(V^*) = \inf\left\{\sum_{n \in \mathbb{N}} \gamma(V_n) : (V_n : n \in \mathbb{N}) \subset \mathfrak{V}, \bigcup_{n \in \mathbb{N}} V_n \supset V^*\right\}. \tag{1}$$

Now $(V^*, \emptyset, \emptyset, \emptyset, \ldots)$ is a covering sequence for V^*. Thus we have

$$\inf\left\{\sum_{n\in\mathbb{N}}\gamma(V_n) : (V_n : n \in \mathbb{N}) \subset \mathfrak{V}, \bigcup_{n\in\mathbb{N}} V_n \supset V^*\right\} \leq \gamma(V^*).$$

Thus we have $\mu_{[\gamma,\mathfrak{V}]}(V^*) \leq \gamma(V^*)$. ∎

Theorem 2.7. **($\mu_{[\gamma,\mathfrak{V}]}$-measurability)** *Let $\mu_{[\gamma,\mathfrak{V}]}$ be an outer measure on a set X generated by a premeasure γ on a covering class $\mathfrak{V}$ for X, that is, for every $E \in \mathfrak{P}(X)$ we have*

$$\mu_{[\gamma,\mathfrak{V}]}(E) = \inf\left\{\sum_{n\in\mathbb{N}}\gamma(V_n) : (V_n : n \in \mathbb{N}) \subset \mathfrak{V}, \bigcup_{n\in\mathbb{N}} V_n \supset E\right\}.$$

Then the following two conditions, the first of which is the Carathéodory condition for the $\mu_{[\gamma,\mathfrak{V}]}$-measurability of E, are equivalent:

1° $\mu_{[\gamma,\mathfrak{V}]}(A) = \mu_{[\gamma,\mathfrak{V}]}(A \cap E) + \mu_{[\gamma,\mathfrak{V}]}(A \cap E^c)$ *for every $A \in \mathfrak{P}(X)$*
and
2° $\mu_{[\gamma,\mathfrak{V}]}(V) = \mu_{[\gamma,\mathfrak{V}]}(V \cap E) + \mu_{[\gamma,\mathfrak{V}]}(V \cap E^c)$ *for every $V \in \mathfrak{V}$.*

Proof. Since 1° implies 2°, it remains to show that 2° implies 1°. Let us assume 2°. Let $E \in \mathfrak{P}(X)$. For an arbitrary $A \in \mathfrak{P}(X)$, let $(V_n : n \in \mathbb{N})$ be an arbitrary sequence in $\mathfrak{V}$ such that $\bigcup_{n\in\mathbb{N}} V_n \supset A$. By 2° we have for every $n \in \mathbb{N}$

$$\mu_{[\gamma,\mathfrak{V}]}(V_n) = \mu_{[\gamma,\mathfrak{V}]}(V_n \cap E) + \mu_{[\gamma,\mathfrak{V}]}(V_n \cap E^c). \tag{1}$$

Summing over $n \in \mathbb{N}$, we have

$$\begin{aligned}\sum_{n\in\mathbb{N}}\mu_{[\gamma,\mathfrak{V}]}(V_n) &= \sum_{n\in\mathbb{N}}\mu_{[\gamma,\mathfrak{V}]}(V_n \cap E) + \sum_{n\in\mathbb{N}}\mu_{[\gamma,\mathfrak{V}]}(V_n \cap E^c)\\ &\geq \mu_{[\gamma,\mathfrak{V}]}\left(\bigcup_{n\in\mathbb{N}}(V_n \cap E)\right) + \mu_{[\gamma,\mathfrak{V}]}\left(\bigcup_{n\in\mathbb{N}}(V_n \cap E^c)\right)\\ &= \mu_{[\gamma,\mathfrak{V}]}\left(\left(\bigcup_{n\in\mathbb{N}}V_n\right) \cap E\right) + \mu_{[\gamma,\mathfrak{V}]}\left(\left(\bigcup_{n\in\mathbb{N}}V_n\right) \cap E^c\right)\\ &\geq \mu_{[\gamma,\mathfrak{V}]}(A \cap E) + \mu_{[\gamma,\mathfrak{V}]}(A \cap E^c),\end{aligned} \tag{2}$$

where the first inequality is by the countable subadditivity of the outer measure $\mu_{[\gamma,\mathfrak{V}]}$ and the second inequality is by the monotonicity of $\mu_{[\gamma,\mathfrak{V}]}$. According to Proposition 2.6, we have $\mu_{[\gamma,\mathfrak{V}]}(V_n) \leq \gamma(V_n)$ for every $n \in \mathbb{N}$. This implies that

$$\sum_{n\in\mathbb{N}}\gamma(V_n) \geq \sum_{n\in\mathbb{N}}\mu_{[\gamma,\mathfrak{V}]}(V_n) \geq \mu_{[\gamma,\mathfrak{V}]}(A \cap E) + \mu_{[\gamma,\mathfrak{V}]}(A \cap E^c). \tag{3}$$

This shows that $\mu_{[\gamma,\mathfrak{V}]}(A \cap E) + \mu_{[\gamma,\mathfrak{V}]}(A \cap E^c)$ is a lower bound for the collection of nonnegative extended real numbers $\left\{\sum_{n\in\mathbb{N}}\gamma(V_n) : (V_n : n \in \mathbb{N}) \subset \mathfrak{V}, \bigcup_{n\in\mathbb{N}} V_n \supset A\right\}$. By the definition of $\mu_{[\gamma,\mathfrak{V}]}(A)$, $\mu_{[\gamma,\mathfrak{V}]}(A)$ is the infimum, that is, the greatest lower bound, of this collection. Thus we have

$$\mu_{[\gamma,\mathfrak{V}]}(A) \geq \mu_{[\gamma,\mathfrak{V}]}(A \cap E) + \mu_{[\gamma,\mathfrak{V}]}(A \cap E^c). \tag{4}$$

On the other hand, the subadditivity of the outer measure $\mu_{[\gamma,\mathfrak{V}]}$ implies

$$\mu_{[\gamma,\mathfrak{V}]}(A) \leq \mu_{[\gamma,\mathfrak{V}]}(A \cap E) + \mu_{[\gamma,\mathfrak{V}]}(A \cap E^c). \tag{5}$$

With (4) and (5), we have 1°. This completes the proof that 2° implies 1°. ■

Next we compare two outer measures $\mu_{[\gamma,\mathfrak{V}]}$ and $\mu_{[\gamma,\mathfrak{W}]}$ on a set X by comparing the covering classes $\mathfrak{V}$ and $\mathfrak{W}$.

Proposition 2.8. *Let $\mathfrak{V}$ and $\mathfrak{W}$ be two covering classes for a set X such that $\mathfrak{V} \subset \mathfrak{W}$. Let γ be a premeasure on the covering class $\mathfrak{W}$. Then for the two outer measures $\mu_{[\gamma,\mathfrak{V}]}$ and $\mu_{[\gamma,\mathfrak{W}]}$ we have*

$$\mu_{[\gamma,\mathfrak{W}]}(E) \leq \mu_{[\gamma,\mathfrak{V}]}(E) \quad \textit{for every } E \in \mathfrak{P}(X).$$

Proof. Let $E \in \mathfrak{P}(X)$. By Definition 2.5, we have

$$\mu_{[\gamma,\mathfrak{V}]}(E) := \inf\left\{\sum_{n\in\mathbb{N}} \gamma(V_n) : (V_n : n \in \mathbb{N}) \subset \mathfrak{V}, \bigcup_{n\in\mathbb{N}} V_n \supset E\right\} \tag{1}$$

and

$$\mu_{[\gamma,\mathfrak{W}]}(E) := \inf\left\{\sum_{n\in\mathbb{N}} \gamma(V_n) : (V_n : n \in \mathbb{N}) \subset \mathfrak{W}, \bigcup_{n\in\mathbb{N}} V_n \supset E\right\}. \tag{2}$$

Then since $\mathfrak{W} \supset \mathfrak{V}$, the collection of nonnegative extended real numbers on which we take the infimum in (1) is contained in the collection of nonnegative extended real numbers in (2). This implies that the infimum in (1) is greater than the infimum in (2). Thus $\mu_{[\gamma,\mathfrak{W}]} \leq \mu_{[\gamma,\mathfrak{V}]}$. ■

We show next that a monotone system of outer measures generates an outer measure.

Theorem 2.9. *Consider a collection of outer measures on a set X, $\{\mu_\xi : \xi \in (\alpha, \beta)\}$, where $\alpha \in [-\infty, \infty)$ and $\beta \in (-\infty, \infty]$.*
(a) *Suppose $\mu_\xi(E) \uparrow$ as $\xi \downarrow \alpha$ for every $E \in \mathfrak{P}(X)$. Let $\mu(E) = \lim_{\xi\downarrow\alpha} \mu_\xi(E)$ for $E \in \mathfrak{P}(X)$. Then μ is an outer measure on X.*
(b) *Suppose $\mu_\xi(E) \uparrow$ as $\xi \uparrow \beta$ for every $E \in \mathfrak{P}(X)$. Let $\mu(E) = \lim_{\xi\uparrow\beta} \mu_\xi(E)$ for $E \in \mathfrak{P}(X)$. Then μ is an outer measure on X.*

Proof. 1. Let us prove (a). To show that μ is an outer measure on X, we show that the set function μ on $\mathfrak{P}(X)$ is nonnegative extended real-valued, $\mu(\emptyset) = 0$, μ is monotone on $\mathfrak{P}(X)$, and countably subadditive on $\mathfrak{P}(X)$.

Now since $\mu_\xi(E) \geq 0$ and $\mu_\xi(E) \uparrow$ as $\xi \downarrow \alpha$, we have

$$\mu(E) = \lim_{\xi \downarrow \alpha} \mu_\xi(E) \in [0, \infty].$$

This shows that μ is nonnegative extended real-valued on $\mathfrak{P}(X)$.

Since $\mu_\xi(\emptyset) = 0$ for every $\xi \in (\alpha, \beta)$, we have

$$\mu(\emptyset) = \lim_{\xi \downarrow \alpha} \mu_\xi(\emptyset) = \lim_{\xi \downarrow \alpha} 0 = 0.$$

To show the monotonicity of μ on $\mathfrak{P}(X)$, let $E, F \in \mathfrak{P}(X)$ and $E \subset F$. By the monotonicity of μ_ξ on $\mathfrak{P}(X)$ for every $\xi \in (\alpha, \beta)$ we have $\mu_\xi(E) \leq \mu_\xi(F)$. Then we have

$$\mu(E) = \lim_{\xi \downarrow \alpha} \mu_\xi(E) \leq \lim_{\xi \downarrow \alpha} \mu_\xi(F) = \mu(F).$$

This proves the monotonicity of μ on $\mathfrak{P}(X)$.

To show that μ is countably subadditive on $\mathfrak{P}(X)$, let $(E_n : n \in \mathbb{N}) \subset \mathfrak{P}(X)$. Now since μ_ξ is an outer measure on X, μ_ξ is countably subadditive on $\mathfrak{P}(X)$ and thus

$$\mu_\xi\left(\bigcup_{n \in \mathbb{N}} E_n\right) \leq \sum_{n \in \mathbb{N}} \mu_\xi(E_n). \tag{1}$$

By the definition of μ we have $\mu_\xi(E_n) \uparrow \mu(E_n)$ as $\xi \downarrow \alpha$ so that $\mu_\xi(E_n) \leq \mu(E_n)$ for every $\xi \in (\alpha, \beta)$ and every $n \in \mathbb{N}$. Thus we have for every $\xi \in (\alpha, \beta)$

$$\sum_{n \in \mathbb{N}} \mu_\xi(E_n) \leq \sum_{n \in \mathbb{N}} \mu(E_n). \tag{2}$$

By (1) and (2), we have for every $\xi \in (\alpha, \beta)$

$$\mu_\xi\left(\bigcup_{n \in \mathbb{N}} E_n\right) \leq \sum_{n \in \mathbb{N}} \mu(E_n). \tag{3}$$

Then by (3) we have

$$\mu\left(\bigcup_{n \in \mathbb{N}} E_n\right) = \lim_{\xi \downarrow \alpha} \mu_\xi\left(\bigcup_{n \in \mathbb{N}} E_n\right) \leq \sum_{n \in \mathbb{N}} \mu(E_n).$$

This shows the countable subadditivity of μ and completes the proof that μ is an outer measure on X.

2. (b) is proved by the same argument as above. ■

We show next that for every outer measure ν on a set X there exists an outer measure $\mu_{[\gamma,\mathfrak{V}]}$ on X such that $\nu = \mu_{[\gamma,\mathfrak{V}]}$. Thus an arbitrary outer measure on X can be represented as $\mu_{[\gamma,\mathfrak{V}]}$ with some premeasure γ on a covering class $\mathfrak{V}$ for X.

Theorem 2.10. (Characterization of Outer Measures) *Let ν be an outer measure on a set X. Then there exists a premeasure γ on a covering class $\mathfrak{V}$ for X such that $\nu = \mu_{[\gamma,\mathfrak{V}]}$.*

Proof. An outer measure ν on X is certainly a premeasure on a covering class $\mathfrak{V}$ for X. Note that by Observation 2.2, $\mathfrak{P}(X)$ is a covering class for X. Consider the outer measure $\mu_{[\nu,\mathfrak{P}(X)]}$ on X generated by the premeasure ν on the covering class $\mathfrak{P}(X)$. Then for every $E \in \mathfrak{P}(X)$ we have by Definition 2.5

$$(1) \qquad \mu_{[\nu,\mathfrak{P}(X)]}(E) := \inf\left\{\sum_{n\in\mathbb{N}} \nu(V_n) : (V_n : n \in \mathbb{N}) \subset \mathfrak{P}(X),\ \bigcup_{n\in\mathbb{N}} V_n \supset E\right\}.$$

Selecting a sequence in the covering class $\mathfrak{P}(X)$, $(V_n : n \in \mathbb{N}) := (E, \emptyset, \emptyset, \emptyset, \ldots)$, we have $\bigcup_{n\in\mathbb{N}} V_n = E$ and $\sum_{n\in\mathbb{N}} \nu(V_n) = \nu(E)$ and then

$$(2) \qquad \inf\left\{\sum_{n\in\mathbb{N}} \nu(V_n) : (V_n : n \in \mathbb{N}) \subset \mathfrak{P}(X),\ \bigcup_{n\in\mathbb{N}} V_n \supset E\right\} = \nu(E).$$

Then (1) and (2) imply that $\mu_{[\nu,\mathfrak{P}(X)]}(E) = \nu(E)$. This proves that $\nu = \mu_{[\gamma,\mathfrak{V}]}$. ∎

Observation 2.11. An outer measure $\mu_{[\gamma,\mathfrak{V}]}$ as constructed in Theorem 2.4 may or may not be a regular outer measure. We show next that a regular outer measure of the type $\mu_{[\gamma,\mathfrak{V}]}$ exists. This allows us to say "a regular outer measure $\mu_{[\gamma,\mathfrak{V}]}$" from the outset.

Theorem 2.12. (Existence and Characterization of Regular Outer Measure)
(a) *There exists a regular outer measure $\mu_{[\gamma,\mathfrak{V}]}$ on an arbitrary set X.*
(b) *Every regular outer measure on a set X is an outer measure of the type $\mu_{[\gamma,\mathfrak{V}]}$.*

Proof. 1. Let us prove (a). Let ν be an outer measure on X. We obtain a regular outer measure $\widetilde{\nu}$ by Theorem 1.41. Then according to Theorem 2.10, there exist a premeasure γ on X and a covering class $\mathfrak{V}$ for X such that $\widetilde{\nu} = \mu_{[\gamma,\mathfrak{V}]}$.

2. Let us prove (b). Every regular outer measure is certainly an outer measure. Every outer measure on X is an outer measure of the type $\mu_{[\gamma,\mathfrak{V}]}$ according to Theorem 2.10. ∎

[II] Construction of Outer Measures on a Metric Space

Review. (Topological Spaces)

• To fix our terminology let us review definitions of some topological concepts. Let X be a set. A collection $\mathfrak{O}$ of subsets of X is called a topology on X if it satisfies the following axioms:

I $\emptyset \in \mathfrak{O}$,

II $X \in \mathfrak{O}$,

III $\{E_\alpha : \alpha \in A\} \subset \mathfrak{O} \Rightarrow \bigcup_{\alpha \in A} E_\alpha \in \mathfrak{O}$,

IV $E_1, E_2 \in \mathfrak{O} \Rightarrow E_1 \bigcap E_2 \in \mathfrak{O}$.

The pair $(X, \mathfrak{O})$ is called a topological space. The members of $\mathfrak{O}$ are called the open sets of the topological space. For brevity, we often write X for $(X, \mathfrak{O})$.

• A subset E of X is called a closed set if its complement E^c is an open set. Thus X is both an open set and a closed and so is $\emptyset$.

• An arbitrary union of open sets is an open set and a finite intersection of open sets is an open set. An arbitrary intersection of closed sets is a closed set and a finite union of closed sets is a closed set.

• The interior E° of a subset E of X is defined as the union of all open sets contained in E. Thus it is the greatest open set contained in E.

• The closure $\overline{E}$ of E is defined as the intersection of all closed sets containing E. It is the smallest closed set containing E.

• The boundary ∂E of E is defined by $\partial E = \big(E^\circ \cup (E^c)^\circ\big)^c$.

• A subset E of X is called a compact set if for every collection $\mathfrak{V}$ of open sets such that $E \subset \bigcup_{V \in \mathfrak{V}} V$ there exists a finite subcollection $\{V_1, \ldots, V_N\}$ such that $E \subset \bigcup_{n=1}^N V_n$.

• A subset E of X is called a G_δ-set if it is the intersection of countably many open sets.

• A subset E of X is called an F_σ-set if it is the union of countably many closed sets.

• Given a topological space $(X, \mathfrak{O})$. We write $\sigma(\mathfrak{O})$ for the smallest σ-algebra of subsets of X containing the collection $\mathfrak{O}$ of all open sets. We call $\sigma(\mathfrak{O})$ the Borel σ-algebra of subsets of X and write $\mathfrak{B}_X$ for it.

• Let $\mathfrak{C}$ be the collection of all closed sets in the topological space $(X, \mathfrak{O})$ and write $\sigma(\mathfrak{C})$ for the smallest σ-algebra of subsets of X containing the collection $\mathfrak{C}$. It follows immediately that $\sigma(\mathfrak{C}) = \sigma(\mathfrak{O})$. Thus we have $\mathfrak{B}_X = \sigma(\mathfrak{C})$.

Review. (Metric Spaces)

• Let X be an arbitrary set. A function ρ on $X \times X$ is called a metric on X if it satisfies the following conditions:

1° $\rho(x, y) \in [0, \infty)$ for $x, y \in X$,

2° $\rho(x, y) = 0 \Leftrightarrow x = y$,

3° $\rho(x, y) = \rho(y, x)$ for $x, y \in X$,

4° triangle inequality: $\rho(x, y) \le \rho(x, z) + \rho(z, y)$ for $x, y, z \in X$.

The pair (X, ρ) is called a metric space.

• In $\mathbb{R}^n$, if we define $\rho(x, y) = |x - y| = \{\sum_{k=1}^n (x_k - y_k)^2\}^{1/2}$ for $x = (x_1, \ldots, x_n)$ and

$y = (y_1, \ldots, y_n)$ in $\mathbb{R}^n$, then ρ satisfies conditions 1°, 2°, 3°, and 4° above and is thus a metric. This metric on $\mathbb{R}^n$ is called the Euclidean metric.

• In a metric space (X, ρ), if $x_0 \in X$ and $r > 0$ the set $B(x_0, r) = \{x \in X : \rho(x, x_0) < r\}$ is called an open ball with center x_0 and radius r. A subset E of X is called an open set if for each $x \in E$ there exists $r > 0$ such that $B(x, r) \subset E$. An open ball is indeed an open set in the sense defined above. The collection of all open sets in a metric space satisfies the axioms I, II, III, and IV and is thus a topology. We write $\mathfrak{O}(\rho)$ for this topology. Henceforth when we say a metric space (X, ρ) we mean the topological space $(X, \mathfrak{O}(\rho))$.

• A set E in a metric space $(X, \mathfrak{O}(\rho))$ is said to be bounded if there exist $x_0 \in X$ and $r > 0$ such that $E \subset B(x_0, r)$. A set E in $\mathbb{R}^n$ is a compact set if and only if E is a bounded and closed set.

Review. (Distance in a Metric Space)

• Given a metric space $(X, \mathfrak{O}(\rho))$. For $x, y \in X$ we call $\rho(x, y)$ the distance between x and y. Thus the distance between x and y is equal to 0 if and only if $x = y$.

• The distance between a point $x \in X$ and a set $E \in \mathfrak{P}(X)$ is defined by

$$\rho(x, E) = \inf_{y \in E} \rho(x, y).$$

If $x \in E$ then $\rho(x, E) = 0$ but the converse is false. (For instance $\{0\}$ and $(0, \infty)$ in $\mathbb{R}$.) If E is a closed set then $\rho(x, E) = 0$ if and only if $x \in E$.

• The distance between two sets $E, F \in \mathfrak{P}(X)$ is defined by

$$\rho(E, F) = \inf_{x \in E, y \in F} \rho(x, y).$$

If $E \cap F \neq \emptyset$, then $\rho(E, F) = 0$.

If $E \cap F = \emptyset$, $\rho(E, F)$ may still be equal to 0 even if E and F are closed sets. If E is a closed set and F is a compact set, then $\rho(E, F) = 0$ if and only if $E \cap F \neq \emptyset$.

If $E \cap F = \emptyset$, we may still have $\rho(E, F) = 0$. (For instance let E be the closed upper half-plane and F be the lower open half-plane in $\mathbb{R}^2$.)

Definition 2.13. (Diameter of a Set) *Let (X, ρ) be a metric space. The diameter $|E|$ of $E \in \mathfrak{P}(X)$ such that $E \neq \emptyset$ is defined by setting*

$$|E| = \sup\{\rho(x', x'') : x', x'' \in E\}.$$

We define $|\emptyset| = 0$.

Observation 2.14. Given a metric space (X, ρ). Let $E \in \mathfrak{P}(X)$. We have:

(a) $|E| \in [0, \infty]$.

(b) $|E| < \infty$ if and only if E is a bounded set in X.

(c) If $E \neq \emptyset$, then $|E| = 0$ if and only if E is a singleton.
(d) If E consists of two distinct points x and y only, then $|E| = \rho(x, y) > 0$.
(e) monotonicity: If $E \subset F$ then $|E| \leq |F|$.
(f) non-subadditivity: Subadditivity $|E \cup F| \leq |E| + |F|$ does not hold in general. There can exist $\{E, F\} \subset \mathfrak{P}(X)$ such that $|E \cup F| > |E| + |F|$.
(g) Let $\overline{B}(x_0, r)$ be a closed ball in a metric space (X, ρ). Then $|\overline{B}(x_0, r)| \leq 2r$.
(h) Let $\overline{B}(x_0, r)$ be a closed ball in the metric space $(\mathbb{R}^n, \rho_e)$. Then $|\overline{B}(x_0, r)| = 2r$.
(i) There exists a metric space (X, ρ) such that for a closed ball $\overline{B}(x_0, r)$ in it we have actually $|\overline{B}(x_0, r)| < 2r$.

Proof. (a), (b), (c), (d) and (e) are immediate from Definition 2.13.

1. Let us prove (f). Consider the metric space $(\mathbb{R}, \rho_e)$. Let $a, b, c \in \mathbb{R}$ and $a < b < c$. Let $E = (a, b)$ and $F = \{c\}$. Then we have $|E| = b - a$, $|F| = 0$ and $|E \cup F| = c - a > b - a = |E| + |F|$.

2. Let us prove (g). Let $x', x'' \in \overline{B}(x_0, r)$. Then we have

$$\rho_e(x', x'') \leq \rho_e(x', x_0) + \rho_e(x_0, x'') \leq r + r = 2r. \tag{1}$$

This implies then

$$|\overline{B}(x_0, r)| = \sup\left\{\rho_e(x', x'') : x', x'' \in \overline{B}(x_0, r)\right\} \leq 2r. \tag{2}$$

3. Let us prove (h). Let $\overline{B}(x_0, r)$ be a closed ball in the metric space $(\mathbb{R}^n, \rho_e)$. Then (2) above holds. Let L be a line in $\mathbb{R}^n$ passing through x_0. Then the intersection $L \cap \overline{B}(x_0, r)$ is a closed interval I in L. For the endpoints x' and x'' of I we have distance $\rho_e(x', x'') = 2r$. This implies that

$$\sup\left\{\rho_e(x', x'') : x', x'' \in \overline{B}(x_0, r)\right\} = 2r.$$

Thus we have $|\overline{B}(x_0, r)| = 2r$.

4. Let us prove (i). Let (X, ρ) be this first quadrant of $(\mathbb{R}^2, \rho_e)$ and consider the closed ball $\overline{B}(0, r)$ in it. Then we have $|\overline{B}(0, r)| = \sqrt{2}r < 2r$. ■

Lemma 2.15. *Let (X, ρ) be a metric space.*
(a) *For every $E \in \mathfrak{P}(X)$, we have $\left|\overline{E}\right| = |E|$.*
(b) *For every $E \in \mathfrak{P}(X)$ and $\varepsilon > 0$, there exists an open set G such that $G \supset E$ and $|G| \leq |E| + \varepsilon$.*
(c) *For every $E \in \mathfrak{P}(X)$ and $\varepsilon > 0$, let $G = \bigcup_{x \in E} B(x, \varepsilon)$. Then $|G| \leq |E| + 2\varepsilon$.*
(d) *For every $E \in \mathfrak{P}(X)$ such that $|E| \in (0, \infty)$ there exists a closed ball $\overline{B}$ such that $\overline{B} \supset E$ and $|\overline{B}| \leq 2|E|$. Indeed, for any $x_0 \in E$, $\overline{B}(x_0, |E|)$ is such a closed ball.*

Proof. 1. Let us prove (a). Since $E \subset \overline{E}$, we have $|E| \leq \left|\overline{E}\right|$ by Observation 2.14. Let $\varepsilon > 0$. If $x, y \in \overline{E}$ then there exist $x_0, y_0 \in E$ such that $\rho(x, x_0) < \varepsilon$ and $\rho(y, y_0) < \varepsilon$. Then $\rho(x, y) \leq \rho(x, x_0) + \rho(x_0, y_0) + \rho(y_0, y) \leq |E| + 2\varepsilon$. Thus $\left|\overline{E}\right| \leq |E| + 2\varepsilon$. Since this holds for every $\varepsilon > 0$, we have $\left|\overline{E}\right| \leq |E|$. Therefore we have $\left|\overline{E}\right| = |E|$.

2. Let us prove (b). Let $\varepsilon > 0$. Let $G = \bigcup_{x \in E} B\big(x, \frac{\varepsilon}{2}\big)$. Then G is an open set and $G \supset E$. Let $x, y \in G$. Then $x \in B\big(x_0, \frac{\varepsilon}{2}\big)$ and $y \in B\big(y_0, \frac{\varepsilon}{2}\big)$ for some $x_0, y_0 \in E$. Now we have $\rho(x,y) \le \rho(x,x_0) + \rho(x_0,y_0) + \rho(y_0,y) \le |E| + \varepsilon$. Then $|G| \le |E| + \varepsilon$.

3. Let us prove (c). For $E \in \mathfrak{P}(X)$ and $\varepsilon > 0$, let $G = \bigcup_{x \in E} B(x, \varepsilon)$. We have $|G| = \sup\big\{\rho(y', y'') : y', y'' \in G\big\}$. Now if $y' \in G$ then $y' \in B(x', \varepsilon)$ for some $x' \in E$ and if $y'' \in G$ then $y'' \in B(x'', \varepsilon)$ for some $x'' \in E$. Then

$$\rho(y', y'') \le \rho(y', x') + \rho(x', x'') + \rho(x'', y'') \le \varepsilon + |E| + \varepsilon.$$

Then $|G| = \sup\big\{\rho(y', y'') : y', y'' \in G\big\} \le |E| + 2\varepsilon$.

4. Let us prove (d). Suppose $E \in \mathfrak{P}(X)$ and $|E| \in (0, \infty)$. Let $x_0 \in E$. Then for any $x \in E$ we have $\rho(x_0, x) \le |E|$ so that $x \in \overline{B}(x_0, |E|)$. Thus we have $E \subset \overline{B}(x_0, |E|)$. Moreover we have $|\overline{B}(x_0, |E|)| \le 2|E|$ by (g) of Observation 2.14. ∎

Review. (Separable Metric Spaces)

• A set E in a topological space X is said to be dense in X if $\overline{E} = X$.

• A set E in a topological space X is dense in X if and only if every non-empty open set in X contains some point of E.

• A topological space X is said to be separable if it has a countable dense subset, that is, there exists a countable subset E of X such that $\overline{E} = X$.

• A topological space X is said to satisfy the Second Axiom of Countability if it has a countable open base, that is, there exists a countable collection of open sets such that every open set in X is a union of members of this collection.

• If a topological space X satisfies the Second Axiom of Countability then it is a separable topological space.

• If a metric space (X, ρ) is separable then it satisfies the Second Axiom of Countability. (Indeed, if $\{a_n : n \in \mathbb{N}\}$ is a countable dense subset of X and if $\{r_k : k \in \mathbb{N}\}$ is the collection of all positive rational numbers, then the collection of open balls $\big\{B_{r_k}(a_n) : n \in \mathbb{N}, k \in \mathbb{N}\big\}$ is a countable open base for X.)

Proposition 2.16. *Let (X, ρ) be a separable metric space.*
(a) *Every open set in X is an F_σ-set.*
(b) *Every closed set in X is a G_δ-set.*

Proof. 1. Let us prove (a). Let us show first that the open set $\emptyset$ is an F_σ-set. Now $\emptyset$ is an open set and it also is a closed set. We have $\emptyset = \bigcup_{i \in \mathbb{N}} \emptyset_i$. Thus the open set $\emptyset$ is a countable union of closed sets $\emptyset_i$ and is thus an F_σ-set.

Let G be a non-empty open set in X. Since X is a separable metric space, there exists a countable dense subset $D = \{\xi_i : i \in \mathbb{N}\}$ of X. Let $\{r_j : j \in \mathbb{N}\}$ be the set of all positive rational numbers. For each $i \in \mathbb{N}$ and $j \in \mathbb{N}$, construct a closed ball

$$\overline{B}_{i,j} = \overline{B}(\xi_i, r_j) = \big\{x \in X : \rho(\xi_i, x) \le r_j\big\}.$$

Let $\mathcal{B} = \{\overline{B}_{i,j} : i \in \mathbb{N}, j \in \mathbb{N}\}$, a countable collection of closed balls. Then let $\mathcal{B}_0$ be the subcollection of $\mathcal{B}$ consisting of all those members of $\mathcal{B}$ that are contained in the open set G. Let F be the union of all members of $\mathcal{B}_0$. Then F is an F_σ-set and $F \subset G$. If we show that $G \subset F$ then $G = F$ and we are done.

Let us show that $G \subset F$. Let $x \in G$ be arbitrarily chosen. Now since G is an open set, there exists $r > 0$ such that

$$(1) \qquad B(x, r) \subset G.$$

Consider the open ball $B(x, \frac{r}{4})$. Since D is a dense subset of X, every open set in X contains some point of D. Thus there exists $i_0 \in \mathbb{N}$ such that $\xi_{i_0} \in B(x, \frac{r}{4})$. Let $j_0 \in \mathbb{N}$ be such that

$$(2) \qquad \frac{r}{4} < r_{j_0} < \frac{r}{2}.$$

Since $\xi_{i_0} \in B(x, \frac{r}{4})$, we have $\rho(x, \xi_{i_0}) < \frac{r}{4} < r_{j_0}$. Then we have

$$(3) \qquad x \in B(\xi_{i_0}, r_{J_0}) \subset \overline{B}(\xi_{i_0}, r_{J_0}).$$

Now $\overline{B}(\xi_{i_0}, r_{J_0}) \in \mathcal{B}$. Let us show that we actually have

$$(4) \qquad \overline{B}(\xi_{i_0}, r_{J_0}) \in \mathcal{B}_0.$$

To prove (4), we show that $\overline{B}(\xi_{i_0}, r_{J_0}) \subset G$. Thus let $y \in \overline{B}(\xi_{i_0}, r_{J_0})$. Then we have

$$\rho(x, y) \leq \rho(x, \xi_{i_0}) + \rho(\xi_{i_0}, y) < \frac{r}{4} + r_{j_0} < \frac{r}{4} + \frac{r}{2} < r.$$

Thus we have $y \in B(x, r)$. Since this holds for an arbitrary $y \in \overline{B}(\xi_{i_0}, r_{J_0})$, we have $\overline{B}(\xi_{i_0}, r_{J_0}) \subset B(x, r) \subset G$. This proves (4). According to (3) and (4), if $x \in G$ then x is contained in a member of $\mathcal{B}_0$. Then since F is the union of all members of $\mathcal{B}_0$, x is contained in F. Therefore we have $G \subset F$.

2. Let us prove (b). Now if F is a closed set in X then F^c is an open set in X and is therefore an F_σ-set by (a). Then F is the complement of an F_σ-set and is thus a G_δ-set. ∎

Remark 2.17. Converses of statements (a) and (b) in Proposition 2.16 are false. We show this below by constructing counter-examples.

Example 1. Consider the separable metric space $(\mathbb{R}, \rho_e)$. Then $\{n\} \subset \mathbb{R}$ is a closed set in $\mathbb{R}$ for every $n \in \mathbb{N}$ and thus $\mathbb{N} = \bigcup_{n \in \mathbb{N}} \{n\}$ is an F_σ-set in $\mathbb{R}$. But $\mathbb{N}$ is not an open set in $\mathbb{R}$.

Example 2. Consider the separable metric space $(\mathbb{R}^2, \rho_e)$. Then $(-1, 1) \times (0, 1 + \frac{1}{n})$ is an open set in $\mathbb{R}^2$ for every $n \in \mathbb{N}$. The G_δ-set $\bigcap_{n \in \mathbb{N}} (-1, 1) \times (0, 1 + \frac{1}{n}) = (-1, 1) \times (0, 1]$ is not a closed set in $\mathbb{R}^2$. ∎

Proposition 2.18. (Generation of Open Sets by a Countable Collection of Closed Balls) *Given a separable metric space (X, ρ). There exists a countable collection $\mathfrak{K}$ of closed balls in X such that every non-empty open set in X is a union of members of $\mathfrak{K}$.*

Proof. Since the metric space (X, ρ) is separable there exists a countable dense subset A of X. Let A be represented as $A = \{a_n : n \in \mathbb{N}\}$. Let $\{r_k : k \in \mathbb{N}\}$ be the collection of all positive rational numbers. Let

$$\overline{B}(a_n, r_k) = \{x \in X : \rho(x, a_n) \leq r_k\},$$

that is, a closed ball in X with center a_n and radius r_k. Let

$$\mathfrak{K} = \{\overline{B}(a_n, r_k) : n \in \mathbb{N}, k \in \mathbb{N}\}.$$

Thus defined, $\mathfrak{K}$ is a countable collection of closed balls in X.

Let G be a non-empty open set in X. Let us show that G is a union of members of $\mathfrak{K}$. Let $x \in G$. Let r_k be so small that $B(x, 4r_k) \subset G$. Since A is dense in X, every non-empty open set in X contains some point of A. Thus there exists $n \in \mathbb{N}$ such that $a_n \in B(x, r_k)$. Now $a_n \in B(x, r_k)$ implies $x \in B(a_n, r_k)$. Then we have

$$x \in B(a_n, r_k) \subset \overline{B}(a_n, r_k) \subset B(a_n, 2r_k).$$

Then for $x \in B(a_n, 2r_k)$ we have $\rho(x, a_n) < 2r_k < 4r_k$ so that we have

$$B(a_n, 2r_k) \subset B(x, 4r_k) \subset G$$

and then we have

$$x \in \overline{B}(a_n, r_k) \subset B(a_n, 2r_k) \subset B(x, 4r_k) \subset G.$$

Then we have

$$G = \bigcup_{x \in G} \{x\} \subset \bigcup_{x \in G} \overline{B}(a_n, r_k) \subset G.$$

This implies that we have

$$G = \bigcup_{x \in G} \overline{B}(a_n, r_k).$$

This completes the proof. ■

Definition 2.19. (Fine-covering Class) *Let (X, ρ) be a metric space. A covering class $\mathfrak{V}$ for X is called a fine-covering class if it satisfies the additional condition that for every $\delta \in (0, \infty]$ there exists a sequence $(V_n : n \in \mathbb{N}) \subset \mathfrak{V}$ such that $|V_n| \leq \delta$ for $n \in \mathbb{N}$ and $\bigcup_{n \in \mathbb{N}} V_n = X$.*
For an arbitrary $E \in \mathfrak{P}(X)$, a sequence $(V_n : n \in \mathbb{N}) \subset \mathfrak{V}$ such that $\bigcup_{n \in \mathbb{N}} V_n \supset E$ and $|V_n| \leq \delta$ for $n \in \mathbb{N}$ is called a δ-covering sequence for E.

Observation 2.20. Let (X, ρ) be a metric space. If $\mathfrak{V}$ is a fine-covering class for X and $\mathfrak{W}$ is a collection of subsets of X such that $\mathfrak{W} \supset \mathfrak{V}$, then $\mathfrak{W}$ is a fine-covering class for X.

We show next that for a separable metric space a fine-covering class always exists.

Lemma 2.21. *Let (X, ρ) be a separable metric space. Then for every $\delta > 0$ there exists $\{x_n : n \in \mathbb{N}\} \subset X$ such that $\bigcup_{i \in \mathbb{N}} B(x_n, \delta) = X$.*

Proof. If (X, ρ) is a separable metric space, then there exists a countable dense subset $D = \{x_n : n \in \mathbb{N}\}$ of X. With an arbitrary $\delta > 0$, consider a collection of open balls in X given by $\{B(x_n, \delta) : n \in \mathbb{N}\}$. Let us show that $\bigcup_{i \in \mathbb{N}} B(x_n, \delta) = X$.

Let $x \in X$ be arbitrarily chosen. Since D is a dense subset of X, every non-empty open set in X contains some point of D. Thus there exists $n_0 \in \mathbb{N}$ such that $x_{n_0} \in B(x, \delta)$. Then $\rho(x_{n_0}, x) < \delta$ and hence $x \in B(x_{n_0}, \delta)$ and then $x \in \bigcup_{n \in \mathbb{N}} B(x_n, \delta)$. Since this holds for an arbitrary $x \in X$, we have $X \subset \bigcup_{n \in \mathbb{N}} B(x_n, \delta)$ and then $\bigcup_{n \in \mathbb{N}} B(x_n, \delta) = X$. ∎

Theorem 2.22. (Existence of Fine-covering Class) *Let (X, ρ) be a separable metric space.*
(a) *Let $\mathfrak{S}_X^o$ be the collection of all open balls in X and $\emptyset$. Then $\mathfrak{S}_X^o$ is a fine-covering class.*
(b) *Let $\mathfrak{S}_X^c$ be the collection of all closed balls in X and $\emptyset$. Then $\mathfrak{S}_X^c$ is a fine-covering class for X.*

Proof. 1. To show that $\mathfrak{S}_X^o$ is a fine-covering class for X, we show that for every $\delta > 0$ there exists a sequence of open balls, $\big(B(x_n, r_n) : n \in \mathbb{N}\big)$, such that $\bigcup_{n \in \mathbb{N}} B(x_n, r_n) = X$ and $|B(x_n, r_n)| \le \delta$ for every $n \in \mathbb{N}$.

Let $\delta > 0$. Now according to Lemma 2.21, for every $r > 0$ there exists $\{x_n : n \in \mathbb{N}\} \subset X$ such that $\bigcup_{i \in \mathbb{N}} B(x_n, r) = X$. Then for our $\frac{\delta}{2} > 0$ there exists $\{x_n : n \in \mathbb{N}\} \subset X$ such that $\bigcup_{i \in \mathbb{N}} B\big(x_n, \frac{\delta}{2}\big) = X$. By (f) of Observation 2.14, we have $|B\big(x_n, \frac{\delta}{2}\big)| \le \delta$ for every $n \in \mathbb{N}$. This completes the proof that $\mathfrak{S}_X^o$ is a fine-covering class for X.

2. We show that $\mathfrak{S}_X^c$ is a fine-covering class for X by the same argument as for $\mathfrak{S}_X^o$. We need only observe that $|\overline{B}\big(x_n, \frac{\delta}{2}\big)| = |B\big(x_n, \frac{\delta}{2}\big)|$ according to (a) of Lemma 2.15. ∎

Theorem 2.23. *Let (X, ρ) be a separable metric space.*
(a) *The collection $\mathfrak{O}_X$ of all open sets in X is a fine-covering class for X.*
(b) *The collection $\mathfrak{C}_X$ of all closed sets in X is a fine-covering class for X.*
(c) *$\mathfrak{B}_X$ is a fine-covering class for X.*
(d) *$\mathfrak{P}(X)$ is a fine-covering class for X.*

Proof. 1. Since (X, ρ) is a separable metric space, the collection $\mathfrak{S}_X^o$ of all open balls in X and $\emptyset$ is a fine covering class for X according to Theorem 2.22. Then since $\mathfrak{S}_X^o \subset \mathfrak{O}_X$, $\mathfrak{O}_X$ is a fine-covering class for X by Observation 2.20.

2. The collection $\mathfrak{S}_X^c$ of all closed balls in X and $\emptyset$ is a fine-covering class for X according to Theorem 2.22. Then since $\mathfrak{S}_X^c \subset \mathfrak{C}_X$, $\mathfrak{C}_X$ is a fine-covering class for X by Observation 2.20.

3. By (a), $\mathfrak{O}_X$ is a fine-covering class for X. Then since $\mathfrak{O}_X \subset \mathfrak{B}_X$, $\mathfrak{B}_X$ is a fine-covering class for X by Observation 2.20.

4. Since $\mathfrak{O}_X \subset \mathfrak{P}(X)$ and $\mathfrak{O}_X$ is a fine-covering class for X by (a), $\mathfrak{P}(X)$ is a fine-covering class for X by Observation 2.20. ∎

Definition 2.24. (Truncation of a Fine-covering Class by $\delta \in (0, \infty]$) *Let (X, ρ) be a metric space and let $\mathfrak{V}$ be a fine-covering class for X. For an arbitrary $\delta \in (0, \infty]$, let $\mathfrak{V}|_\delta$ be the subcollection of $\mathfrak{V}$ consisting of all $V \in \mathfrak{V}$ with $|V| \le \delta$. We call $\mathfrak{V}|_\delta$ a truncation of the fine-covering class $\mathfrak{V}$ by$\delta > 0$.* (Observe that $\mathfrak{V}|_\infty = \mathfrak{V}$.)

We show next that a truncation of a fine-covering class remains a fine-covering class.

Proposition 2.25. *Let (X, ρ) be a metric space and let $\mathfrak{V}$ be a fine-covering class for X. For every $\delta \in (0, \infty]$, the truncation $\mathfrak{V}|_\delta$ of the fine-covering class $\mathfrak{V}$ is a fine-covering class for X.*

Proof. To show that $\mathfrak{V}|_\delta$ is a fine-covering class for X, we show that for every $\eta > 0$ there exists a sequence $(V_n : n \in \mathbb{N}) \subset \mathfrak{V}|_\delta$ such that $\bigcup_{n\in\mathbb{N}} V_n = X$ and $|V_n| \le \eta$ for every $n \in \mathbb{N}$. Now since $\mathfrak{V}$ is a fine-covering class for X, for $\min\{\eta, \delta\} > 0$ there exists a sequence $(V_n : n \in \mathbb{N}) \subset \mathfrak{V}$ such that $\bigcup_{n\in\mathbb{N}} V_n = X$ and $|V_n| \le \min\{\eta, \delta\}$ for every $n \in \mathbb{N}$. Then $|V_n| \le \min\{\eta, \delta\} \le \delta$ and thus $V_n \in \mathfrak{V}|_\delta$. We also have $|V_n| \le \min\{\eta, \delta\} \le \eta$. Thus we have a sequence $(V_n : n \in \mathbb{N}) \subset \mathfrak{V}|_\delta$ such that $\bigcup_{n\in\mathbb{N}} V_n = X$ and $|V_n| \le \eta$ for every $n \in \mathbb{N}$. This proves that $\mathfrak{V}|_\delta$ is a fine-covering class for X. ∎

Proposition 2.26. *If $0 < \delta' < \delta'' \le \infty$, then $\mathfrak{V}|_{\delta'} \subset \mathfrak{V}|_{\delta''}$.*

Proof. Note that if $V \in \mathfrak{V}|_{\delta'}$ then $|V| \le \delta'$ so that $|V| \le \delta''$ and then $V \in \mathfrak{V}|_{\delta''}$. ∎

Observation 2.27. Consider an outer measure $\mu_{[\gamma, \mathfrak{V}|_\delta]}$ on a separable metric space (X, ρ) generated by a premeasure γ on a truncated fine-covering class $\mathfrak{V}|_\delta$ for X. Recall that according to Definition 2.5, $\mu_{[\gamma,\mathfrak{V}|_\delta]}$ is defined by setting for every $E \in \mathfrak{P}(X)$

$$\mu_{[\gamma,\mathfrak{V}|_\delta]}(E) := \inf\left\{\sum_{n\in\mathbb{N}} \gamma(V_n) : (V_n : n \in \mathbb{N}) \subset \mathfrak{V}|_\delta, \bigcup_{n\in\mathbb{N}} V_n \supset E\right\}.$$

Theorem 2.28. *For $\delta \in (0, \infty]$, consider outer measures $\mu_{[\gamma,\mathfrak{V}|_\delta]}$ on a separable metric space (X, ρ). Then we have*

(1) $$\mu_{[\gamma,\mathfrak{V}|_\delta]}(E) \uparrow \quad \text{as } \delta \downarrow 0 \text{ for every } E \in \mathfrak{P}(X)$$

and

(2) $$\lim_{\delta\downarrow 0} \mu_{[\gamma,\mathfrak{V}|_\delta]}(E) \in [0, \infty] \quad \text{for every } E \in \mathfrak{P}(X).$$

Let us define a set function ν on $\mathfrak{P}(X)$ by setting

(3) $$\nu(E) = \lim_{\delta\downarrow 0} \mu_{[\gamma,\mathfrak{V}|_\delta]}(E) \quad \text{for every } E \in \mathfrak{P}(X).$$

Then ν is an outer measure on X.

Proof. Let $0 < \delta_1 < \delta_2 \leq \infty$. By Observation 2.27, for every $E \in \mathfrak{P}(X)$ we have

(4) $$\mu_{[\gamma,\mathfrak{V}|_{\delta_1}]}(E) = \inf\left\{\textstyle\sum_{i\in\mathbb{N}} \gamma(V_i) : (V_i : i \in \mathbb{N}) \subset \mathfrak{V}|_{\delta_1}, \bigcup_{i\in\mathbb{N}} V_i \supset E\right\}.$$

(5) $$\mu_{[\gamma,\mathfrak{V}|_{\delta_2}]}(E) = \inf\left\{\textstyle\sum_{i\in\mathbb{N}} \gamma(V_i) : (V_i : i \in \mathbb{N}) \subset \mathfrak{V}|_{\delta_2}, \bigcup_{i\in\mathbb{N}} V_i \supset E\right\}.$$

By Proposition 2.26, we have $\mathfrak{V}_{\delta_1} \subset \mathfrak{V}_{\delta_2}$. Then the set of nonnegative extended real numbers on the right side of (4) is a subset of the set of nonnegative extended real numbers on the right side of (5). Thus the infimum of the former is greater than or equal to that of the latter, that is, we have

(6) $$\mu_{[\gamma,\mathfrak{V}|_{\delta_1}]}(E) \geq \mu_{[\gamma,\mathfrak{V}|_{\delta_2}]}(E).$$

This proves (1) and (2). Then by (a) of Theorem 2.9, ν is an outer measure on X. ∎

[III] Construction of Regular Outer Measures

Theorem 2.29. (Construction of Regular Outer Measures) *Let μ be an outer measure on a set X generated by a premeasure γ on a covering class $\mathfrak{V}$ for X. Define a set function $\widetilde{\mu}$ on $\mathfrak{P}(X)$ by setting for every $A \in \mathfrak{P}(X)$*

$$\widetilde{\mu}(A) = \inf \left\{ \mu(E) : A \subset E \in \mathfrak{M}(\mu) \right\}.$$

Then $\widetilde{\mu}$ is a regular outer measure on X.

Proof. With the outer measure μ we have a measure space $(X, \mathfrak{M}(\mu), \mu)$. Then our theorem is a particular case of Theorem 1.38. ∎

Definition 2.30. (Regular Outer Measure Derived from Premeasure on Covering Class) *We call the regular outer measure $\widetilde{\mu}$ constructed in Theorem 2.29 a regular outer measure derived from the premeasure γ on the covering class $\mathfrak{V}$.*

Remark 2.31. Theorem 2.29 shows that a regular outer measure on a set is easily obtained and thus a regular outer measure is not a rarity. Observe also that a regular outer measure is still an outer measure and thus according to Theorem 2.10 it is an outer measure generated by some premeasure on some covering class for the set.

Proposition 2.32. (Sufficient Condition for Regularity) *Let μ be an outer measure on a set X generated by a premeasure γ on a covering class $\mathfrak{V}$ for X. Consider $\mathfrak{M}(\mu)$, the σ-algebra of μ-measurable subsets of X. If $\mathfrak{M}(\mu) \supset \mathfrak{V}$, then μ is a regular outer measure.*

Proof. Assume that $\mathfrak{M}(\mu) \supset \mathfrak{V}$. To show that μ is a regular outer measure, we verify that for every $A \in \mathfrak{P}(X)$ there exists $E \in \mathfrak{M}(\mu)$ such that $E \supset A$ and $\mu(E) = \mu(A)$.

Now let $A \in \mathfrak{P}(X)$. If $\mu(A) = \infty$, then with $X \in \mathfrak{M}(\mu)$ we have $A \subset X$ and then $\mu(A) \leq \mu(X)$ so that $\mu(X) = \infty = \mu(A)$.

Consider the case that $\mu(A) < \infty$. For each $m \in \mathbb{N}$, let a sequence $(V_{m,n} : n \in \mathbb{N}) \subset \mathfrak{V}$ be such that

(1) $$\bigcup_{n \in \mathbb{N}} V_{m,n} \supset A$$

and

(2) $$\sum_{n \in \mathbb{N}} \gamma(V_{m,n}) < \mu(A) + \frac{1}{m}.$$

Then let

(3) $$W_m = \bigcup_{n \in \mathbb{N}} V_{m,n}$$

and

(4) $$E = \bigcap_{m \in \mathbb{N}} W_m.$$

Since $\mathfrak{V} \subset \mathfrak{M}(\mu)$, we have $V_{m,n} \in \mathfrak{M}(\mu)$. Then since $\mathfrak{M}(\mu)$ is a σ-algebra, (3) implies that $W_m \in \mathfrak{M}(\mu)$ and then (4) implies that $E \in \mathfrak{M}(\mu)$. Note also that (1), (3) and (4) imply that $A \subset E$ and hence we have

(5) $$\mu(A) \leq \mu(E).$$

It remains to prove the reverse inequality. For each $m \in \mathbb{N}$ we have by the countable subadditivity of the outer measure μ

(6) $$\begin{aligned} \mu(W_m) &= \mu\left(\bigcup_{n \in \mathbb{N}} V_{m,n}\right) \leq \sum_{n \in \mathbb{N}} \mu(V_{m,n}) \\ &\leq \sum_{n \in \mathbb{N}} \gamma((V_{m,n}) \quad \text{by Proposition 2.6} \\ &< \mu(A) + \frac{1}{m} \quad \text{by (2).} \end{aligned}$$

For each $m \in \mathbb{N}$ we have $E \subset W_m$ by (4) and then

(7) $$\mu(E) \leq \mu(W_m) < \mu(A) + \frac{1}{m}.$$

Since this holds for every $m \in \mathbb{N}$, we have $\mu(E) \leq \mu(A)$. This and (5) imply that we have $\mu(A) = \mu(E)$. ■

[IV] Countable Open Bases for the Separable Metric Space $(\mathbb{R}^n, \rho_e)$

We construct a countable open base for the metric space $(\mathbb{R}^n, \rho_e)$ consisting of open balls in $\mathbb{R}^n$. Existence of such an open base in $\mathbb{R}^n$ facilitates our subsequent discussion of outer measures.

Let $\mathbb{Q}$ be the set of all rational numbers in $\mathbb{R}$. Now $\mathbb{Q}$ is a countable set and its cardinality is $\aleph_0$. Let us call a point in $\mathbb{R}^n$ a rational point if its n coordinates are all rational numbers. Then the n-fold product $\mathbb{Q} \times \mathbb{Q} \times \cdots \times \mathbb{Q}$ is the collection of all rational points in $\mathbb{R}^n$ and its cardinality is n-fold product $\aleph_0 \times \aleph_0 \times \cdots \times \aleph_0 = \aleph_0$.
Observe that the set of all rational points in $\mathbb{R}^n$ is dense in $\mathbb{R}^n$.
Let us call an open ball $B(x, r)$ in $\mathbb{R}^n$ a rational open ball if its center $x \in \mathbb{R}^n$ is a rational point and its radius $r > 0$ is a rational number. The cardinality of the set of all positive rational number is $\aleph_0$.
Let us write $\mathcal{B}(\mathbb{R}^n)$ for the collection of all rational open balls in $\mathbb{R}^n$. Then the cardinality of $\mathcal{B}(\mathbb{R}^n)$ is equal to $\aleph_0 \times \aleph_0 = \aleph_0$.

Theorem 2.33. *The collection of all rational open balls $\mathcal{B}(\mathbb{R}^n)$ constitutes a countable open base for $(\mathbb{R}^n, \rho_e)$, that is, every open set in $\mathbb{R}^n$ is a union of members of the countable set $\mathcal{B}(\mathbb{R}^n)$.*

Proof. Let O be a non-empty open set in $\mathbb{R}^n$. Since the set of all rational points in $\mathbb{R}^n$ is dense in $\mathbb{R}^n$, the open set O contains a rational point x. Then with sufficiently small rational radius $r > 0$ we have a rational open ball $B(x, r) \subset O$.

Consider all the rational open balls contained in O. Their union is contained in O. We show next that O is actually equal to the union by showing that every point in O is contained in some constituent of the union.

Let $x \in O$. Since O is an open set, there exists a rational number $r > 0$ such that $B(x, r) \subset O$. Consider the open ball $B(x, \frac{r}{2})$ where the radius $\frac{r}{2} > 0$ is still a rational number. Let y be a rational point contained in the open ball $B(x, \frac{r}{2})$. Then we have $x \in B(y, \frac{r}{2}) \subset B(x, r) \subset O$. Now $B(y, \frac{r}{2})$ is a constituent of the union. Thus we have shown that every point in O is contained in some constituent of the union. ∎

An open rectangle in $\mathbb{R}^n$, $\prod_{i=1}^n (\alpha_i, \beta_i)$, is called a rational open rectangle if $\alpha_1, \ldots, \alpha_n$ and $\beta_1, \ldots \beta_n$ are all rational numbers.
Observe that the center of a rational open rectangle is a rational point in $\mathbb{R}^n$.
A rational open rectangle $\prod_{i=1}^n (\alpha_i, \beta_i)$ is determined by $n+1$ of its 2^n vertices. Since the cardinality of the set of all rational points in $\mathbb{R}^n$ is equal to $\aleph_0$, there exist $n+1$-fold product $\aleph_0 \times \aleph_0 \times \cdots \times \aleph_0 = \aleph_0$ rational open rectangles.
Let $\mathcal{R}(\mathbb{R}^n)$ be the collection of all rational open rectangles. The cardinality of $\mathcal{R}(\mathbb{R}^n)$ is equal to $\aleph_0$.
Let us observe that every rational open ball contains a concentric rational open rectangle and conversely every rational open rectangle contains a concentric rational open ball.

Theorem 2.34. *The collection of all rational open rectangles* $\mathcal{R}(\mathbb{R}^n)$ *constitutes a countable open base for* $(\mathbb{R}^n, \rho_e)$, *that is, every open set in* $\mathbb{R}^n$ *is a union of members of the countable set* $\mathcal{R}(\mathbb{R}^n)$.

Proof. This theorem is proved by the same argument as in the proof of Theorem 2.33. ■

§3 Metric Outer Measures and Borel Outer Measures

[I] Metric Outer Measure on a Metric Space

Definition 3.1. (Positive Separation) *Given a metric space (X, ρ). We say that two sets $E, F \in \mathfrak{P}(X)$ are positively separated if the distance between the two sets $\rho(E, F) > 0$.*

Definition 3.2. (Metric Outer Measure) *Consider a metric space (X, ρ). Let μ be an outer measure on X. We say that μ is a metric outer measure if for every pair $\{E, F\}$ of positively separated subsets of X we have the equality $\mu(E \cup F) = \mu(E) + \mu(F)$.*

To show that a metric outer measure as defined above exists, we construct a metric outer measure on an arbitrary metric space.

Lemma 3.3. *Let (X, ρ) be a metric space and let A, B and C be subsets of X. If $\rho(A, B) > |C|$, then we cannot have both $A \cap C \neq \emptyset$ and $B \cap C \neq \emptyset$.*

Proof. Recall that $\rho(A, B) = \inf\{\rho(x, y) : x \in A, y \in B\}$ and $|C| = \sup\{\rho(x, y) : x, y \in C\}$ by Definition 2.13.

Now assume that $\rho(A, B) > |C|$. Suppose we have both $A \cap C \neq \emptyset$ and $B \cap C \neq \emptyset$. Let $x \in A \cap C$ and $y \in B \cap C$. Then since $x, y \in C$, we have

$$|C| \geq \rho(x, y) \geq \inf\left\{\rho(x', y') : x' \in A, y' \in B\right\} = \rho(A, B).$$

This contradicts the assumption that $\rho(A, B) > |C|$. This shows that we cannot have both $A \cap C \neq \emptyset$ and $B \cap C \neq \emptyset$. ■

Theorem 3.4. (Construction of Metric Outer Measure) *Let (X, ρ) be a metric space. Let γ be a premeasure on a covering class $\mathfrak{V}$ for X. For every $n \in \mathbb{N}$, let $\mathfrak{V}_n = \mathfrak{V}|_{\frac{1}{n}}$ and let $\mu_n = \mu_{[\gamma, \mathfrak{V}_n]}$ be the outer measure on X generated by the premeasure γ on the covering class $\mathfrak{V}_n$ for X. Then for every $E \in \mathfrak{P}(X)$, we have $\mu_n(E) \uparrow$ as $n \to \infty$. Define a set function μ on $\mathfrak{P}(X)$ by setting*

$$\mu(E) = \lim_{n \to \infty} \mu_n(E).$$

Then μ is a metric outer measure on X.

Proof. Observe that $\mathfrak{V}_n \downarrow$ as $n \to \infty$ and thus $\mu_n(E) \uparrow$ for every $E \in \mathfrak{P}(X)$ by Proposition 2.8. Then μ as defined above is an outer measure on X by Theorem 2.9.

It remains to show that μ is a metric outer measure. Thus we are to show that if $E, F \in \mathfrak{P}(X)$ and $\rho(E, F) > 0$ then

$$\mu(E \cup F) = \mu(E) + \mu(F). \tag{1}$$

Now the subadditivity of the outer measure μ implies that $\mu(E \cup F) \leq \mu(E) + \mu(F)$. Thus it remains to prove the reverse inequality, that is,

$$\mu(E \cup F) \geq \mu(E) + \mu(F). \tag{2}$$

If $\mu(E \cup F) = \infty$ then (2) is trivially true. Thus assume that $\mu(E \cup F) < \infty$. Now we have $\rho(E, F) > 0$. Choose $N \in \mathbb{N}$ so large that $\rho(E, F) > \frac{1}{N}$.

Let $\varepsilon > 0$. For every $n \in \mathbb{N}$, there exists a sequence $(V_{n,k}; k \in \mathbb{N}) \subset \mathfrak{V}_n$ such that $\bigcup_{k\in\mathbb{N}}(V_{n,k}) \supset E \cup F$ and

$$\sum_{k\in\mathbb{N}} \gamma(V_{n,k}) \leq \mu_n(E \cup F) + \varepsilon.$$

Let us assume that any $V_{n,k}$ that is disjoint from $E \cup F$ has been dropped. Now for $n \geq N$ and $k \in \mathbb{N}$, we have $|V_{n,k}| \leq \frac{1}{n} \leq \frac{1}{N} < \rho(E, F)$ so that $V_{n,k}$ cannot intersect both E and F according to Lemma 3.3 and thus we have either $V_{n,k} \cap E \neq \emptyset$ or $V_{n,k} \cap F \neq \emptyset$. Let

$$\mathbb{N}_1 = \{k \in \mathbb{N} : V_{n,k} \cap E \neq \emptyset\}.$$

$$\mathbb{N}_2 = \{k \in \mathbb{N} : V_{n,k} \cap F \neq \emptyset\}.$$

We have $\mathbb{N}_1 \cap \mathbb{N}_2 = \emptyset$ and $\mathbb{N}_1 \cup \mathbb{N}_2 = \mathbb{N}$. We also have $\mu_n(E) \leq \sum_{k\in\mathbb{N}_1} \gamma(V_{n,k})$ and $\mu_n(F) \leq \sum_{k\in\mathbb{N}_2} \gamma(V_{n,k})$. Thus we have

$$\begin{aligned}\mu_n(E) + \mu_n(F) &\leq \sum_{k\in\mathbb{N}_1} \gamma(V_{n,k}) + \sum_{k\in\mathbb{N}_2} \gamma(V_{n,k}) \\ &= \sum_{k\in\mathbb{N}} \gamma(V_{n,k}) \leq \mu_n(E \cup F) + \varepsilon.\end{aligned}$$

Since this holds for every $\varepsilon > 0$, we have

$$\mu_n(E) + \mu_n(F) \leq \mu_n(E \cup F). \tag{3}$$

Then letting $n \to \infty$, we have

$$\mu(E) + \mu(F) \leq \mu(E \cup F),$$

which is (2). This completes the proof. ■

Let $(X, \mathfrak{A}, \mu)$ be a measure space. Let $(A_n : n \in \mathbb{N}) \subset \mathfrak{A}$ be an increasing sequence and let $A = \bigcup_{n\in\mathbb{N}} A_n = \lim_{n\to\infty} A_n$. Then according to Theorem 1.9, we have

$$\mu(A) = \lim_{n\to\infty} \mu(A_n). \tag{1}$$

This is a consequence of the countable additivity of μ on the σ-algebra $\mathfrak{A}$.

More generally, let μ be an outer measure on a set X. Let $(A_n : n \in \mathbb{N}) \subset \mathfrak{P}(X)$ be an increasing sequence and let $A = \bigcup_{n\in\mathbb{N}} A_n = \lim_{n\to\infty} A_n$. Then the monotonicity of μ on $\mathfrak{P}(X)$ implies

$$\mu(A) \geq \lim_{n\to\infty} \mu(A_n). \tag{2}$$

We show below that if μ is a metric outer measure on a metric space (X, ρ) then for a certain type of increasing sequence $(A_n : n \in \mathbb{N}) \subset \mathfrak{P}(X)$, the equality (1) holds.

Proposition 3.5. *Let (X, ρ) be a metric space. Let μ be a metric outer measure on X. Let G be an open set in X such that $G \neq \emptyset$ and $G \neq X$. Let $A \in \mathfrak{P}(X)$ be such that $A \subset G$. For $n \in \mathbb{N}$, let*

$$A_n = \left\{x \in A : \rho(x, G^c) \geq \tfrac{1}{n}\right\}. \tag{1}$$

Then $(A_n : n \in \mathbb{N})$ is an increasing sequence of subsets of A and

$$A = \bigcup_{n \in \mathbb{N}} A_n = \lim_{n \to \infty} A_n. \tag{2}$$

Moreover, we have

$$\mu(A) = \lim_{n \to \infty} \mu(A_n). \tag{3}$$

Proof. 1. Clearly $(A_n : n \in \mathbb{N})$ is an increasing sequence of subsets of A. Observe that G^c is a closed set and $G^c \neq \emptyset$ and $G^c \neq X$. Since G^c is a closed set, $\rho(x, G^c) = 0$ if and only if $x \in G^c$. Thus $\rho(x, G^c) > 0$ for every $x \in G$ and in particular $\rho(x, G^c) > 0$ for every $x \in A$. Then for every $x \in A$ we have $\rho(x, G^c) \geq \frac{1}{n}$ for sufficiently large $n \in \mathbb{N}$ and consequently $x \in A_n$ for such n. This shows that every x in A is in A_n for some $n \in \mathbb{N}$. This shows that $A = \bigcup_{n \in \mathbb{N}} A_n$ and proves (2).

2. Let us prove (3). Since $A_n \uparrow$ as $n \to \infty$, we have $\mu(A_n) \uparrow$ as $n \to \infty$ by the monotonicity of the outer measure μ. Thus $\lim_{n \to \infty} \mu(A_n)$ exists in $[0, \infty]$. Now $A \supset A_n$ implies $\mu(A) \geq \mu(A_n)$ by the monotonicity of μ. Then we have

$$\mu(A) \geq \lim_{n \to \infty} \mu(A_n). \tag{4}$$

Thus it remains to show

$$\mu(A) \leq \lim_{n \to \infty} \mu(A_n). \tag{5}$$

Now we showed above that $(A_n : n \in \mathbb{N})$ is an increasing sequence and moreover we have $A = \bigcup_{n \in \mathbb{N}} A_n$. Then for each $n \in \mathbb{N}$, let

$$B_n = A_{n+1} \setminus A_n = \left\{x \in A : \tfrac{1}{n+1} \leq \rho(x, G^c) < \tfrac{1}{n}\right\}. \tag{6}$$

Then we have

$$A = A_{2n} \cup \left(\bigcup_{k \geq 2n} B_k \right) = A_{2n} \cup \left(\bigcup_{k \geq n} B_{2k} \right) \cup \left(\bigcup_{k \geq n} B_{2k+1} \right).$$

Then by the countable subadditivity of the outer measure μ, we have

$$\mu(A) \leq \mu(A_{2n}) + \sum_{k \geq n} \mu(B_{2k}) + \sum_{k \geq n} \mu(B_{2k+1}). \tag{7}$$

Regarding the convergence or divergence of the two series on the right side of (7), there are three possible cases to treat. These are:

Case 1. $\sum_{k\in\mathbb{N}} \mu(B_{2k}) < \infty$ and $\sum_{k\in\mathbb{N}} \mu(B_{2k+1}) < \infty$.

Case 2. $\sum_{k\in\mathbb{N}} \mu(B_{2k}) = \infty$.

Case 3. $\sum_{k\in\mathbb{N}} \mu(B_{2k+1}) = \infty$.

Consider Case 1. In this case convergence of the two series implies

$$\lim_{n\to\infty} \sum_{k\geq n} \mu(B_{2k}) = 0 \quad \text{and} \quad \lim_{n\to\infty} \sum_{k\geq n} \mu(B_{2k+1}) = 0.$$

Then letting $n \to \infty$ in (7), we have

$$\mu(A) \leq \lim_{n\to\infty} \mu(A_{2n}) + 0 + 0 = \lim_{n\to\infty} \mu(A_n).$$

This proves (5).

Next consider Case 2. Now from the definition of B_n by (6), we have

$$B_{2k} = \left\{ x \in A : \frac{1}{2k+1} \leq \rho(x, G^c) < \frac{1}{2k} \right\}.$$

$$B_{2k+2} = \left\{ x \in A : \frac{1}{2k+3} \leq \rho(x, G^c) < \frac{1}{2k+2} \right\}.$$

This implies that $\rho(B_{2k}, B_{2k+2}) \geq \frac{1}{2k+1} - \frac{1}{2k+2} > 0$, that is, B_{2k} and B_{2k+2} are positively separated. Then since μ is a metric outer measure, we have

$$\mu\left(\bigcup_{k=1}^{n-1} B_{2k}\right) = \sum_{k=1}^{n-1} \mu(B_{2k}).$$

But $A_{2n} \supset \bigcup_{k=1}^{n-1} B_{2k}$ and this implies

$$\mu(A_{2n}) \geq \mu\left(\bigcup_{k=1}^{n-1} B_{2k}\right) = \sum_{k=1}^{n-1} \mu(B_{2k}).$$

Letting $n \to \infty$ in the last inequality, we obtain

$$\lim_{n\to\infty} \mu(A_n) = \lim_{n\to\infty} \mu(A_{2n}) \geq \sum_{k\in\mathbb{N}} \mu(B_{2k}) = \infty \geq \mu(A).$$

This proves (5). Finally for Case 3, we prove (5) by the same argument as for Case 2. ∎

[II] Borel Outer Measure on a Topological Space

Definition 3.6. (Borel Outer Measure) *Given a topological space X. An outer measure μ on X is called a Borel outer measure if we have $\mathfrak{B}_X \subset \mathfrak{M}(\mu)$.*

Definition 3.7. (Borel Measure Space) *Given a topological space X. A measure space $(X, \mathfrak{A}, \mu)$ is called a Borel measure space if $\mathfrak{B}_X \subset \mathfrak{A}$.*

Observation 3.8. Let μ be a Borel outer measure on a topological space X. Then $\mathfrak{B}_X \subset \mathfrak{M}(\mu)$ and this implies that $(X, \mathfrak{M}(\mu), \mu)$ is a Borel measure space.

The concept of a Borel outer measure on an arbitrary topological space is defined by Definition 3.6 above. An immediate question is: Does such an outer measure exist? We answer this question in the affirmative by constructing a counting outer measure on an arbitrary topological space and then showing that a counting outer measure is a Borel outer measure.

Theorem 3.9. (Counting Outer Measure) *Let X be an arbitrary set and let μ be a set function on $\mathfrak{P}(X)$ defined by setting $\mu(E)$ to be equal to the number of elements in E for every $E \in \mathfrak{P}(X)$. Then μ is an outer measure on X. We call this outer measure a counting outer measure.*

Proof. Our set function μ on $\mathfrak{P}(X)$ satisfies conditions $1°$, $2°$, $3°$ and $4°$ of Definition 1.14 and is thus an outer measure on X.

Theorem 3.10. *Let μ be a counting outer measure on an arbitrary set X. Then we have $\mathfrak{M}(\mu) = \mathfrak{P}(X)$.*

Proof. Since μ is a counting outer measure, for every $E \in \mathfrak{P}(X)$ we have

$$\mu(A) = \mu(A \cap E) + \mu(A \cap E^c) \quad \text{for every } A \in \mathfrak{P}(X).$$

This shows that every $E \in \mathfrak{P}(X)$ is μ-measurable and thus $\mathfrak{M}(\mu) = \mathfrak{P}(X)$. ∎

Theorem 3.11. *Let μ be a counting outer measure on an arbitrary topological space X. Then μ is a Borel outer measure on X.*

Proof. We have $\mathfrak{B}_X \subset \mathfrak{P}(X) = \mathfrak{M}(\mu)$ by Theorem 3.10. Thus we have $\mathfrak{B}_X \subset \mathfrak{M}(\mu)$. This shows that μ is a Borel outer measure. ∎

Next we present some criteria for an outer measure on a topological space to be a Borel outer measure.

Theorem 3.12. (Criterion for Borel Outer Measure) *Given a topological space $(X, \mathfrak{O})$. Let μ be an outer measure on X. Then μ is a Borel outer measure if and only if $\mathfrak{O} \subset \mathfrak{M}(\mu)$. (Thus μ is a Borel outer measure if and only if every open set in X is μ-measurable.)*

Proof. Consider the Borel σ-algebra $\mathfrak{B}_X = \sigma(\mathfrak{O})$.

1. If μ is a Borel outer measure, then we have $\mathfrak{B}_X \subset \mathfrak{M}(\mu)$, that is, $\sigma(\mathfrak{O}) \subset \mathfrak{M}(\mu)$. This implies $\mathfrak{O} \subset \mathfrak{M}(\mu)$.

2. Conversely suppose $\mathfrak{O} \subset \mathfrak{M}(\mu)$. Then since $\mathfrak{M}(\mu)$ is a σ-algebra of subsets of X containing $\mathfrak{O}$ and $\sigma(\mathfrak{O})$ is the smallest σ-algebra of subsets of X containing $\mathfrak{O}$, we have $\sigma(\mathfrak{O}) \subset \mathfrak{M}(\mu)$, that is, $\mathfrak{B}_X \subset \mathfrak{M}(\mu)$. ■

Corollary 3.13. *Given a topological space $(X, \mathfrak{O})$. Let $\mathfrak{C}$ be the collection of all closed sets in X. Let μ be an outer measure on X. Then μ is a Borel outer measure if and only if $\mathfrak{C} \subset \mathfrak{M}(\mu)$. (Thus μ is a Borel outer measure if and only if every closed set in X is μ-measurable.)*

Proof. Immediate from Theorem 3.12 since $\sigma(\mathfrak{C}) = \sigma(\mathfrak{O}) = \mathfrak{B}_X$. ■

In particular when a topological space X is actually a metric space (X, ρ), we have the following criterion for an outer measure to be a Borel outer measure.

Theorem 3.14. (Equivalence of Borel Outer Measure to Metric Outer Measure) *Let μ be an outer measure on a metric space (X, ρ). Then μ is a Borel outer measure if and only if μ is a metric outer measure.*

Proof. 1. Suppose μ is a metric outer measure. To show that μ is a Borel outer measure, we show that $\mathfrak{B}_X \subset \mathfrak{M}(\mu)$. Since $\mathfrak{B}_X$ is the smallest σ-algebra of subsets of X containing all the closed sets, it suffices to show that every closed set F is contained in $\mathfrak{M}(\mu)$, that is, F in μ-measurable, that is, F satisfies the condition:

$$\mu(E) = \mu(E \cap F) + \mu(E \setminus F) \quad \text{for every } E \in \mathfrak{P}(X). \tag{1}$$

Now if $F = \emptyset$ then (1) reduces to $\mu(E) = \mu(E)$ which is trivially true. Similarly if $F = X$ then (1) reduces to $\mu(E) = \mu(E)$ which is trivially true.

Thus it remains to prove (1) for a closed set F such that $F \neq \emptyset$ and $F \neq X$. Let $G = F^c$. Then G is an open set such that $G \neq X$ and $G \neq \emptyset$. Now let $E \in \mathfrak{P}(X)$ and let

$$A = E \setminus F = E \cap F^c = E \cap G \subset G. \tag{2}$$

$$A_n = \left\{x \in A : \rho(x, G^c) \geq \tfrac{1}{n}\right\} \quad \text{for } n \in \mathbb{N}. \tag{3}$$

Then we have

$$\rho(A_n, F) = \rho(A_n, G^c) \geq \tfrac{1}{n} \quad \text{for } n \in \mathbb{N}. \tag{4}$$

Our sets G, A and A_n satisfy the conditions in Proposition 3.5 and thus we have

(5) $$\lim_{n\to\infty} \mu(A_n) = \mu(A) = \mu(E \setminus F).$$

Now (4) shows that A_n and F are positively separated. Then since μ is a metric outer measure we have

(6) $$\mu\big((E \cap F) \cup A_n\big) = \mu(E \cap F) + \mu(A_n).$$

Observe that we have

$$E = (E \cap F) \cup (E \setminus F) = (E \cap F) \cup A \supset (E \cap F) \cup A_n.$$

Then (6) implies that we have

$$\mu(E) \geq \mu\big((E \cap F) \cup A_n\big) = \mu(E \cap F) + \mu(A_n).$$

Letting $n \to \infty$ and applying (5), we obtain

(7) $$\mu(E) \geq \mu(E \cap F) + \mu(E \setminus F).$$

On the other hand, by the subadditivity of the outer measure μ we have

(8) $$\mu(E) = \mu\big((E \cap F) \cup (E \setminus F)\big) \leq \mu(E \cap F) = \mu(E \setminus F).$$

With (7) and (8), we have (1).

2. Conversely suppose $\mathfrak{B}_X \subset \mathfrak{M}(\mu)$. To show that μ is a metric outer measure, we show that if $A_1, A_2 \in \mathfrak{P}(X)$ and $\rho(A_1, A_2) > 0$ then $\mu(A_1 \cup A_2) = \mu(A_1) + \mu(A_2)$.

Let $A_1, A_2 \in \mathfrak{P}(X)$ and $\rho(A_1, A_2) > 0$, say $\rho(A_1, A_2) = c > 0$. For each $x \in A_1$, let

$$G(x) = \left\{y \in X : \rho(x, y) < \tfrac{c}{2}\right\} \quad \text{and} \quad G = \textstyle\bigcup_{x \in A_1} G(x).$$

Then G is an open set, $A_1 \subset G$ and $G \cap A_2 = \emptyset$. Since G is an open set, we have $G \in \mathfrak{B}_X \subset \mathfrak{M}(\mu)$ and thus G is μ-measurable. This implies that for $A_1 \cup A_2 \in \mathfrak{P}(X)$ we have

(9) $$\mu(A_1 \cup A_2) = \mu\big((A_1 \cup A_2) \cap G\big) + \mu\big((A_1 \cup A_2) \cap G^c\big).$$

But $A_1 \subset G$ and $G \cap A_2 = \emptyset$, implying that $(A_1 \cup A_2) \cap G = A_1$ and $(A_1 \cup A_2) \cap G^c = A_2$. Thus (9) reduces to $\mu(A_1 \cup A_2) = \mu(A_1) + \mu(A_2)$, proving that μ is a metric outer measure. ■

The last theorem shows the equivalence of a Borel outer measure to a metric outer measure on a metric space. Next we derive from this equivalence a criterion for an outer measure on a metric space to be a metric outer measure.

Theorem 3.15. (Criterion for Metric Outer Measure) *Let μ be an outer measure on a metric space (X,ρ). Then μ is a metric outer measure if and only if every open set in X is μ-measurable.*

Proof. Let μ be an outer measure on a metric space (X,ρ). According to Theorem 3.14, μ is a metric outer measure if and only if μ is a Borel outer measure. Then according to Theorem 3.12, μ is a Borel outer measure if and only if every open set in X is μ-measurable. Thus μ is a metric outer measure if and only if every open set in X is μ-measurable. ∎

Corollary 3.16. *Let μ be an outer measure on a metric space (X,ρ). Then μ is a metric outer measure if and only if every closed set in X is μ-measurable.*

Proof. The Corollary follows from Theorem 3.14 and Corollary 3.13. ∎

[III] Borel-regularity of Borel Outer Measure

According to Definition 1.35, an outer measure μ on a set X is called a regular outer measure if for every $E \in \mathfrak{P}(X)$ there exists $F \in \mathfrak{M}(\mu)$ such that $F \supset E$ and $\mu(F) = \mu(E)$. For Borel outer measures on a topological space we have an even stronger regularity condition as follows.

Definition 3.17. (Borel-regular Borel Outer Measure) *A Borel outer measure μ on a topological space X is called a Borel-regular Borel outer measure if it satisfies the condition that for every $A \in \mathfrak{P}(X)$ there exists $B \in \mathfrak{B}_X$ such that $B \supset A$ and $\mu(B) = \mu(A)$.*

Let μ be a Borel outer measure on a topological space X. Let $(A_n : n \in \mathbb{N}) \subset \mathfrak{P}(X)$ be an increasing sequence. Then $\lim_{n\to\infty} A_n$ exists and indeed we have $\bigcup_{n\in\mathbb{N}} A_n = \lim_{n\to\infty} A_n$. By the monotonicity of the outer measure μ, we have

$$\lim_{n\to\infty} \mu(A_n) \le \mu\left(\bigcup_{n\in\mathbb{N}} A_n\right) = \mu\left(\lim_{n\to\infty} A_n\right).$$

We show in the next theorem that if μ is Borel-regular then we have the equality

$$\lim_{n\to\infty} \mu(A_n) = \mu\left(\bigcup_{n\in\mathbb{N}} A_n\right) = \mu\left(\lim_{n\to\infty} A_n\right).$$

Theorem 3.18. *Let μ be a Borel-regular Borel outer measure on a topological space X. Let $(A_n : n \in \mathbb{N}) \subset \mathfrak{P}(X)$ be an increasing sequence. Then we have*

$$\lim_{n\to\infty} \mu(A_n) = \mu\left(\lim_{n\to\infty} A_n\right).$$

Proof. Let $(A_n : n \in \mathbb{N}) \subset \mathfrak{P}(X)$ be an increasing sequence. We showed in Theorem 1.36 that we have $\lim_{n\to\infty} \mu(A_n) \leq \mu(\lim_{n\to\infty} A_n)$. Thus it remains to prove

$$(1) \qquad \mu\left(\lim_{n\to\infty} A_n\right) \leq \lim_{n\to\infty} \mu(A_n).$$

Since μ is Borel-regular, for every $n \in \mathbb{N}$ there exists $B_n \in \mathfrak{B}_X$ such that $A_n \subset B_n$ and $\mu(A_n) = \mu(B_n)$. Consider the measure space $(X, \mathfrak{B}_X, \mu)$ and the sequence $(B_n : n \in \mathbb{N}) \subset \mathfrak{B}_X$. According to Theorem 1.11, we have

$$(2) \qquad \mu\left(\liminf_{n\to\infty} B_n\right) \leq \liminf_{n\to\infty} \mu(B_n).$$

Returning to the increasing sequence $(A_n : n \in \mathbb{N}) \subset \mathfrak{P}(X)$, the existence of $\lim_{n\to\infty} A_n$ implies that $\lim_{n\to\infty} A_n = \liminf_{n\to\infty} A_n$. Then we have

$$\begin{aligned}
(3) \qquad \mu\left(\lim_{n\to\infty} A_n\right) &= \mu\left(\liminf_{n\to\infty} A_n\right) \\
&\leq \mu\left(\liminf_{n\to\infty} B_n\right) \quad \text{by } A_n \subset B_n \text{ and monotonicity of } \mu \\
&\leq \liminf_{n\to\infty} \mu(B_n) \quad \text{by (2)} \\
&= \liminf_{n\to\infty} \mu(A_n) \quad \text{since } \mu(A_n) = \mu(A_n) \\
&= \lim_{n\to\infty} \mu(A_n) \quad \text{since } \lim_{n\to\infty} \mu(A_n) \text{ exists.}
\end{aligned}$$

This proves (1) and completes the proof. ∎

Proposition 3.19. (Sufficient Condition for Borel-regularity of Borel Outer Measure) *Let μ be a Borel outer measure on a topological space X. Let $\mathfrak{O}_X$ be the collection of all open sets in X. Suppose μ satisfies the condition that for every $A \in \mathfrak{P}(X)$ we have*

$$\mu(A) = \inf\left\{\mu(O) : A \subset O \in \mathfrak{O}_X\right\}.$$

Then μ is a Borel-regular Borel outer measure.

Proof. Suppose μ satisfies the condition. Let $A \in \mathfrak{P}(X)$. Then for every $k \in \mathbb{N}$ there exists an open set O_k such that $A \subset O_k$ and $\mu(O_k) \leq \mu(A) + \frac{1}{k}$. Consider the sequence $(O_k : k \in \mathbb{N})$ of open sets containing A. Let $V_k = \bigcap_{i=1}^k O_i$ for $k \in \mathbb{N}$. Then $(V_k : k \in \mathbb{N})$ is a decreasing sequence of open sets containing A. Since $V_k \subset O_k$, we have $\mu(V_k) \leq \mu(O_k) \leq \mu(A) + \frac{1}{k}$ for every $k \in \mathbb{N}$. Since $V_k \in \mathfrak{B}_X$ for every $k \in \mathbb{N}$, we have $B := \bigcap_{k\in\mathbb{N}} V_k \in \mathfrak{B}_X$. Since $A \subset V_k$ for every $k \in \mathbb{N}$, we have $A \subset B$ and $\mu(A) \leq \mu(B)$. Now since $B \subset V_k$ for every $k \in \mathbb{N}$, we have $\mu(B) \leq \mu(V_k) \leq \mu(A) + \frac{1}{k}$ for every $k \in \mathbb{N}$. This implies that $\mu(B) \leq \mu(A)$. Then we have $\mu(A) = \mu(B)$. Thus we have shown that for every $A \in \mathfrak{P}(X)$ there exists $B \in \mathfrak{B}_X$ such that $A \subset B$ and $\mu(A) = \mu(B)$. This proves that μ is a Borel regular Borel outer measure. ∎

Theorem 3.20. (Borel-regularization of Borel Outer Measure) *Let μ be a Borel outer measure on a topological space X. There exists a Borel-regular Borel outer measure $\widetilde{\mu}$ on X such that $\widetilde{\mu} = \mu$ on $\mathfrak{B}_X$.*

Proof. Since μ is a Borel outer measure on X, we have $\mathfrak{B}_X \subset \mathfrak{M}(\mu)$. Now $(X, \mathfrak{M}(\mu), \mu)$ is a measure space. This implies that $(X, \mathfrak{B}_X, \mu)$ is a measure space. Then with the measure space $(X, \mathfrak{B}_X, \mu)$, let us define a set function $\widetilde{\mu}$ on $\mathfrak{P}(X)$ by setting for every $A \in \mathfrak{P}(X)$

$$\widetilde{\mu}(A) := \inf \left\{\mu(E) : A \subset E \in \mathfrak{B}_X\right\}.$$

Then according to Theorem 1.38, we have:
(a) $\widetilde{\mu} = \mu$ on $\mathfrak{B}_X$.
(b) $\widetilde{\mu}$ is an outer measure on X.
(c) For every $A \in \mathfrak{P}(X)$ there exists $E \in \mathfrak{B}_X$ such that $A \subset E$ and $\widetilde{\mu}(A) = \mu(E)$.
Note that (c) shows that $\widetilde{\mu}$ is a Borel-regular Borel outer measure on X. ∎

The concept of a Borel-regular Borel outer measure on a topological space is defined by Definition 3.17. Does such a Borel outer measure exist? To answer this question in the affirmative, we construct a Borel-regular Borel outer measure below. Recall that the existence of a Borel outer measure on a topological space was proved by Theorem 3.11.

We show next that a Borel regular Borel outer measure can be constructed on an arbitrary Borel outer measure.

Theorem 3.21. (Construction of Borel-regular Borel Outer Measure) *Let μ_0 be a Borel outer measure on a topological space X. Define a set function μ on $\mathfrak{P}(X)$ by setting for every $A \in \mathfrak{P}(X)$*

$$\mu(A) = \inf \left\{\mu_0(E) : A \subset E \in \mathfrak{B}_X\right\}.$$

Then μ is a Borel-regular Borel outer measure on X.

Proof. Since μ_0 is a Borel outer measure on X, we have a measure space $(X, \mathfrak{B}_X, \mu_0)$. Let us define a set function μ on $\mathfrak{P}(X)$ by setting $\mu(A) = \inf \left\{\mu_0(E) : A \subset E \in \mathfrak{B}_X\right\}$ for every $A \in \mathfrak{P}(X)$. Then according to Theorem 1.38 we have:
(a) $\mu = \mu_0$ on $\mathfrak{B}_X$.
(b) μ is an outer measure on X.
(c) For every $A \in \mathfrak{P}(X)$ there exists $E \in \mathfrak{B}_X$ such that $A \subset E$ and $\mu(A) = \mu_0(E)$.
(d) $\mathfrak{B}_X \subset \mathfrak{M}(\mu)$.

Then (d) shows that μ is a Borel outer measure and then (c) and (a) show that μ is a Borel-regular Borel outer measure. ∎

We show next that on a metric space (X, ρ) a Borel-regular Borel outer measure can be constructed more directly as follows:
In Theorem 3.14, we showed that an outer measure on a metric space (X, ρ) is a Borel outer measure if and only if it is a metric outer measure.
In Theorem 3.4, we constructed a metric outer measure by starting with an arbitrary premeasure γ on an arbitrary covering class $\mathfrak{V}$ for X.
We show below that if we start with a covering class $\mathfrak{V}$ for X such that $\mathfrak{V} \subset \mathfrak{B}_X$ then the metric outer measure constructed as in Theorem 3.4 is a Borel-regular Borel outer measure.

Theorem 3.22. (Construction of Borel-regular Borel Outer Measure on Metric Space) *Let (X, ρ) be a metric space. Let $\mathfrak{V}$ be a covering class for X such that $\mathfrak{V} \subset \mathfrak{B}_X$ and let γ be a premeasure on $\mathfrak{V}$. For $n \in \mathbb{N}$, let $\mathfrak{V}_n = \mathfrak{V}|_{\frac{1}{n}}$ and let μ_n be the outer measure on X generated by the premeasure γ on the covering class $\mathfrak{V}_n$ for X, that is,*

(1)
$$\mu_n = \mu_{[\gamma, \mathfrak{V}_n]}$$

and for every $E \in \mathfrak{P}(X)$ we have

(2)
$$\mu_{[\gamma, \mathfrak{V}_n]}(E) = \inf \left\{ \sum_{i \in \mathbb{N}} \gamma(V_i) : (V_i : i \in \mathbb{N}) \subset \mathfrak{V}_n, \bigcup_{i \in \mathbb{N}} V_i \supset E \right\}.$$

Define a set function μ on $\mathfrak{P}(X)$ by setting for every $E \in \mathfrak{P}(X)$

(3)
$$\mu(E) = \lim_{n \to \infty} \mu_n(E).$$

Then μ is a Borel-regular Borel outer measure on (X, ρ).

Proof. According to Theorem 3.14, μ is a metric outer measure on (X, ρ) and hence μ is Borel outer measure on (X, ρ) by Theorem 3.14. Let us show that μ is Borel-regular, that is, for every $E \in \mathfrak{P}(X)$ there exists $B \in \mathfrak{B}_X$ such that $B \supset E$ and $\mu(B) = \mu(E)$.

Let $E \in \mathfrak{P}(X)$ be arbitrarily given. Then for every $n \in \mathbb{N}$ we have by (1) and (2)

(4)
$$\mu_n(E) = \inf \left\{ \sum_{i \in \mathbb{N}} \gamma(V_i) : (V_i : i \in \mathbb{N}) \subset \mathfrak{V}_n, \bigcup_{i \in \mathbb{N}} V_i \supset E \right\}.$$

Select $(W_{n,i} : i \in \mathbb{N}) \subset \mathfrak{V}_n$ such that $\bigcup_{i \in \mathbb{N}} W_{n,i} \supset E$ and

(5)
$$\sum_{i \in \mathbb{N}} \gamma(W_{n,i}) \leq \mu_n(E) + \frac{1}{n}.$$

Let

(6)
$$B = \bigcap_{n \in \mathbb{N}} \left(\bigcup_{i \in \mathbb{N}} W_{n,i} \right) \supset E.$$

Then since $W_{n,i} \in \mathfrak{V}_n \subset \mathfrak{V} \subset \mathfrak{B}_x$ and $\mathfrak{B}_X$ is a σ-algebra of subsets of X, definition of B by (6) implies that $B \in \mathfrak{B}_X$. It remains to show that $\mu(B) = \mu(E)$. Now since $B \supset E$, monotonicity of the outer measure μ implies that $\mu(B) \geq \mu(E)$. Thus it remains to show that $\mu(B) \leq \mu(E)$.

Now for every $n \in \mathbb{N}$ we have

$$\begin{aligned}
(7) \qquad \mu_n(B) &= \mu_{[\gamma,\mathfrak{V}_n]}(B) \\
&= \inf\left\{\sum_{i\in\mathbb{N}} \gamma(V_i) : (V_i : i \in \mathbb{N}) \subset \mathfrak{V}_n, \bigcup_{i\in\mathbb{N}} V_i \supset B\right\} for \\
&\leq \sum_{i\in\mathbb{N}} \gamma(W_{n,i}) \leq \mu_n(E) + \frac{1}{n} \quad \text{by (5).}
\end{aligned}$$

Then we have, by (3), (7) and (3) again

$$\mu(B) = \lim_{n\to\infty} \mu_n(B) \leq \lim_{n\to\infty} \left\{\mu_n(E) + \frac{1}{n}\right\} \leq \mu(E).$$

This completes the proof that $\mu(B) = \mu(E)$. ∎

[IV] Truncation of Borel Outer Measure

Theorem 3.23. (Truncation of Borel Outer Measure) *Let μ be a Borel outer measure on a topological space X. Then for every $A \in \mathfrak{P}(X)$ the truncated outer measure $\mu|_A$ is a Borel outer measure on X.*

Proof. By Proposition 1.32 and Definition 1.33, $\mu|_A$ is an outer measure on X. We also have $\mathfrak{M}(\mu) \subset \mathfrak{M}(\mu|_A)$ by Theorem 1.34. Then since μ is a Borel outer measure we have $\mathfrak{B}_X \subset \mathfrak{M}(\mu) \subset \mathfrak{M}(\mu|_A)$. This shows that $\mu|_A$ is a Borel outer measure on X. ■

Theorem 3.24. (Truncation of Borel-regular Borel Outer Measure) *Let μ be a Borel-regular Borel outer measure on a topological space X. Then for every $B \in \mathfrak{B}_X$ the truncated outer measure $\mu|_B$ is a Borel-regular Borel outer measure on X.*

Proof. Let $E \in \mathfrak{P}(X)$. We have $\mu|_B(E) = \mu(E \cap B)$. Since μ is a Borel-regular Borel outer measure, for our $E \cap B \in \mathfrak{P}(X)$ there exists $G \in \mathfrak{B}_X$ such that $E \cap B \subset G$ and $\mu(E \cap B) = \mu(G)$. Let $F = G \cup B^c$. Since $B, G \in \mathfrak{B}_X$, we have $F \in \mathfrak{B}_X$. Moreover we have

$$E \subset (E \cap B) \cup B^c \subset G \cup B^c = F, \tag{1}$$

$$F \cap B = (G \cup B^c) \cap B = G \cap B. \tag{2}$$

Applying the monotonicity of the outer measure $\mu|_B$ to (1), we have

$$\mu|_B(E) \leq \mu|_B(F). \tag{3}$$

On the other hand,

$$\begin{aligned} \mu|_B(F) &= \mu(F \cap B) = \mu(G \cap B) \quad \text{by (2)} \\ &\leq \mu(G) = \mu(E \cap B) = \mu|_B(E). \end{aligned} \tag{4}$$

Then (3) and (4) imply $\mu|_B(E) = \mu|_B(F)$. This completes the proof. ■

Theorem 3.25. *Let μ be a Borel-regular Borel outer measure on a topological space X. Let $A \in \mathfrak{M}(\mu)$ with $\mu(A) < \infty$. Then there exists $B \in \mathfrak{B}_X$ such that $\mu|_A = \mu|_B$.*

Proof. 1. Since μ is a Borel regular Borel outer measure, for every $C \in \mathfrak{P}(X)$ there exists $B \in \mathfrak{B}_X$ such that $B \supset C$ and $\mu(B) = \mu(C)$. In particular for our $A \in \mathfrak{M}(\mu) \subset \mathfrak{P}(X)$ there exists $B \in \mathfrak{B}_X$ such that $B \supset A$ and $\mu(B) = \mu(A)$. Since $\mu(A) < \infty$, we have $\mu(B \setminus A) = \mu(B) - \mu(A) = 0$. According to Theorem 1.22, $(X, \mathfrak{M}(\mu), \mu)$ is a complete measure space. Thus every subset C of the null set $B \setminus A$ is in $\mathfrak{M}(\mu)$ with $\mu(C) = 0$.

2. Let $E \in \mathfrak{P}(X)$. Now

$$E \cap B = (E \cap A) \cup [E \cap (B \setminus A)]$$

and then

$$\mu(E \cap B) = \mu((E \cap A) \cup [E \cap (B \setminus A)]) \leq \mu(E \cap A) + \mu([E \cap (B \setminus A)]) = \mu(E \cap A).$$

But $B \supset A$ implies $\mu(E \cap B) \geq \mu(E \cap A)$. Thus we have $\mu(E \cap A) = \mu(E \cap B)$, that is, $\mu|_A(E) = \mu|_B(E)$ for every $E \in \mathfrak{P}(X)$. ∎

[V] Approximation of Borel Sets by Borel Outer Measure

Consider a Borel outer measure μ on a metric space (X, ρ). Let us inquire about the possibility of approximating a Borel set internally by closed sets and externally by open sets in terms of the outer measure μ.

Definition 3.26. *Let μ be an outer measure on a metric space (X, ρ).*
(a) *We call μ locally finite if for every $x \in X$ there exists $r > 0$ such that $\mu\big(B_r(x)\big) < \infty$.*
(b) *We say that μ is boundedly finite if for every bounded set E in X we have $\mu(E) < \infty$.*
(Observe that if μ is boundedly finite we may still have $\mu(X) = \infty$.)

Proposition 3.27. *An outer measure μ on a metric space (X, ρ) is boundedly finite if and only if for every ball B in X we have $\mu(B) < \infty$.*

Proof. 1. Suppose μ is boundedly finite. Then since every ball B in X is a bounded set we have $\mu(B) < \infty$.

2. Conversely suppose that for every ball B in X we have $\mu(B) < \infty$. Let E be a bounded set in X. Then there exists a ball B such that $E \subset B$. This implies by the monotonicity of the outer measure μ that $\mu(E) \le \mu(B) < \infty$. ∎

We require the following lemmas in our discussion of approximation.

Lemma 3.28. *Let $(A_i : i \in \mathbb{N})$ and $(B_i : i \in \mathbb{N})$ be two sequences of subsets of a set X. Then*

(1)
$$\bigcup_{i\in\mathbb{N}} A_i \setminus \bigcup_{i\in\mathbb{N}} B_i \subset \bigcup_{i\in\mathbb{N}} (A_i \setminus B_i),$$

(2)
$$\bigcap_{i\in\mathbb{N}} A_i \setminus \bigcap_{i\in\mathbb{N}} B_i \subset \bigcup_{i\in\mathbb{N}} (A_i \setminus B_i).$$

Proof. 1. The equality (1) is immediate. Indeed we have

$$\bigcup_{i\in\mathbb{N}} A_i \setminus \bigcup_{i\in\mathbb{N}} B_i = \bigcup_{i\in\mathbb{N}} \left(A_i \setminus \bigcup_{i\in\mathbb{N}} B_i \right) \subset \bigcup_{i\in\mathbb{N}} (A_i \setminus B_i).$$

2. Let us prove (2). Now we have

$$\bigcap_{i\in\mathbb{N}} A_i \setminus \bigcap_{i\in\mathbb{N}} B_i = \left(\bigcap_{i\in\mathbb{N}} A_i\right) \cap \left(\bigcap_{i\in\mathbb{N}} B_i\right)^c = \left(\bigcap_{i\in\mathbb{N}} A_i\right) \cap \left(\bigcup_{i\in\mathbb{N}} B_i^c\right).$$

Thus if $x \in \bigcap_{i\in\mathbb{N}} A_i \setminus \bigcap_{i\in\mathbb{N}} B_i$, then $x \in A_i$ for every $i \in \mathbb{N}$ and $x \in B_i^c$ for some $i \in \mathbb{N}$ and hence $x \in A_i \cap B_i^c = A_i \setminus B_i$ for some $i \in \mathbb{N}$ and therefore $x \in \bigcup_{i\in\mathbb{N}}(A_i \setminus B_i)$. This proves (2). ∎

Lemma 3.29. *For any three sets A, B, and C, we have $(A \setminus B) \setminus C = (A \setminus C) \setminus B$.*

Proof. Observe that

$$(A \setminus B) \setminus C = (A \cap B^c) \cap C^c = A \cap B^c \cap C^c$$
$$= (A \cap C^c) \cap B^c = (A \setminus C) \setminus B. \blacksquare$$

Theorem 3.30. (Internal Approximation of Borel Sets by Closed Sets) *Let μ be a Borel outer measure on a separable metric space (X, ρ). Let $E \in \mathfrak{B}_X$ with $\mu(E) < \infty$. Then for every $\varepsilon > 0$ there exists a closed set C in X such that $C \subset E$ and $\mu(E \setminus C) < \varepsilon$.*

Proof. 1. Let $\mathfrak{O}_X$ be the collection of all open sets in X and let $\mathfrak{C}_X$ be the collection of all closed sets in X. Let $E \in \mathfrak{B}_X$ with $\mu(E) < \infty$. Let $\nu = \mu|_E$. By Theorem 3.23, ν is a Borel outer measure on X. Note also that $\mathfrak{M}(\mu) \subset \mathfrak{M}(\nu)$ by Theorem 1.34 and $\nu(X) = \mu|_E(X) = \mu(X \cup E) = \mu(E) < \infty$. Let

$$\text{(1)} \qquad \mathfrak{H} = \{A \in \mathfrak{M}(\mu) : \forall \varepsilon > 0, \exists C \in \mathfrak{C}_X \ni \cdot C \subset A, \nu(A \setminus C) < \varepsilon\}.$$

Let us show that
1° $\mathfrak{C}_X \subset \mathfrak{H}$.
2° $\mathfrak{H}$ is closed under countable intersections.
3° $\mathfrak{H}$ is closed under countable unions.
4° $\mathfrak{O}_X \subset \mathfrak{H}$.

Now if $C \in \mathfrak{C}_X$ then $C \in \mathfrak{B}_X \subset \mathfrak{M}(\mu)$. Thus $C \in \mathfrak{M}(\mu)$. Then $C \in \mathfrak{H}$ trivially. Thus $\mathfrak{C}_X \subset \mathfrak{H}$. This verifies 1°.

To verify 2°, let $(A_i : i \in \mathbb{N}) \subset \mathfrak{H}$ and let $A = \bigcap_{i \in \mathbb{N}} A_i$. Since $A_i \in \mathfrak{M}(\mu)$ for every $i \in \mathbb{N}$, we have $A \in \mathfrak{M}(\mu)$. Let $\varepsilon > 0$ be arbitrarily given. Since $A_i \in \mathfrak{H}$ there exists $C_i \in \mathfrak{C}_X$ such that $C_i \subset A_i$ and $\nu(A_i \setminus C_i) < \frac{\varepsilon}{2^i}$. Let $C = \bigcap_{i \in \mathbb{N}} C_i$. Then $C \in \mathfrak{C}_X$ and

$$\nu(A \setminus C) = \nu\left(\bigcap_{i \in \mathbb{N}} A_i \setminus \bigcap_{i \in \mathbb{N}} C_i\right) \leq \nu\left(\bigcup_{i \in \mathbb{N}} (A_i \setminus C_i)\right)$$
$$\leq \sum_{i \in \mathbb{N}} \nu(A_i \setminus C_i) < \sum_{i \in \mathbb{N}} \frac{\varepsilon}{2^i} = \varepsilon,$$

where the first inequality is by (2) of Lemma 3.28. This shows that $A \in \mathfrak{H}$.

To verify 3°, let $(A_i : i \in \mathbb{N}) \subset \mathfrak{H}$ and let $A = \bigcup_{i \in \mathbb{N}} A_i$. Since $A_i \in \mathfrak{M}(\mu)$ for every $i \in \mathbb{N}$, we have $A \in \mathfrak{M}(\mu)$. Let $\varepsilon > 0$ be arbitrarily given. Since $A_i \in \mathfrak{H}$ there exists $C_i \in \mathfrak{C}_X$ such that $C_i \subset A_i$ and $\nu(A_i \setminus C_i) < \frac{\varepsilon}{2^i}$. Now $\left(A \setminus \bigcup_{i=1}^m C_i : m \in \mathbb{N}\right)$ is a decreasing sequence in $\mathfrak{M}(\mu)$. Also $A \setminus \bigcup_{i=1}^m C_i \subset A$ and $\nu(A) \leq \nu(X) < \infty$. Thus we have

$$\begin{aligned}\lim_{m\to\infty}\nu\left(A\setminus\bigcup_{i=1}^{m}C_i\right)&=\nu\left(\lim_{m\to\infty}\left(A\setminus\bigcup_{i=1}^{m}C_i\right)\right)=\nu\left(A\setminus\bigcup_{i\in\mathbb{N}}C_i\right)\\&=\nu\left(\bigcup_{i\in\mathbb{N}}A_i\setminus\bigcup_{i\in\mathbb{N}}C_i\right)\le\nu\left(\bigcup_{i\in\mathbb{N}}(A_i\setminus C_i)\right)\\&\le\sum_{i\in\mathbb{N}}\nu(A_i\setminus C_i)<\sum_{i\in\mathbb{N}}\frac{\varepsilon}{2^i}=\varepsilon,\end{aligned}$$

where the first inequality is by (1) of Lemma 3.28. Thus there exists $N\in\mathbb{N}$ such that $\nu\big(A\setminus\bigcup_{i=1}^{N}C_i\big)<\varepsilon$. Let $C:=\bigcup_{i=1}^{N}C_i\in\mathfrak{C}_X$. This shows that $A\in\mathfrak{H}$.

To verify 4°, note first that $\emptyset\in\mathfrak{O}_X$ and $\emptyset\in\mathfrak{C}_X$ also. Then by 1° we have $\emptyset\in\mathfrak{H}$. Let G be a non-empty open set. By Proposition 2.18, G is a union of countably many closed sets. Thus $G\in\mathfrak{H}$ by 1° and 3°. This shows that $\mathfrak{O}_X\subset\mathfrak{H}$.

2. Let

$$\mathfrak{F}=\{A\in\mathfrak{H}:A^c\in\mathfrak{H}\}\subset\mathfrak{H}.\tag{2}$$

Let us show that

5° $\mathfrak{O}_X\subset\mathfrak{F}$.

6° $\mathfrak{F}$ is closed under complementations.

7° $\mathfrak{F}$ is closed under countable unions.

To verify 5°, let $O\in\mathfrak{O}_X$. Then $O\in\mathfrak{H}$ by 4° and $O^c\in\mathfrak{C}_X\subset\mathfrak{H}$ by 1°. Thus $O\in\mathfrak{F}$. This shows that $\mathfrak{O}_X\subset\mathfrak{F}$.

To verify 6°, let $A\in\mathfrak{F}$. Then we have $A\in\mathfrak{H}$ and $A^c\in\mathfrak{H}$. Then $A^c\in\mathfrak{H}$ and $(A^c)^c=A\in\mathfrak{H}$ and thus $A^c\in\mathfrak{F}$. This shows that $\mathfrak{F}$ is closed under complementations.

To verify 7°, let $(A_i:n\in\mathbb{N})\subset\mathfrak{F}$ and let $A=\bigcup_{i\in\mathbb{N}}A_i$. To show that $A\in\mathfrak{F}$ we show that $A\in\mathfrak{H}$ and $A^c\in\mathfrak{H}$ also. Now since $A_i\in\mathfrak{F}\subset\mathfrak{H}$ and $\mathfrak{H}$ is closed under countable unions by 3°, we have $A\in\mathfrak{H}$. Since $A_i\in\mathfrak{F}$ and $\mathfrak{F}$ is closed under complementations by 6°, we have $A_i^c\in\mathfrak{F}\subset\mathfrak{H}$. Then $A^c=\big(\bigcup_{i\in\mathbb{N}}A_i\big)^c=\bigcap_{i\in\mathbb{N}}A_i^c\in\mathfrak{H}$ since $\mathfrak{H}$ is closed under countable intersections by 2°. Thus we have $A\in\mathfrak{H}$ and $A^c\in\mathfrak{H}$ so that $A\in\mathfrak{F}$.

3. By 5°, 6° and 7°, $\mathfrak{F}$ is a σ-algebra of subsets of X. By 5° we have $\mathfrak{O}_X\subset\mathfrak{F}$ and hence $\mathfrak{B}_X=\sigma(\mathfrak{O}_X)\subset\mathfrak{F}$. Thus if $E\in\mathfrak{B}_X\subset\mathfrak{F}\subset\mathfrak{H}$ then $E\in\mathfrak{H}$ so that by (1) for every $\varepsilon>0$ there exists $C\in\mathfrak{C}_X$ such that $C\subset E$ and $\nu(E\setminus C)<\varepsilon$, that is, $\mu\big((E\setminus C)\cap E\big)<\varepsilon$, that is, $\mu(E\setminus C)<\varepsilon$. ∎

Theorem 3.31. (External Approximation of Borel Sets by Open Sets) *Let μ be a boundedly finite Borel outer measure on a separable metric space (X,ρ). Let $E\in\mathfrak{B}_X$. Then for every $\varepsilon>0$ there exists an open set O such that $O\supset E$ and $\mu(O\setminus E)<\varepsilon$.*

Proof. With $x_0\in X$ arbitrarily chosen, consider the open balls $B_m(x_0)$ with center x_0 and radius m for $m\in\mathbb{N}$. For brevity let us write $O_m:=B_m(x_0)$ for $m\in\mathbb{N}$. Then

$(O_m : m \in \mathbb{N})$ is an increasing sequence of open sets such that $\bigcup_{m\in\mathbb{N}} O_m = X$ and $\mu(O_m) < \infty$ for $m \in \mathbb{N}$ since μ is a boundedly finite outer measure.

Let $E \in \mathfrak{B}_X$. For each $m \in \mathbb{N}$, we have $O_m \setminus E \in \mathfrak{B}_X$ and $\mu(O_m \setminus E) \leq \mu(O_m) < \infty$. Thus by Theorem 3.30 for every $\varepsilon > 0$ there exists a closed sets C_m such that

$$C_m \subset O_m \setminus E \tag{1}$$

and

$$\mu((O_m \setminus E) \setminus C_m) < \frac{\varepsilon}{2^m}.$$

Now $(O_m \setminus E) \setminus C_m = (O_m \setminus C_m) \setminus E$ by Lemma 3.29 so that we have

$$\mu((O_m \setminus C_m) \setminus E) < \frac{\varepsilon}{2^m}. \tag{2}$$

Since O_m is an open set and C_m is a closed set, $O_m \setminus C_m = O_m \cap C_m^c$ is an open set. Let

$$O = \bigcup_{m\in\mathbb{N}} (O_m \setminus C_m). \tag{3}$$

Then O is an open set. By (1), $C_m \subset O_m \cap E^c$ so that $C_m^c \supset \left(O_m \cap E^c\right)^c = O_m^c \cup E \supset E$ and then we have

$$O_m \cap C_m^c \supset O_m \cap E. \tag{4}$$

Thus by (3) and (4), we have

$$\begin{aligned} E = X \cap E = \left(\bigcup_{m\in\mathbb{N}} O_m\right) \cap E = \bigcup_{m\in\mathbb{N}} (O_m \cap E) \\ \subset \bigcup_{m\in\mathbb{N}} (O_m \cap C_m^c) = \bigcup_{m\in\mathbb{N}} (O_m \setminus C_m) = O. \end{aligned}$$

Moreover we have

$$\begin{aligned} \mu(O \setminus E) &= \mu\left(\bigcup_{m\in\mathbb{N}} (O_m \setminus C_m) \setminus E\right) = \mu\left(\bigcup_{m\in\mathbb{N}} ((O_m \setminus C_m) \setminus E)\right) \\ &\leq \sum_{m\in\mathbb{N}} \mu((O_m \setminus C_m) \setminus E) < \sum_{m\in\mathbb{N}} \frac{\varepsilon}{2^m} = \varepsilon, \end{aligned}$$

where the second inequality is by (2). ∎

The last two theorems show internal approximation by closed sets and external approximation by open sets for a Borel set with respect to a Borel outer measure. We show next that assumption of Borel-regularity of a Borel outer measure enables us to approximate an arbitrary set externally by open sets.

Theorem 3.32. (External Approximation of Arbitrary Sets by Open Sets) *Let (X,ρ) be a separable metric space. Let μ be a boundedly finite Borel-regular Borel outer measure on X. Let $\mathfrak{O}_X$ be the collection of open sets in X. Then for every $A \subset X$, we have*

$$\mu(A) = \inf \{\mu(O) : A \subset O, O \in \mathfrak{O}_X\}. \tag{1}$$

Proof. 1. Consider first the case $\mu(A) = \infty$. There exists at least one open set containing A, namely X. Let O be an arbitrary open set containing A. By the monotonicity of the outer measure μ we have $\mu(O) \geq \mu(A) = \infty$ and thus $\mu(O) = \infty$. Then (1) holds trivially.

2. Consider the case $\mu(A) < \infty$. To prove (1) we show that for every $\varepsilon > 0$ there exists an open set O such that $A \subset O$ and

$$\mu(O) < \mu(A) + \varepsilon. \tag{2}$$

Let us assume that $A \in \mathfrak{B}_X$. Then by Theorem 3.31, for every $\varepsilon > 0$ there exists an open set O such that $A \subset O$ and $\mu(O \setminus A) < \varepsilon$. Now the outer measure μ on X is a measure on the σ-algebra $\mathfrak{M}(\mu)$. Then since $\mathfrak{B}_X \subset \mathfrak{M}(\mu)$, μ is a measure on the σ-algebra $\mathfrak{B}_X$. Then since $A, O \in \mathfrak{B}_X$, $A \subset O$, and $\mu(A) < \infty$, we have $\mu(O \setminus A) = \mu(O) - \mu(A)$. Thus $\mu(O) = \mu(A) + \mu(O \setminus A) < \mu(A) + \varepsilon$. This proves (2) for $A \in \mathfrak{B}_X$.

Let A be an arbitrary subset of X. Since μ is a Borel-regular Borel outer measure on X there exists $E \in \mathfrak{B}_X$ such that $A \subset E$ and $\mu(A) = \mu(E)$. Since $E \in \mathfrak{B}_X$, by our result above we have

$$\mu(A) = \mu(E) = \inf \{\mu(O) : E \subset O, O \in \mathfrak{O}_X\}. \tag{3}$$

Since $A \subset E$, the collection of all open sets containing A is greater than the collection of all open set containing E and hence

$$\{\mu(O) : E \subset O, O \in \mathfrak{O}_X\} \subset \{\mu(O) : A \subset O, O \in \mathfrak{O}_X\}$$

and then

$$\inf \{\mu(O) : E \subset O, O \in \mathfrak{O}_X\} \geq \inf \{\mu(O) : A \subset O, O \in \mathfrak{O}_X\}. \tag{4}$$

Then by (3) and (4) we have

$$\mu(A) \geq \inf \{\mu(O) : A \subset O, O \in \mathfrak{O}_X\}. \tag{5}$$

On the other hand from the monotonicity of the outer measure μ, we have

$$\mu(A) \leq \inf \{\mu(O) : A \subset O, O \in \mathfrak{O}_X\}. \tag{6}$$

Then by (5) and (6) we have (1). ∎

To compare a Borel outer measure with a Radon outer measure which we treat next, we prepare a modification of Theorem 3.30 as follows.

Theorem 3.33. (Internal Approximation of Borel Sets by Bounded Closed Sets) *Let μ be a Borel outer measure on a separable metric space (X, ρ). Let $E \in \mathfrak{B}_X$ with $\mu(E) < \infty$. Then for every $\varepsilon > 0$ there exists a bounded closed set K in X such that $K \subset E$ and $\mu(E \setminus K) < \varepsilon$.*

Proof. Let $E \in \mathfrak{B}_X$ with $\mu(E) < \infty$. Let $\varepsilon > 0$ be arbitrarily given. According to Theorem 3.30, there exists a closed set C in X such that

$$C \subset E \quad \text{and} \quad \mu(E \setminus C) < \frac{\varepsilon}{2}. \tag{1}$$

Let $x_0 \in C$ be arbitrarily chosen and fixed. Consider closed balls in X, $\overline{B}(x_0, n)$ for $n \in \mathbb{N}$. Then $\left(\overline{B}(x_0, n) : n \in \mathbb{N}\right)$ is an increasing sequence of bounded closed sets in X and

$$\lim_{n\to\infty} \overline{B}(x_0, n) = \bigcup_{n\in\mathbb{N}} \overline{B}(x_0, n) = X.$$

Let $K_n = \overline{B}(x_0, n) \cap C$ for $n \in \mathbb{N}$. Then $(K_n : n \in \mathbb{N})$ is an increasing sequence of bounded closed sets contained in C and

$$\lim_{n\to\infty} K_n = \bigcup_{n\in\mathbb{N}} K_n = \left(\bigcup_{n\in\mathbb{N}} \overline{B}(x_0, n)\right) \cap C = X \cap C = C. \tag{2}$$

Consider the measure space $(X, \mathfrak{B}_X, \mu)$. Now we have $K_n \in \mathfrak{B}_X$ for every $n \in \mathbb{N}$ and $C \in \mathfrak{B}_X$. Thus (2) implies

$$\lim_{n\to\infty} \mu(K_n) = \mu\Big(\lim_{n\to\infty} K_n\Big) = \mu(C). \tag{3}$$

This implies that there exists $N \in \mathbb{N}$ such that $\mu(C) - \mu(K_N) < \frac{\varepsilon}{2}$, that is,

$$\mu(C \setminus K_N) < \frac{\varepsilon}{2}. \tag{4}$$

Now we have $K_N \subset C \subset E$ so that $E \setminus K_N = (E \setminus C) \cup (C \setminus K_N)$. This then implies by (1) and (4)

$$\mu(E \setminus K_N) = \mu(E \setminus C) + \mu(C \setminus K_N) < \frac{\varepsilon}{2} + \frac{\varepsilon}{2} = \varepsilon.$$

This completes the proof. ∎

Theorem 3.34. *Let μ be a Borel outer measure on a separable metric space (X, ρ). Let $\mathfrak{C}_X^b$ be the collection of all bounded closed sets in X. Then for every $E \in \mathfrak{B}_X$ with $\mu(E) < \infty$, we have*

$$\mu(E) = \sup\left\{\mu(K) : K \subset E, K \in \mathfrak{C}_X^b\right\}. \tag{1}$$

Proof. Observe that the statement (1) is equivalent to the statement that for every $E \in \mathfrak{B}_X$ with $\mu(E) < \infty$, for every $\varepsilon > 0$ there exists $K \in \mathfrak{C}_X^b$ such that

$$K \subset E \quad \text{and} \quad \mu(E) - \mu(K) < \varepsilon. \tag{2}$$

Now according to Theorem 3.33, if $E \in \mathfrak{B}_X$ with $\mu(E) < \infty$, then for every $\varepsilon > 0$ there exists $K \in \mathfrak{C}_b$ such that

$$K \subset E \quad \text{and} \quad \mu(E \setminus K) < \varepsilon. \tag{3}$$

Now $E \in \mathfrak{B}_X$ and $K \in \mathfrak{C}_X^b \subset \mathfrak{B}_X$. Consider the measure space $(X, \mathfrak{B}_X, \mu)$. Then the statement $\mu(E \setminus K) < \varepsilon$ is equivalent to the statement $\mu(E) - \mu(K) < \varepsilon$. Thus (3) proves (2) and hence (1). ■

[VI] Radon Outer Measure

A Radon outer measure is a refinement of a Borel outer measure having additional approximation properties. Before stating a formal definition for a Radon outer measure we review the concept of a compact set in a topological space.

Review. (Compact Sets)
• A set K in a topological space $(X, \mathfrak{O})$ is called a compact set if for every $\{O_\alpha : \alpha \in A\} \subset \mathfrak{O}$ such that $\bigcup_{\alpha \in A} O_\alpha \supset K$ there exists a finite subcollection $\{O_{\alpha_1}, \dots, O_{\alpha_m}\} \subset \{O_\alpha : \alpha \in A\}$ such that $\bigcup_{i=1}^m O_{\alpha_i} \supset K$.
• If $K_1, \dots, K_m$ are compact sets in $(X, \mathfrak{O})$, then $\bigcup_{i=1}^m K_i$ is a compact set.
• If K is a compact set and C is a closed set in $(X, \mathfrak{O})$, then $K \cap C$ is a compact set.
• A topological space $(X, \mathfrak{O})$ is called a Hausdorff space if for every $x_1, x_2 \in X$ such that $x_1 \neq x_2$ there exist $O_1, O_2 \in \mathfrak{O}$ such that $x_1 \in O_1$, $x_2 \in O_2$ and $O_1 \cap O_2 = \emptyset$.
• Every metric space (X, ρ) is a Hausdorff space.
• Every compact set in a Hausdorff space is a closed set.
• In particular, every compact set in a metric space (X, ρ) is a closed set.
• If $\{K_\alpha : \alpha \in A\}$ is an arbitrary collection of compact sets in a Hausdorff space, then $\bigcap_{\alpha \in A} K_\alpha$ is a compact set.

Proposition 3.35. Every compact set K in a metric space (X, ρ) is a bounded closed set.

Proof. 1. Every compact set in a Hausdorff space is a closed set. Since a metric space is a Hausdorff space, every compact set in a metric space is a closed set.

2. Let us show that if K is a compact set in a metric space (X, ρ) then K is a bounded set. For each $x \in K$, construct an open ball $B(x, 1)$, that is, an open ball with center at x and with radius 1. The collection of open balls, $\{B(x, 1) : x \in K\}$, is an open cover of the compact set K and hence there exists a finite subcover, that is, there exists $\{x_1, \dots, x_p\} \subset K$ such that $K \subset \bigcup_{i=1}^p B(x_i, 1)$. This shows that K is a bounded set. ∎

Definition 3.36. (Bounded-closed Compact Metric Space) *Let us call a metric space (X, ρ) a bounded-closed compact metric space if a set in X is compact if and only if it is a bounded closed set.*

Example. $(\mathbb{R}^n, \rho_e)$ is an example of bounded-closed compact metric space.

Definition 3.37. (Radon Outer Measure) *Let X be a topological space. Let $\mathfrak{O}$ be the collection of all open sets in X and let $\mathfrak{K}$ be the collection of all compact sets in X. An outer measure μ on X is called a Radon outer measure if it satisfies the following conditions:*

1° μ *is a Borel outer measure on X, that is,* $\mathfrak{B}_X \subset \mathfrak{M}(\mu)$.
2° $\mu(K) < \infty$ *for every* $K \in \mathfrak{K}$.
3° $\mu(O) = \sup\big\{\mu(K) : K \subset O, K \in \mathfrak{K}\big\}$ *for every* $O \in \mathfrak{O}$ *with* $\mu(O) < \infty$.
4° $\mu(E) = \inf\big\{\mu(O) : E \subset O, O \in \mathfrak{O}\big\}$ *for every* $E \subset X$.

Proposition 3.38. *Let μ be a Radon outer measure on a bounded-closed compact metric space (X, ρ). Then we have:*
(a) *μ is locally finite.*
(b) *μ is boundedly finite.*

Proof. 1. Let us show that μ is locally finite. Let $x_0 \in X$. With $r > 0$, consider the closed ball with center x_0 and radius r

$$\overline{B}(x_0, r) = \{x \in X : \rho(x_0, x) \leq r\}.$$

This is a bounded closed set in X. Since our metric space (X, ρ) is a bounded-closed compact metric space, the bounded closed set $\overline{B}(x_0, r)$ is a compact set. Then since μ is a Radon outer measure we have $\mu(\overline{B}(x_0, r)) < \infty$. This implies that $\mu(B(x_0, r)) < \infty$. This shows that μ is locally finite.

2. Let us show that μ is boundedly finite. Let E be a bounded set in X. Then $\overline{E}$ is a bounded closed set. Then since (X, ρ) is a bounded-closed compact metric space, $\overline{E}$ is a compact set. Then since μ is a Radon outer measure we have $\mu(\overline{E}) < \infty$. This implies that $\mu(E) < \infty$. This shows that μ is boundedly finite. ∎

Theorem 3.39. (Sufficient Conditions for Radon Outer Measure) *Let (X, ρ) be a bounded-closed compact and separable metric space. Then every boundedly finite Borel-regular Borel outer measure on X is a Radon outer measure. In particular, every finite Borel-regular Borel outer measure on X is a Radon outer measure.*

Proof. 1. Let μ be a boundedly finite Borel-regular Borel outer measure on X. To show that μ is a Radon outer measure, we verify that μ satisfies conditions 2°, 3° and 4° of Definition 3.3. Let $\mathfrak{O}$ be the collection of all open sets in (X, ρ) and let $\mathfrak{K}$ be the collection of all compact sets in (X, ρ).

2. Let K be a compact set in (X, ρ). Then since (X, ρ) is a bounded-closed compact metric space, K is a bounded closed set. Then since μ is a boundedly finite outer measure, we have $\mu(K) < \infty$. This shows that μ satisfies condition 2°.

3. According to Theorem 3.34, we have for every $E \in \mathfrak{B}_X$ with $\mu(E) < \infty$

$$\mu(E) = \sup \{\mu(K) : K \subset E, K \in \mathfrak{K}\}.$$

Then in particular for every $O \in \mathfrak{O} \subset \mathfrak{B}_X$ with $\mu(O) < \infty$, we have

$$\mu(O) = \sup \{\mu(K) : K \subset O, K \in \mathfrak{K}\}.$$

This shows that μ satisfies condition 3°.

4. According to Theorem 3.32, we have for every $A \in \mathfrak{P}(X)$

$$\mu(A) = \inf \{\mu(O) : A \subset O, O \in \mathfrak{O}\}.$$

Then in particular for every $E \in \mathfrak{B}_X \subset \mathfrak{P}(X)$, we have

$$\mu(E) = \inf \{\mu(O) : E \subset O, O \in \mathfrak{O}\}.$$

This shows that μ satisfies condition 4°. ∎

Corollary 3.40. *Let (X, ρ) be a bounded-closed compact and separable metric space. Let μ be a Borel-regular Borel outer measure on X. Let $A \in \mathfrak{B}_X$ with $\mu(A) < \infty$. Then the truncated outer measure $\mu|_A$ is a Radon outer measure on X.*

Proof. According to Theorem 3.24, the truncated outer measure $\mu|_A$ is still a Borel-regular Borel outer measure on X. Then condition $\mu(A) < \infty$ implies that $\mu|_A(X) = \mu(X \cap A) = \mu(A) < \infty$, that is, $\mu|_A$ is a finite outer measure. Then by Theorem 3.39, $\mu|_A$ is a Radon outer measure on X. Then every finite Borel-regular Borel outer measure on X is a Radon outer measure. ∎

Theorem 3.41. *Let μ be a Borel-regular Borel outer measure on the metric space $(\mathbb{R}^n, \rho_e)$. Suppose that $\mu(K) < \infty$ for every compact set $K \subset \mathbb{R}^n$. Then μ is a Radon outer measure on $\mathbb{R}^n$.*

Proof. Let E be a bounded set in $\mathbb{R}^n$. Then $\overline{E}$ is a bounded closed set in $\mathbb{R}^n$ and hence $\overline{E}$ is a compact set in $\mathbb{R}^n$ and then $\mu(\overline{E}) < \infty$ by our assumption. This implies that $\mu(E) < \infty$ and hence the outer measure μ is a finite outer measure. Then by Theorem 3.39, μ is a Radon outer measure on $\mathbb{R}^n$. ∎

Observation 3.42. Let μ be a Borel outer measure on a topological space X. If μ satisfies the condition that

$$\mu(A) = \inf\{\mu(V) : A \subset V, V \text{ is open}\} \text{ for every } A \subset X,$$

then μ is a Borel-regular outer measure.

Proof. Suppose μ satisfies the condition. Let $A \subset X$. Then for every $k \in \mathbb{N}$ there exists an open set V_k such that $A \subset V_k$ and $\mu(V_k) \leq \mu(A) + 1/k$. Consider the sequence $(V_k : k \in \mathbb{N})$ of open sets containing A. Let $O_k = \cap_{i=1}^k V_i$ for $k \in \mathbb{N}$. Then $(O_k : k \in \mathbb{N})$ is a decreasing sequence of open sets containing A. Since $O_k \subset V_k$, we have $\mu(O_k) \leq \mu(V_k) \leq \mu(A) + 1/k$ for every $k \in \mathbb{N}$. Since O_k is a Borel set for every $k \in \mathbb{N}$, we have $B := \cap_{k \in \mathbb{N}} O_k$ is a Borel set. Since $A \subset O_k$ for every $k \in \mathbb{N}$, we have $A \subset B$ and $\mu(A) \leq \mu(B)$. Now since $B \subset O_k$ for every $k \in \mathbb{N}$, we have $\mu(B) \leq \mu(O_k) \leq \mu(A) + 1/k$ for every $k \in \mathbb{N}$. Then $\mu(B) = \mu(A)$. ∎

Observation 3.43. Every Radon outer measure μ on a topological space X is a Borel-regular outer measure on X. This follows from the fact that condition 2° in Definition 3.37 implies Borel-regularity of μ according to Observation 3.42.

[VII] Lebesgue Outer Measure

We construct the Lebesgue outer measure on $(\mathbb{R}^n, \rho_e)$ and show that it is an example of the Borel-regular Borel outer measure and the Radon outer measure.

Definition 3.44. *Consider the separable metric space $(\mathbb{R}^n, \rho_e)$.*
(a) *We call a set $\prod_{i=1}^n (\alpha_i, \beta_i] \subset \mathbb{R}^n$ an open-closed rectangle. We call a set $\prod_{i=1}^n (\alpha_i, \beta_i) \subset \mathbb{R}^n$ an open rectangle.*
(b) *Let $\mathfrak{V}_{oc}$ be the collection of all open-closed rectangles in $\mathbb{R}^n$ and $\emptyset$. Let $\mathfrak{V}_o$ be the collection of all open rectangles in $\mathbb{R}^n$ and $\emptyset$.*
($\mathfrak{V}_{oc}$ and $\mathfrak{V}_o$ are then fine-covering classes for $\mathbb{R}^n$ according to Definition 2.19.)
(c) *Let v be a set function on $\mathfrak{V}_{oc}$ defined by $v(R) =$ volume of R for $R \in \mathfrak{V}_{oc}$ and $v(\emptyset) = 0$. We also write v for a set function on $\mathfrak{V}_o$ defined by $v(R) =$ volume of R for $R \in \mathfrak{V}_o$ and $v(\emptyset) = 0$.*
(v is then a premeasure on the covering classes $\mathfrak{V}_{oc}$ and $\mathfrak{V}_o$ for $\mathbb{R}^n$ by Definition 2.3.)

Observation 3.45. Observe that the advantage of an open-closed rectangle in $\mathbb{R}^n$ is that every open-closed rectangle can be decomposed into finitely many disjoint open-closed rectangles.

Definition 3.46. (Lebesgue Outer Measure μ_L^n) *Consider the separable metric space $(\mathbb{R}^n, \rho_e)$. We call the outer measure $\mu_{[v,\mathfrak{V}_{oc}]}$ on $\mathbb{R}^n$ generated by the premeasure v on the covering class $\mathfrak{V}_{oc}$ for $\mathbb{R}^n$ the Lebesgue outer measure on $\mathbb{R}^n$.*
We write μ_L^n for $\mu_{[v,\mathfrak{V}_{oc}]}$.

Proposition 3.47. (Equivalent Construction of Lebesgue Outer Measure) *Consider the Lebesgue outer measure $\mu_L^n = \mu_{[v,\mathfrak{V}_{oc}]}$ on $(\mathbb{R}^n, \rho_e)$. Let $\nu = \mu_{[v,\mathfrak{V}_o]}$ on $(\mathbb{R}^n, \rho_e)$. Then $\nu = \mu_L^n$ on $(\mathbb{R}^n, \rho_e)$.*

Proof. Let $E \in \mathfrak{P}(\mathbb{R}^n)$. Then we have

$$(1) \quad \mu_L^n(E) = \mu_{[v,\mathfrak{V}_{oc}]}(E) = \inf\Big\{\sum_{k\in\mathbb{N}} v(R_k) : (R_k : k \in \mathbb{N}) \subset \mathfrak{V}_{oc}, \bigcup_{k\in\mathbb{N}} R_k \supset E\Big\}$$

and

$$(2) \quad \nu(E) = \mu_{[v,\mathfrak{V}_o]}(E) = \inf\Big\{\sum_{k\in\mathbb{N}} v(O_k) : (O_k : k \in \mathbb{N}) \subset \mathfrak{V}_o, \bigcup_{k\in\mathbb{N}} O_k \supset E\Big\}.$$

1. Let us show first that $\nu(E) \le \mu_L^n(E)$. Let $\varepsilon > 0$. Let $(R_k : k \in \mathbb{N})$ be an arbitrary sequence in $\mathfrak{V}_{oc}$ such that $\bigcup_{k\in\mathbb{N}} R_k \supset E$. Then let $(O_k : k \in \mathbb{N})$ be an arbitrary sequence in $\mathfrak{V}_o$ such that $O_k \supset R_k$ and $v(O_k) \le v(R_k) + \frac{\varepsilon}{2^k}$ for every $k \in \mathbb{N}$. Then we have $\sum_{k\in\mathbb{N}} v(O_k) \le \sum_{k\in\mathbb{N}} v(R_k) + \varepsilon$. This implies that

$$\inf\Big\{\sum_{k\in\mathbb{N}} v(O_k) : (O_k : k \in \mathbb{N}) \subset \mathfrak{V}_o, \bigcup_{k\in\mathbb{N}} O_k \supset E\Big\} \le \sum_{k\in\mathbb{N}} v(R_k) + \varepsilon.$$

Since this holds for every sequence $(R_k : k \in \mathbb{N})$ in $\mathfrak{V}_{oc}$ such that $\bigcup_{k\in\mathbb{N}} R_k \supset E$, we have

$$\inf\left\{\sum_{k\in\mathbb{N}} v(O_k) : (O_k : k \in \mathbb{N}) \subset \mathfrak{V}_o, \bigcup_{k\in\mathbb{N}} O_k \supset E\right\}$$

$$\leq \inf\left\{\sum_{k\in\mathbb{N}} v(R_k) : (R_k : k \in \mathbb{N}) \subset \mathfrak{V}_{oc}, \bigcup_{k\in\mathbb{N}} R_k \supset E\right\} + \varepsilon,$$

that is, we have $\nu(E) \leq \mu_L^n(E) + \varepsilon$. Then since this holds for every $\varepsilon > 0$, we have $\nu(E) \leq \mu_L^n(E)$.

2. We show that $\mu_L^n(E) \leq \nu(E)$ by interchanging the roles of $\mathfrak{V}_{oc}$ and $\mathfrak{V}_o$ in the argument above. ■

Proposition 3.48. **(a)** *For the Lebesgue outer measure $\mu_L^n = \mu_{[v,\mathfrak{V}_{oc}]}$ on $\mathbb{R}^n$, we have*

(1) $$\mu_{[v,\mathfrak{V}_{oc}]}(R) = v(R) \quad \textit{for every } R \in \mathfrak{V}_{oc}.$$

(b) *For the outer measure $\nu = \mu_{[v,\mathfrak{V}_o]}$ on $\mathbb{R}^n$, we have*

(2) $$\mu_{[v,\mathfrak{V}_o]}(O) = v(O) \quad \textit{for every } O \in \mathfrak{V}_o.$$

Proof. 1. Let us prove (1). Let $R \in \mathfrak{V}_{oc}$. Then by Definition 2.5 for $\mu_{[v,\mathfrak{V}_{oc}]}$, we have

(1) $$\mu_{[v,\mathfrak{V}_{oc}]}(R) = \inf\left\{\sum_{k\in\mathbb{N}} v(R_k) : (R_k : n \in \mathbb{N}) \subset \mathfrak{V}_{oc}, \bigcup_{k\in\mathbb{N}} R_k \supset R\right\}.$$

Let Γ be a set of nonnegative extended real numbers defined by setting

(2) $$\Gamma = \left\{\sum_{k\in\mathbb{N}} v(R_k) : (R_k : k \in \mathbb{N}) \subset \mathfrak{V}_{oc}, \bigcup_{k\in\mathbb{N}} R_k \supset R\right\}.$$

Now for every sequence $(R_k : k \in \mathbb{N}) \subset \mathfrak{V}_{oc}$ such that $\bigcup_{k\in\mathbb{N}} R_k \supset R$, we have

(3) $$v(R) \leq \sum_{k\in\mathbb{N}} v(R_k).$$

Then (3) implies that the set Γ is bounded below by $v(R)$. On the other hand, with the one-term sequence $(R_k : k \in \mathbb{N}) := (R) \subset \mathfrak{V}_{oc}$, we have $\bigcup_{k\in\mathbb{N}} R_k \supset R$ and

(4) $$\sum_{k\in\mathbb{N}} v(R_k) = v(R).$$

This shows that $v(R)$ is not only a lower bound of Γ but is actually the minimum of Γ. Now the existence of $\min \Gamma$ implies that $\inf \Gamma = \min \Gamma$. Thus we have

$$\begin{aligned} v(R) &= \min \Gamma = \inf \Gamma \\ &= \inf\left\{\sum_{k\in\mathbb{N}} v(R_k) : (R_k : k \in \mathbb{N}) \subset \mathfrak{V}_{oc}, \bigcup_{k\in\mathbb{N}} R_k \supset R\right\} \quad \text{by (2)} \\ &= \mu_{[v,\mathfrak{V}_{oc}]}(R) \quad \text{by (1)}. \end{aligned}$$

This completes the proof of (1).

2. (2) is proved by the same argument as for (1). ■

Theorem 3.49. *Consider the Lebesgue outer measure* $\mu_L^n = \mu_{[v,\mathfrak{V}_{oc}]}$ *on the separable metric space* $(\mathbb{R}^n, \rho_e)$. *We have:*
(a) *The Lebesgue outer measure* $\mu_L^n = \mu_{[v,\mathfrak{V}_{oc}]}$ *is a metric outer measure on* $(\mathbb{R}^n, \rho_e)$.
(b) *The Lebesgue outer measure* $\mu_L^n = \mu_{[v,\mathfrak{V}_{oc}]}$ *is a Borel outer measure on* $(\mathbb{R}^n, \rho_e)$.

Proof. 1. To show that μ_L^n is a metric outer measure we show that if $E_1, E_2 \in \mathfrak{P}(\mathbb{R}^n)$ are such that $\rho_e(E_1, E_2) > 0$, then $\mu_L^n(E_1 \cup E_2) = \mu_L^n(E_1) + \mu_L^n(E_2)$. Since the outer measure μ_L^n is subadditive, it suffices to show

(1) $$\mu_L^n(E_1 \cup E_2) \geq \mu_L^n(E_1) + \mu_L^n(E_2).$$

Let us show that for every sequence $(R_k : k \in \mathbb{N}) \subset \mathfrak{V}_r$ such that $\bigcup_{k\in\mathbb{N}} R_k \supset E_1 \cup E_2$, we have

(2) $$\sum_{k\in\mathbb{N}} v(R_k) \geq \mu_L^n(E_1) + \mu_L^n(E_2).$$

Let $\delta = \rho_E(E_1, E_2) > 0$. Now for each $k \in \mathbb{N}$, R_k can be decomposed into a finite collection $\{R_{k,1}, \ldots, R_{k,p_k}\} \subset \mathfrak{V}_r$ in such a way that

1° $\{R_{k,1}, \ldots, R_{k,p_k}\}$ is a disjoint collection and $R_k = R_{k,1} \cup \cdots \cup R_{k,p_k}$,

2° the lengths of the edges of $R_{k,1}, \ldots, R_{k,p_k}$ are all bounded above by $\delta/(2\sqrt{n})$.

Note that 1° implies that $v(R_k) = v(R_{k,1}) + \cdots + v(R_{k,p_k})$. Let the countable collection $\mathfrak{R} = \{R_{k,j_k} : j_k = 1, \ldots, p_k; k \in \mathbb{N}\}$ be arbitrarily enumerated as $\mathfrak{R} = \{F_m : m \in \mathbb{N}\}$. We have $\bigcup_{m\in\mathbb{N}} F_m \supset E_1 \cup E_2$ and

(3) $$\sum_{m\in\mathbb{N}} v(F_m) = \sum_{k\in\mathbb{N}} \sum_{j_k=1}^{p_k} v(R_{k,j_k}) = \sum_{k\in\mathbb{N}} v(R_k).$$

Let us show that no member of $\mathfrak{R} = \{F_m : m \in \mathbb{N}\}$ can intersect both E_1 and E_2. Assume the contrary, that is, for some $m \in \mathbb{N}$ the set F_m intersects both E_1 and E_2. Then $F_m \cap E_1 \neq \emptyset$ and $F_m \cap E_2 \neq \emptyset$. Let $x_1 \in F_m \cap E_1$ and $x_2 \in F_m \cap E_2$. Then $\delta = \rho_e(E_1, E_2) \leq \rho_e(x_1, x_2)$. Now since $x_1, x_2 \in F_m$ and every edge of F_m has length $\leq \delta/(2\sqrt{n})$, we have $\rho_e(x_1, x_2) \leq \left[n\left(\delta/(2\sqrt{n})\right)^2\right]^{1/2} = \frac{\delta}{2}$. Thus we have $\delta \leq \rho_e(x_1, x_2) \leq \delta/2$, a contradiction. This shows that no member of $\mathfrak{R} = \{F_m : m \in \mathbb{N}\}$ can intersect both E_1 and E_2. Let us classify $\{F_m : m \in \mathbb{N}\}$ into three subclasses: $\{F'_m : m \in \mathbb{N}\}$ consisting of those F_m which intersect E_1, $\{F''_m : m \in \mathbb{N}\}$ consisting of those F_m which intersect E_2, and $\{F'''_m : m \in \mathbb{N}\}$ consisting of those F_m which are disjoint from both E_1 and E_2. Then we have $E_1 \subset \bigcup_{m\in\mathbb{N}} F'_m$ and

$E_2 \subset \bigcup_{m \in \mathbb{N}} F''_m$ and therefore

(4)
$$\sum_{m \in \mathbb{N}} v(F_m) = \sum_{m \in \mathbb{N}} v(F'_m) + \sum_{m \in \mathbb{N}} v(F''_m) + \sum_{m \in \mathbb{N}} v(F'''_m)$$
$$\geq sum_{m \in \mathbb{N}} v(F'_m) + \sum_{m \in \mathbb{N}} v(F''_m)$$
$$\geq \mu^n_L(E_1) + \mu^n_L(E_2).$$

By (3) and (4), we have (2). Now since (2) holds for every sequence $(R_k : k \in \mathbb{N})$ in $\mathfrak{I}^n_{oc}$ such that $\bigcup_{k \in \mathbb{N}} R_k \supset E_1 \cup E_2$, we have $\mu^n_L(E_1 \cup E_2) \geq \mu^n_L(E_1) + \mu^n_L(E_2)$. This proves (1) and completes the proof that μ^n_L is a metric outer measure.

2. Let us show that μ^n_L is a Borel outer measure. We showed above that μ^n_L is a metric outer measure. According to Theorem 3.14, an outer measure on $(\mathbb{R}^n, \rho_e)$ is Borel outer measure if and only if it is a metric outer measure. Thus μ^n_L is a Borel outer measure. ∎

Theorem 3.50. (Borel-regularity of the Lebesgue Outer Measure on $\mathbb{R}^n$)
The Lebesgue outer measure $\mu^n_L = \mu_{[v, \mathfrak{V}_{oc}]}$ on $(\mathbb{R}^n, \rho_e)$ has the following properties.
(a) *For every $E \in \mathfrak{P}(\mathbb{R}^n)$ and $\varepsilon > 0$, there exists an open set O in $\mathbb{R}^n$ such that $O \supset E$ and*
$$\mu^n_L(E) \leq \mu^n_L(O) \leq \mu^n_L(E) + \varepsilon.$$
(b) *For every $E \in \mathfrak{P}(\mathbb{R}^n)$, there exists a G_δ-set G in $\mathbb{R}^n$ such that $G \supset E$ and*
$$\mu^n_L(G) = \mu^n_L(E).$$
(c) *μ^n_L is a Borel-regular Borel outer measure.*

Proof. 1. Let $E \in \mathfrak{P}(\mathbb{R}^n)$. By Definition 3.46 and Proposition 3.47, we have
$$\mu^n_L(E) = \inf \left\{ \sum_{k \in \mathbb{N}} v(O_k) : (O_k : k \in \mathbb{N}) \subset \mathfrak{V}_o, \bigcup_{k \in \mathbb{N}} O_k \supset E \right\}.$$
Thus for every $\varepsilon > 0$ there exists $(O_k : k \in \mathbb{N}) \subset \mathfrak{V}_o$ such that $\bigcup_{k \in \mathbb{N}} O_k \supset E$ and

(1)
$$\mu^n_L(E) \leq \sum_{k \in \mathbb{N}} v(O_k) \leq \mu^n_L(E) + \varepsilon.$$

Let $O = \bigcup_{k \in \mathbb{N}} O_k$. Then O is an open set and $E \subset \bigcup_{k \in \mathbb{N}} O_k = O$. Thus by the monotonicity of the outer measure μ^n_L we have

(2)
$$\mu^n_L(E) \leq \mu^n_L(O).$$

Also, by the countable subadditivity of the outer measure μ^n_L we have

(3)
$$\mu^n_L(O) = \mu^n_L\left(\bigcup_{k \in \mathbb{N}} O_k\right) \leq \sum_{k \in \mathbb{N}} \mu^n_L(O_k)$$
$$= \sum_{k \in \mathbb{N}} v(O_k) \leq \mu^n_L(E) + \varepsilon,$$

where the second equality is by (2) of Proposition 3.48 and the last inequality is by (1). By (2) and (3) we have (a).

2. Let $E \in \mathfrak{P}(\mathbb{R}^n)$. By (a), for every $k \in \mathbb{N}$ there exists an open set O_k such that $O_k \supset E$ and $\mu_L^n(E) \leq \mu_L^n(O_k) \leq \mu_L^n(E) + \frac{1}{k}$. Let $G = \bigcap_{k \in \mathbb{N}} O_k$. Then G is a G_δ-set and $G \supset E$. Since $G \subset O_k$ for every $k \in \mathbb{N}$, we have $\mu_L^n(E) \leq \mu_L^n(G) \leq \mu_L^n(O_k) \leq \mu_L^n(E) + \frac{1}{k}$ by the monotonicity of the outer measure μ_L^n. Since this holds for every $k \in \mathbb{N}$, we have $\mu_L^n(E) \leq \mu_L^n(G) \leq \mu_L^n(E)$ and therefore $\mu_L^n(E) = \mu_L^n(G)$.

3. According to Theorem 3.49, μ_L^n is a Borel outer measure on $\mathbb{R}^n$. Now if G is a G_δ-set in $\mathbb{R}^n$ then $G \in \mathfrak{B}_{\mathbb{R}^n}$. Then according to (b), for every $E \in \mathfrak{P}(\mathbb{R}^n)$ there exists $G \in \mathfrak{B}_{\mathbb{R}^n}$ such that $G \supset E$ and $\mu_L^n(G) = \mu_L^n(E)$. This shows that μ_L^n is a Borel-regular Borel outer measure according to Definition 3.17. ∎

Theorem 3.51. *Consider the Lebesgue outer measure $\mu_L^n = \mu_{[v, \mathfrak{V}_{oc}]}$ on $(\mathbb{R}^n, \rho_e)$. We have:*
(a) *$\mu_L^n = \mu_{[\gamma_v, \mathfrak{V}_{oc}]}$ is a boundedly finite outer measure.*
(b) *$\mu_L^n = \mu_{[\gamma_v, \mathfrak{V}_{oc}]}$ is a Radon outer measure.*

Proof. 1. Let us prove (a). Thus we are to show that if E is a bounded set in $\mathbb{R}^n$ then $\mu_{[v, \mathfrak{V}_{oc}]}(E) < \infty$. Let $E \in \mathfrak{P}(\mathbb{R}^n)$ be a bounded set. Then there exists $r > 0$ such that $E \subset B(0, r)$, an open ball with center $0 \in \mathbb{R}^n$ and radius $r > 0$. Now we have $B(0, r) \subset \prod_{i=1}^n (-r, r] \in \mathfrak{V}_{oc}$. Then we have, applying Proposition 3.48,

$$\mu_{[v, \mathfrak{V}_{oc}]}(E) \leq \mu_{[v, \mathfrak{V}_{oc}]}(B(0, r)) \leq \mu_{[v, \mathfrak{V}_{oc}]}\left(\prod_{i=1}^n (-r, r]\right)$$
$$= v\left(\prod_{i=1}^n (-r, r]\right) = (2r)^n < \infty.$$

This shows that $\mu_{[v, \mathfrak{V}_{oc}]}$ is a boundedly finite outer measure.

2. Our metric space $(\mathbb{R}^n, \rho_e)$ is an example of bounded-closed compact separable metric space. We showed above that μ_L^n is a boundedly finite outer measure. We showed in Theorem 3.50 that μ_L^n is a Borel-regular Borel outer measure. Then by Theorem 3.39, μ_L^n is a Radon outer measure. ∎

§4 Hausdorff Measures

[I] Hausdorff Outer Measure

Definition 2.1 (Covering Class) requires that a covering class $\mathfrak{V}$ of subsets of a set X include $\emptyset$ as a member.
Definition 2.3 (Premeasure) defines premeasure γ as a set function on a covering class $\mathfrak{V}$ satisfying conditions:
1° $\gamma(V) \in [0, \infty]$ for every $V \in \mathfrak{V}$.
2° $\gamma(\emptyset) = 0$.

Definition 4.1. (Hausdorff Premeasure of Dimension $s \in [0, \infty)$**)** *Let (X, ρ) be a metric space. Let $s \in [0, \infty)$. Let γ_s be a set function on a covering class $\mathfrak{V}$ defined by setting*
1° $\gamma_s(V) = |V|^s$ for $V \in \mathfrak{V}$ such that $V \neq \emptyset$.
2° $\gamma_s(\emptyset) = 0$.
We call γ_s an s-dimensional Hausdorff premeasure on $\mathfrak{V}$.

Remark 4.2. Condition 2° in Definition 4.1 is explicitly stipulated since condition 1° for every $V \in \mathfrak{V}$ would imply in the case $s = 0$ that $\gamma_0(\emptyset) = |\emptyset|^0 = 0^0 = 1$, contradicting condition 2° of Definition 2.3.

Let (X, ρ) be a separable metric space. Then $\mathfrak{P}(X)$ is a fine-covering class for X according to Theorem 2.22 and for every $\delta \in (0, \infty]$ the truncation $\mathfrak{P}(X)|_\delta$ is a fine-covering class for X according to Proposition 2.24.
Let $s \in [0, \infty)$. With an s-dimensional Hausdorff premeasure γ_s on a fine-covering class $\mathfrak{P}(X)|_\delta$ for X, consider the outer measure on X generated by the premeasure γ_s on the covering class $\mathfrak{P}(X)|_\delta$ for X, that is, for every $E \in \mathfrak{P}(X)$,

$$\mu_{[\gamma_s, \mathfrak{P}(X)|_\delta]}(E) := \inf \left\{ \sum_{n \in \mathbb{N}} \gamma_s(V_n) : (V_n : n \in \mathbb{N}) \subset \mathfrak{P}(X)|_\delta, \bigcup_{n \in \mathbb{N}} V_n \supset E \right\}.$$

Convention 4.3. Consider a sequence $(V_n : n \in \mathbb{N})$ in a covering class $\mathfrak{V}$ for X. If $V_k = \emptyset$ then V_k does not cover any point in X and besides we have $\gamma_s(V_k) = 0$ by Definition 4.1 and thus $\gamma_s(V_k)$ has no contribution to the sum $\sum_{n \in \mathbb{N}} \gamma_s(V_n)$ and hence existence of V_k in the sequence $(V_n : n \in \mathbb{N})$ has no effect on $\mu_{[\gamma_s, \mathfrak{V}]}(E)$. Thus in defining $\mu_{[\gamma_s, \mathfrak{V}]}(E)$, we assume that every sequence $(V_n : n \in \mathbb{N})$ in a covering class $\mathfrak{V}$ is such that $V_n \neq \emptyset$ for every $n \in \mathbb{N}$. This convention/assumption results in having $\gamma_s(V_n) = |V_n|^s$ for every $n \in \mathbb{N}$.

Definition 4.4. *Let (X, ρ) be a separable metric space. Let $s \in [0, \infty)$ and $\delta \in (0, \infty]$. Let $\mathcal{H}_\delta^s$ be an outer measure on X defined by setting for every $E \in \mathfrak{P}(X)$*

$$\mathcal{H}_\delta^s(E) = \mu_{[\gamma_s, \mathfrak{P}(X)|_\delta]}(E) = \inf \left\{ \sum_{n \in \mathbb{N}} \gamma_s(V_n) : (V_n : n \in \mathbb{N}) \subset \mathfrak{P}(X)|_\delta, \bigcup_{n \in \mathbb{N}} V_n \supset E \right\}.$$

Let us call $\mathcal{H}_\delta^s$ an s-dimensional δ-level Hausdorff outer measure.

Consider the collection $\{\mathcal{H}_\delta^s : s \in [0,\infty), \delta \in (0,\infty]\}$ of outer measures on a separable metric space (X,ρ). Now by Definition 4.4, we have for every $E \in \mathfrak{P}(X)$

$$\mathcal{H}_\delta^s(E) = \mu_{[\gamma_s,\mathfrak{P}(X)|\delta]}(E).$$

According to Theorem 2.27, we have $\mu_{[\gamma_s,\mathfrak{P}(X)|\delta]}(E) \uparrow$ as $\delta \downarrow 0$ and $\lim_{\delta\downarrow 0} \mu_{[\gamma_s,\mathfrak{P}(X)|\delta]}(E) \in [0,\infty]$ and if we define a set function ν on $\mathfrak{P}(X)$ by setting for every $E \in \mathfrak{P}(X)$

$$\nu(E) = \lim_{\delta\downarrow 0} \mu_{[\gamma_s,\mathfrak{P}(X)|\delta]}(E),$$

then ν is an outer measure on (X,ρ).

Definition 4.5. **(s-dimensional Hausdorff Outer Measure)** *Let (X,ρ) be a separable metric space. Let $s \in [0,\infty)$ and $\delta \in (0,\infty]$. Consider the outer measure $\mathcal{H}^s$ on X defined by setting for every $E \in \mathfrak{P}(X)$*

$$\mathcal{H}^s(E) := \lim_{\delta\downarrow 0} \mathcal{H}_\delta^s(E).$$

We call $\mathcal{H}^s$ an s-dimensional Hausdorff outer measure on X.

Theorem 4.6. *Consider the s-dimensional Hausdorff outer measure $\mathcal{H}^s$ on a separable metric space (X,ρ). We have:*
(a) *$\mathcal{H}^s$ is a metric outer measure on (X,ρ).*
(b) *$\mathcal{H}^s$ is a Borel outer measure on (X,ρ).*

Proof. 1. Let $s \in [0,\infty)$ and $\delta \in (0,\infty]$. Then for every $E \in \mathfrak{P}(X)$ we have

$$\mathcal{H}^s(E) = \lim_{\delta\downarrow 0} \mathcal{H}_\delta^s(E).$$

In particular we have

$$\mathcal{H}^s(E) = \lim_{n\to\infty} \mathcal{H}_{\frac{1}{n}}^s(E).$$

Then according to Theorem 3.4, $\mathcal{H}^s$ is a metric outer measure.

2. According to Theorem 3.14, an outer measure on a metric space (X,ρ) is a metric outer measure if and only if it is a Borel outer measure. In (a) we showed that $\mathcal{H}^s$ is a metric outer measure on (X,ρ). Then $\mathcal{H}^s$ is a Borel outer measure on (X,ρ) by Theorem 3.14. ∎

We showed above that $\mathcal{H}^s$ is a Borel outer measure on (X,ρ). We show later that $\mathcal{H}^s$ is a Borel-regular Borel outer measure after an alternate construction of $\mathcal{H}^s$.

Next let us study the family $\{\mathcal{H}_\delta^s : s \in [0,\infty), \delta \in (0,\infty]\}$ of outer measures on (X,ρ).

Lemma 4.7. *Consider $\mathcal{H}^0_\delta(E)$ for $\delta \in (0,\infty]$ and $\mathcal{H}^0(E)$ for $E \in \mathfrak{P}(X)$.*
(a) *If $\mathcal{H}^0_{\delta_0}(E) = 0$ for some $\delta_0 \in (0,\infty]$, then $E = \emptyset$.*
(b) *If $\mathcal{H}^0(E) = 0$, then $E = \emptyset$.*

Proof. 1. Let us prove (a). Let $E \in \mathfrak{P}(X)$. According to Definition 4.4, we have

$$\mathcal{H}^0_{\delta_0}(E) = \inf\Big\{\sum_{n\in\mathbb{N}} \gamma_0(V_n) : (V_n : n \in \mathbb{N}) \subset \mathfrak{P}(X)|_{\delta_0}, \bigcup_{n\in\mathbb{N}} V_n \supset E\Big\}.$$

Then the assumption $\mathcal{H}^0_{\delta_0}(E) = 0$ implies that for every $\varepsilon \in (0,1)$ there exists a sequence $(V_n : n \in \mathbb{N}) \subset \mathfrak{P}(X)|_{\delta_0}$ such that $\bigcup_{n\in\mathbb{N}} V_n \supset E$ and $\sum_{n\in\mathbb{N}} \gamma_0(V_n) < \varepsilon$.

To show that $E = \emptyset$, assume the contrary, that is, assume that $E \neq \emptyset$. Then since $\bigcup_{n\in\mathbb{N}} V_n \supset E$, there exists $n_0 \in \mathbb{N}$ such that $V_{n_0} \neq \emptyset$. Then we have $\gamma_0(V_{n_0}) = |V_{n_0}|^0 = 1$ and consequently $\sum_{n\in\mathbb{N}} \gamma_0(V_n) \geq 1 > \varepsilon$, a contradiction. Therefore we must have $E = \emptyset$.

2. Let us prove (b). Suppose $\mathcal{H}^0(E) = 0$. Since $\mathcal{H}^0(E) \geq \mathcal{H}^0_\delta(E)$ for every $\delta \in (0,\infty]$, we have $\mathcal{H}^0_\delta(E) = 0$ for every $\delta \in (0,\infty]$. Then by (a), we have $E = \emptyset$. ∎

Corollary 4.8. *Consider $\mathcal{H}^0_\delta(E)$ for $\delta \in (0,\infty]$ and $\mathcal{H}^0(E)$ for $E \in \mathfrak{P}(X)$.*
(a) *If $E \neq \emptyset$ then for every $\delta \in (0,\infty]$ we have $\mathcal{H}^0_\delta(E) > 0$.*
(b) *In particular for any $x \in X$ we have $\mathcal{H}^0_\delta(\{x\}) = 1$ for every $\delta \in (0,\infty]$.*
(c) *For any $x \in X$ we have $\mathcal{H}^0(\{x\}) = 1$.*

Proof. 1. (a) is the contra-positive of (a) in Lemma 4.7.

2. To prove (b), observe that we have by Definition 4.4

$$(1)\qquad \mathcal{H}^0_\delta(\{x\}) = \inf\Big\{\sum_{n\in\mathbb{N}} \gamma_0(V_n) : (V_n : n \in \mathbb{N}) \subset \mathfrak{P}(X)|_\delta, \bigcup_{n\in\mathbb{N}} V_n \supset \{x\}\Big\}.$$

Now $\bigcup_{n\in\mathbb{N}} V_n \supset \{x\}$ implies that there exists $n_0 \in \mathbb{N}$ such that $x \in V_{n_0}$ and then $V_{n_0} \neq \emptyset$ and then $\gamma_0(V_{n_0}) = |V_{n_0}|^0 = 1$ and then $\sum_{n\in\mathbb{N}} \gamma_0(V_n) \geq 1$. Thus we have

$$(2)\qquad \inf\Big\{\sum_{n\in\mathbb{N}} \gamma_0(V_n) : (V_n : n \in \mathbb{N}) \subset \mathfrak{P}(X)|_\delta, \bigcup_{n\in\mathbb{N}} V_n \supset \{x\}\Big\} \geq 1.$$

Let $V \in \mathfrak{P}(X)|_\delta$ be such that that $\{x\} \subset V$. Then $V \neq \emptyset$ and $\gamma_0(V) = |V|^0 = 1$. Then with the one-term covering sequence $(V_n : n \in \mathbb{N}) := (V)$ for $\{x\}$, we have

$$(3)\qquad \sum_{n\in\mathbb{N}} \gamma_0(V_n) = |V|^0 = 1.$$

Then (2) and (3) imply

$$(4)\qquad \inf\Big\{\sum_{n\in\mathbb{N}} \gamma_0(V_n) : (V_n : n \in \mathbb{N}) \subset \mathfrak{P}(X)|_\delta, \bigcup_{n\in\mathbb{N}} V_n \supset \{x\}\Big\} = 1.$$

Then (1) and (4) imply that $\mathcal{H}^0_\delta(\{x\}) = 1$.

3. Since $\mathcal{H}^0(\{x\}) = \lim_{\delta \downarrow 0} \mathcal{H}^0_\delta(\{x\})$ and $\mathcal{H}^0_\delta(\{x\}) = 1$ for every $\delta \in (0, \infty]$, we have $\mathcal{H}^0(\{x\}) = 1$. ∎

Proposition 4.9. *Consider $\mathcal{H}^s_\delta(E)$ where $s \in [0, \infty)$, $\delta \in (0, \infty]$ and $E \in \mathfrak{P}(X)$. We have:*
(a) *If $\mathcal{H}^s_{\delta_0}(E) > 0$ for some $\delta_0 \in (0, \infty]$, then $\mathcal{H}^s_\delta(E) > 0$ for every $\delta \in (0, \infty]$.*
(b) *If $\mathcal{H}^s_{\delta_0}(E) = 0$ for some $\delta_0 \in (0, \infty]$, then $\mathcal{H}^s_\delta(E) = 0$ for every $\delta \in (0, \infty]$.*

Proof. 1. Let us prove (a). Suppose $\mathcal{H}^s_{\delta_0}(E) > 0$ for some $\delta_0 \in (0, \infty]$. Now $\mathcal{H}^s_\delta(E) \uparrow$ as $\delta \downarrow 0$ and hence $\mathcal{H}^s_\delta(E) \downarrow$ as $\delta \uparrow \infty$. Then to show that $\mathcal{H}^s_\delta(E) > 0$ for every $\delta \in (0, \infty]$, it suffices to show that $\mathcal{H}^s_\infty(E) > 0$.

Case $s \in (0, \infty)$. To show that $\mathcal{H}^s_\infty(E) > 0$, assume the contrary, that is, assume that $\mathcal{H}^s_\infty(E) = 0$. This implies that for an arbitrary $\varepsilon > 0$ there exists a sequence $(V_n : n \in \mathbb{N}) \subset \mathfrak{P}(X)|_\infty$ such that $\bigcup_{n\in\mathbb{N}} V_n \supset E$ and $\sum_{n\in\mathbb{N}} \gamma_s(V_n) < \varepsilon^s$. Then for every $n \in \mathbb{N}$ we have $\gamma_s(V_n) < \varepsilon^s$. If $V_n \neq \emptyset$, then $\gamma_s(V_n) = |V_n|^s$ by Definition 4.1 and thus $|V_n|^s < \varepsilon^s$ and then $|V_n| < \varepsilon$ so that $V_n \in \mathfrak{P}(X)|_\varepsilon$. If $V_n = \emptyset$, then $V_n \in \mathfrak{P}(X)|_\varepsilon$ trivially. This shows that we actually have $(V_n : n \in \mathbb{N}) \subset \mathfrak{P}(X)|_\varepsilon$. This implies

$$\mathcal{H}^s_\varepsilon(E) \le \sum_{n\in\mathbb{N}} \gamma_s(V_n) < \varepsilon^s$$

and then

$$\mathcal{H}^s(E) = \lim_{\varepsilon \downarrow 0} \mathcal{H}^s_\varepsilon(E) \le \lim_{\varepsilon \downarrow 0} \varepsilon^s = 0 \quad \text{since } s > 0.$$

Then since $\mathcal{H}^s_{\delta_0}(E) \le \mathcal{H}^s(E)$, we have $\mathcal{H}^s_{\delta_0}(E) = 0$, contradicting the assumption that $\mathcal{H}^s_{\delta_0}(E) > 0$. Thus $\mathcal{H}^s_\infty(E) > 0$ must hold.

Case $s = 0$. To show that $\mathcal{H}^0_\infty(E) > 0$, assume the contrary, that is, assume that $\mathcal{H}^0_\infty(E) = 0$. Now $\mathcal{H}^0_\infty(E) = 0$ implies that $E = \emptyset$ according to (a) of Lemma 4.7. Then $\mathcal{H}^0_{\delta_0}(E) = \mathcal{H}^0_{\delta_0}(\emptyset) = 0$, contradicting the assumption that $\mathcal{H}^0_{\delta_0}(E) > 0$. Thus $\mathcal{H}^0_\infty(E) > 0$ must hold.

2. Let us prove (b). Suppose $\mathcal{H}^s_{\delta_0}(E) = 0$ for some $\delta_0 \in (0, \infty]$. To show that $\mathcal{H}^s_\delta(E) = 0$ for every $\delta \in (0, \infty]$, assume the contrary, that is, assume that there exists $\delta_0 \in (0, \infty]$ such that $\mathcal{H}^s_{\delta_0}(E) > 0$. Then by (a) we have $\mathcal{H}^s_\delta(E) > 0$ for every $\delta \in (0, \infty]$. This is a contradiction. ∎

Theorem 4.10. *For $s \in [0, \infty)$ and $\delta \in (0, \infty]$, consider the outer measures $\mathcal{H}^s_\delta$ and $\mathcal{H}^s$ on a separable metric space (X, ρ). Let $E \in \mathfrak{P}(X)$.*
(a) *Either $\mathcal{H}^s_\delta(E) > 0$ for all $\delta \in (0, \infty]$ or $\mathcal{H}^s_\delta(E) = 0$ for all $\delta \in (0, \infty]$.*
(b) *If $\mathcal{H}^s(E) > 0$, then $\mathcal{H}^s_\delta(E) > 0$ for all $\delta \in (0, \infty]$.*
(c) *If $\mathcal{H}^s(E) = 0$, then $\mathcal{H}^s_\delta(E) = 0$ for all $\delta \in (0, \infty]$.*

Proof. 1. (a) is just a restatement of Proposition 4.9.

2. Let us prove (b). Suppose $\mathcal{H}^s(E) > 0$. Then since $\mathcal{H}^s_\delta(E) \uparrow \mathcal{H}^s(E)$ as $\delta \downarrow 0$, there exists $\delta_0 \in (0, \infty]$ such that $\mathcal{H}^s_{\delta_0}(E) > 0$. This implies that $\mathcal{H}^s_\delta(E) > 0$ for every $\delta \in (0, \infty]$ according to (a) of Proposition 4.9.

3. Let us prove (c). Suppose $\mathcal{H}^s(E) = 0$. Then since $\mathcal{H}^s_\delta(E) \uparrow \mathcal{H}^s(E)$ as $\delta \downarrow 0$, we have $\mathcal{H}^s_\delta(E) = 0$ for every $\delta \in (0, \infty]$. ∎

[II] Counting Outer Measure and $\mathcal{H}^0$

We defined and treated counting-measure space in Theorem 1.24. Let us give a formal definition of a counting outer measure now.

Theorem 4.11. (Counting Outer Measure) *Let X be an arbitrary set. Let us define a set function $\mathcal{N}$ on $\mathfrak{P}(X)$ by setting for every $E \in \mathfrak{P}(X)$*

$$\mathcal{N}(E) = \begin{cases} \text{number of elements of } E & \text{if } E \text{ is a finite set,} \\ \infty & \text{if } E \text{ is an infinite set.} \end{cases}$$

Then $\mathcal{N}$ is an outer measure on X. We call $\mathcal{N}$ a counting outer measure on X.

Proof. It is easily verified that $\mathcal{N}$ satisfies conditions 1°, 2°, 3° and 4° of Definition 1.14 (Outer Measure). ■

Theorem 4.12. *Consider $\mathcal{H}^0$, the Hausdorff outer measure of dimension 0 on a separable metric space (X, ρ). Let $\mathcal{N}$ be the counting outer measure on X. Then we have $\mathcal{H}^0 = \mathcal{N}$.*

Proof. 1. To show that $\mathcal{H}^0 = \mathcal{N}$ on X, we show that for every $E \in \mathfrak{P}(X)$ we have

$$\mathcal{H}^0(E) = \begin{cases} \text{number of elements of } E & \text{if } E \text{ is a finite set,} \\ \infty & \text{if } E \text{ is an infinite set.} \end{cases}$$

2. Since $\mathcal{H}^0$ is an outer measure, we have $\mathcal{H}^0(\emptyset) = 0$.

3. Let $k \in \mathbb{N}$. Suppose E consists of k distinct elements $x_1, \dots, x_k$. Then $\rho(x_i, x_j) > 0$ for $i, j = 1, \dots, k$ and $i \neq j$ and thus $\min_{i \neq j} \rho(x_i, x_j) > 0$. This implies that

$$\rho(\{x_i\}, \{x_{i+1}, \dots, x_k\}) > 0 \quad \text{for } i = 1, \dots, k-1,$$

that is, the two sets, $\{x_i\}$ and $\{x_{i+1}, \dots, x_k\}$, are positively separated. Now $\mathcal{H}^0$ is a metric outer measure on (X, ρ) according to Theorem 4.6. Thus we have

$$\mathcal{H}^0(\{x_1, \dots, x_k\}) = \mathcal{H}^0(\{x_1\} \cup \{x_2, \dots, x_k\}) = \mathcal{H}^0(\{x_1\}) + \mathcal{H}^0(\{x_2, \dots, x_k\}).$$

Then repeating the argument we have

$$\mathcal{H}^0(\{x_2, \dots, x_k\}) = \mathcal{H}^0(\{x_2\} \cup \{x_3, \dots, x_k\}) = \mathcal{H}^0(\{x_2\}) + \mathcal{H}^0(\{x_3, \dots, x_k\}).$$

Iterating the process we have finally

$$\mathcal{H}^0(\{x_{k-1}, x_k\}) = \mathcal{H}^0(\{x_{k_1}\} \cup \{x_k\}) = \mathcal{H}^0(\{x_{k-1}\}) + \mathcal{H}^0(\{x_k\}).$$

Summarizing the equalities obtained above, we have

$$\mathcal{H}^0(\{x_1, \dots, x_k\}) = \sum_{i=1}^{k} \mathcal{H}^0(\{x_i\}).$$

Now according to (c) of Corollary 4.8 we have $\mathcal{H}^0(\{x_i\}) = 1$. Thus we have

$$\mathcal{H}^0(\{x_1, \ldots, x_k\}) = k.$$

Thus we have shown that if $E \in \mathfrak{P}(X)$ consists of k distinct elements then $\mathcal{H}^0(E) = k$.

4. If E is an infinite set, for each $k \in \mathbb{N}$ let E_{k+1} be a subset of E consisting of $k+1$ distinct elements of E. By the monotonicity of the outer measure $\mathcal{H}^0$, we have $\mathcal{H}^0(E) \geq \mathcal{H}^0(E_{k+1}) = k+1 > k$. Since this holds for every $k \in \mathbb{N}$, we have $\mathcal{H}^0(E) = \infty$. ∎

Theorem 4.12 shows that the Hausdorff outer measure of dimension 0, $\mathcal{H}^0$, is a counting outer measure on a separable metric space (X, ρ). We show next that $\mathcal{H}^s$ with $s > 0$ is not a counting outer measure. We show that for every countable set $E \in \mathfrak{P}(X)$ we have $\mathcal{H}^s(E) = 0$.

Proposition 4.13. *Consider the Hausdorff outer measure $\mathcal{H}^s$ where $s > 0$ on a separable metric space (X, ρ). For every countable subset E of X, we have $\mathcal{H}^s(E) = 0$.*

Proof. Let $E = \{x_n : n \in \mathbb{N}\} \subset X$ be an arbitrary countable subset of X. Since $\mathcal{H}^s(E) = \lim_{\delta \downarrow 0} \mathcal{H}^s_\delta(E)$, to show that $\mathcal{H}^s(E) = 0$ it suffices to show that $\mathcal{H}^s_\delta(E) = 0$ for every $\delta \in (0, \infty]$.

Let $\delta \in (0, \infty]$. Consider the fine-covering class $\mathfrak{P}(X)|_\delta$ for X in Definition 4.4 for $\mathcal{H}^s_\delta$. Let $\varepsilon \in (0,1)$ be arbitrarily chosen. Define a sequence of open balls in X, $(W_n : n \in \mathbb{N})$, by setting

$$W_n = B(x_n, r_n) \quad \text{with} \quad r_n = \frac{1}{2} \cdot 2^{-\frac{n}{s}} \varepsilon \delta.$$

Then we have $|W_n| = 2r_n = 2^{-\frac{n}{s}} \varepsilon \delta \leq \varepsilon \delta < \delta$. This shows that we have $(W_n : n \in \mathbb{N}) \subset \mathfrak{P}(X)|_\delta$. Observe that since $W_n \neq \emptyset$, we have by Definition 4.1

$$\gamma_s(W_n) = |W_n|^s = \left(2^{-\frac{n}{s}} \varepsilon \delta\right)^s = 2^{-n} \varepsilon^s \delta^s.$$

Then we have

$$\begin{aligned}\mathcal{H}^s_\delta(E) &= \inf \left\{ \sum_{n \in \mathbb{N}} \gamma_s(V_n) : (V_n : n \in \mathbb{N}) \subset \mathfrak{P}(X)|_\delta, \bigcup_{n \in \mathbb{N}} V_n \supset E \right\} \\ &\leq \sum_{n \in \mathbb{N}} |W_n|^s = \sum_{n \in \mathbb{N}} \frac{1}{2^n} \varepsilon^s \delta^s = \varepsilon^s \delta^s.\end{aligned}$$

Since this holds for every $\varepsilon \in (0,1)$, we have $\mathcal{H}^s_\delta(E) = 0$. ∎

[III] Alternate Constructions for the Hausdorff Outer Measure

By definition, $\mathcal{H}^s(E) = \lim_{\delta \downarrow 0} \mathcal{H}^s_\delta(E)$ for $E \in \mathfrak{P}(X)$ and $\mathcal{H}^s_\delta$ is an outer measure on a separable metric space (X, ρ) generated by the premeasure γ_s on a fine-covering class $\mathfrak{P}(X)|_\delta$ for X. Let $\mathfrak{O}_X$ be the collection of all open sets in X and let $\mathfrak{C}_X$ be the collection of all closed sets in X. By Theorem 2.22, $\mathfrak{O}_X$ and $\mathfrak{C}_X$ are fine-covering classes for X. We show below that $\mathcal{H}^s$ can be constructed by using the same premeasure γ_s defined on the smaller fine-covering classes $\mathfrak{O}_X|_\delta$ and $\mathfrak{C}_X|_\delta$ than $\mathfrak{P}(X)|_\delta$.

Observation 4.14. Let (X, ρ) be a separable metric space. Let $\mathfrak{O}_X$ be the collection of all open sets in X and let $\mathfrak{C}_X$ be the collection of all closed sets in X. By Theorem 2.22, $\mathfrak{O}_X$ and $\mathfrak{C}_X$ are fine-covering classes for X. With $s \in [0, \infty)$ and $\delta \in (0, \infty]$, define two set functions $\mathcal{F}^s_\delta$ and $\mathcal{G}^s_\delta$ on $\mathfrak{P}(X)$ by setting for every $E \in \mathfrak{P}(X)$

$$\mathcal{F}^s_\delta(E) = \inf \left\{ \sum_{n \in \mathbb{N}} \gamma_s(F_n) : (F_n : n \in \mathbb{N}) \subset \mathfrak{C}_X|_\delta, \bigcup_{n \in \mathbb{N}} F_n \supset E \right\},$$

$$\mathcal{G}^s_\delta(E) = \inf \left\{ \sum_{n \in \mathbb{N}} \gamma_s(G_n) : (G_n : n \in \mathbb{N}) \subset \mathfrak{O}_X|_\delta, \bigcup_{n \in \mathbb{N}} G_n \supset E \right\}.$$

By Theorem 2.4, $\mathcal{F}^s_\delta$ and $\mathcal{G}^s_\delta$ are outer measures on X. Then according to Theorem 2.28, $\mathcal{F}^s_\delta(E) \uparrow$ and $\mathcal{G}^s_\delta(E) \uparrow$ as $\delta \downarrow 0$ and if we define two set functions $\mathcal{F}^s$ and $\mathcal{G}^s$ on $\mathfrak{P}(X)$ by setting for every $E \in \mathfrak{P}(X)$

$$\mathcal{F}^s(E) = \lim_{\delta \downarrow 0} \mathcal{F}^s_\delta(E),$$

$$\mathcal{G}^s(E) = \lim_{\delta \downarrow 0} \mathcal{G}^s_\delta(E),$$

then $\mathcal{F}^s$ and $\mathcal{G}^s$ are outer measures on X. We show below that $\mathcal{F}^s(E) = \mathcal{G}^s(E) = \mathcal{H}^s(E)$, that is, $\mathcal{F}^s = \mathcal{G}^s = \mathcal{H}^s$ on X.

Theorem 4.15. (Alternate Constructions for the Hausdorff Outer Measure) *For the outer measures $\mathcal{F}^s$ and $\mathcal{G}^s$ on a separable metric space (X, ρ) as defined in Observation 4.14, we have*

$$\mathcal{F}^s = \mathcal{G}^s = \mathcal{H}^s \quad \text{on } X.$$

Proof. Let us show first that for $0 < \delta < \delta'$ and $E \in \mathfrak{P}(X)$, we have

$$\mathcal{G}^s_\delta(E) \geq \mathcal{H}^s_\delta(E) = \mathcal{F}^s_\delta(E) \geq \mathcal{G}^s_{\delta'}(E). \tag{1}$$

To prove the first inequality in (1), recall that by Definition 4.4 and Convention 4.3, we have

$$\mathcal{H}^s_\delta(E) = \inf \left\{ \sum_{n \in \mathbb{N}} |V_n|^s : (V_n : n \in \mathbb{N}) \subset \mathfrak{P}(X)|_\delta, \bigcup_{n \in \mathbb{N}} V_n \supset E \right\}.$$

Now $\mathfrak{O}_X \subset \mathfrak{P}(X)$ implies $\mathfrak{O}_X|_\delta \subset \mathfrak{P}(X)|_\delta$ and this implies $\mathcal{G}_\delta^s(E) \geq \mathcal{H}_\delta^s(E)$ by Proposition 2.8. This proves the first inequality in (1).

Similarly $\mathfrak{C}_X \subset \mathfrak{P}(X)$ implies $\mathfrak{C}_X|_\delta \subset \mathfrak{P}(X)|_\delta$ and this implies $\mathcal{F}_\delta^s(E) \geq \mathcal{H}_\delta^s(E)$. Let us prove the reverse inequality, that is, $\mathcal{F}_\delta^s(E) \leq \mathcal{H}_\delta^s(E)$. Now if $\mathcal{H}_\delta^s(E) = \infty$, then $\mathcal{F}_\delta^s(E) \leq \mathcal{H}_\delta^s(E)$ holds trivially. Suppose $\mathcal{H}_\delta^s(E) < \infty$. Then for an arbitrary $\eta > 0$, there exists a sequence $(V_n : n \in \mathbb{N}) \subset \mathfrak{P}(X)|_\delta$ such that $\bigcup_{n\in\mathbb{N}} V_n \supset E$ and $\sum_{n\in\mathbb{N}} \gamma_s(V_n) \leq \mathcal{H}_\delta^s(E) + \eta$. Now $\overline{V_n} \in \mathfrak{C}_X$ and $|\overline{V_n}| = |V_n|$ for $n \in \mathbb{N}$ so that $(\overline{V_n} : n \in \mathbb{N}) \subset \mathfrak{C}_X|_\delta$ and $\bigcup_{n\in\mathbb{N}} \overline{V_n} \supset E$. Then we have $\mathcal{F}_\delta^s(E) \leq \sum_{n\in\mathbb{N}} \gamma_s(\overline{V_n})$. Now if $\overline{V_n} \neq \emptyset$ then $\gamma_s(\overline{V_n}) = |\overline{V_n}|^s = |V_n|^s = \gamma_s(V_n)$ and if $\overline{V_n} = \emptyset$ then $\gamma_s(\overline{V_n}) = 0 = \gamma_s(V_n)$. Substituting these into the last inequality, we have $\mathcal{F}_\delta^s(E) \leq \sum_{n\in\mathbb{N}} \gamma_s(V_n) \leq \mathcal{H}_\delta^s + \eta$. Then the arbitrariness of $\eta > 0$ implies that $\mathcal{F}_\delta^s(E) \leq \mathcal{H}_\delta^s(E)$. With this reverse inequality, we have the equality $\mathcal{H}_\delta^s(E) = \mathcal{F}_\delta^s(E)$.

It remains to prove the last inequality in (1). If $\mathcal{F}_\delta^s(E) = \infty$, then $\mathcal{F}_\delta^s(E) \geq \mathcal{G}_{\delta'}^s(E)$ holds trivially. Assume that $\mathcal{F}_\delta^s(E) < \infty$. Then for an arbitrary $\eta > 0$, there exists a sequence $(F_n : n \in \mathbb{N}) \subset \mathfrak{C}_X|_\delta$ such that $\bigcup_{n\in\mathbb{N}} F_n \supset E$ and

$$\sum_{n\in\mathbb{N}} |F_n|^s \leq \mathcal{F}_\delta^s(E) + \eta. \tag{2}$$

Let us assume that for every $n \in \mathbb{N}$ we select $F_n \neq \emptyset$ so that we have $\gamma_s(F_n) = |F_n|^s$ by Definition 4.1. Since $0 < \delta < \delta'$, for each $n \in \mathbb{N}$ there exists $\varepsilon_n > 0$ such that

$$\begin{cases} \delta + 2\varepsilon_n < \delta'. \\ \{|F_n| + 2\varepsilon_n\}^s < |F_n|^s + \frac{\eta}{2^n}. \end{cases} \tag{3}$$

Let $G_n = \bigcup_{x\in F_n} B(x, \varepsilon_n)$. Then G_n is an open set in X containing F_n and moreover by (c) of Lemma 2.15 we have

$$|G_n| \leq |F_n| + 2\varepsilon_n \leq \delta + 2\varepsilon_n < \delta'. \tag{4}$$

This implies that $(G_n : n \in \mathbb{N}) \subset \mathfrak{O}_X|_{\delta'}$. It also implies

$$|G_n|^s \leq \{|F_n| + 2\varepsilon_n\}^s < |F_n|^s + \frac{\eta}{2^n}. \tag{5}$$

Then we have by (5) and (2)

$$\sum_{n\in\mathbb{N}} |G_n|^s \leq \sum_{n\in\mathbb{N}} \left\{|F_n|^s + \frac{\eta}{2^n}\right\} \leq \mathcal{F}_\delta^s(E) + 2\eta. \tag{6}$$

Then we have

$$\mathcal{G}_{\delta'}^s(E) \leq \sum_{n\in\mathbb{N}} \gamma_s(G_n) = \sum_{n\in\mathbb{N}} |G_n|^s \leq \mathcal{F}_\delta^s(E) + 2\eta. \tag{7}$$

Then by the arbitrariness of $\eta > 0$ we have $\mathcal{G}_{\delta'}^s(E) \leq \mathcal{F}_\delta^s(E)$. This proves the last inequality in (1) and completes the proof of (1).

Now let $\delta \downarrow 0$ in (1). Then we have

(8) $$\mathcal{G}^s(E) \geq \mathcal{H}^s(E) = \mathcal{F}^s(E) \geq \mathcal{G}^s_{\delta'}(E).$$

Then let $\delta \downarrow 0$ in (2) to have

(9) $$\mathcal{G}^s(E) \geq \mathcal{H}^s(E) = \mathcal{F}^s(E) \geq \mathcal{G}^s(E).$$

This completes the proof. ∎

Corollary 4.16. (Borel-regularity of $\mathcal{H}^s$) *The s-dimensional Hausdorff outer measure $\mathcal{H}^s$ on a separable metric space (X, ρ) is a Borel-regular Borel outer measure.*

Proof. According to Theorem 4.6, $\mathcal{H}^s$ is a Borel outer measure on (X, ρ). Thus it remains to show that $\mathcal{H}^s$ is Borel-regular. According to Theorem 4.15, we have $\mathcal{G}^s = \mathcal{H}^s$. Thus to show that $\mathcal{H}^s$ is Borel-regular, we show that $\mathcal{G}^s$ is Borel-regular.

Now by Observation 4.14, we have for every $E \in \mathfrak{P}(X)$

$$\mathcal{G}^s_\delta(E) = \inf\left\{\sum_{n\in\mathbb{N}} \gamma_s(G_n) : (G_n : n \in \mathbb{N}) \subset \mathfrak{O}_X|_\delta, \bigcup_{n\in\mathbb{N}} G_n \supset E\right\}$$

and

$$\mathcal{G}^s(E) = \lim_{\delta\downarrow 0} \mathcal{G}^s_\delta(E).$$

Observe that $\mathfrak{O}_X \subset \mathfrak{B}_X$. Then according to Theorem 3.22, $\mathcal{G}^s$ is a Borel-regular Borel outer measure. This completes the proof that $\mathcal{H}^s$ is a Borel-regular Borel outer measure. ∎

We showed above that the Hausdorff outer measure $\mathcal{H}^s$ on a separable metric space (X, ρ) is Borel-regular, that is, for every $E \in \mathfrak{P}(X)$ there exists $B \in \mathfrak{B}_X$ such that $B \supset E$ and $\mathcal{H}^s(B) = \mathcal{H}^s(E)$. We show next that actually there exists a G_δ-set $G \in \mathfrak{B}_X$ such that $G \supset E$ and $\mathcal{H}^s(G) = \mathcal{H}^s(E)$. Now $\mathcal{H}^s = \mathcal{G}^s$. Thus we show the above regularity property for $\mathcal{G}^s$ rather than for $\mathcal{H}^s$ directly. The reason for the preference of $\mathcal{G}^s$ over $\mathcal{H}^s$ is the fact that $\mathcal{G}^s$ is constructed with $\mathfrak{O}_X$ as a covering class for X.

Proposition 4.17. *Consider the outer measure $\mathcal{G}^s$ on a separable metric space (X, ρ). Let $\mathfrak{O}_X$ be the collection of all open sets in X.*
(a) *For every $E \in \mathfrak{P}(X)$ and $\varepsilon > 0$, there exists $O \in \mathfrak{O}_X$ such that $O \supset E$ and*

$$\mathcal{G}^s(E) \leq \mathcal{G}^s(O) \leq \mathcal{G}^s(E) + \varepsilon.$$

(b) *For every $E \in \mathfrak{P}(X)$, there exists a G_δ-set G such that $G \supset E$ and $\mathcal{G}^s(G) = \mathcal{G}^s(E)$.*

Proof. 1. Let us prove (a). Let $E \in \mathfrak{P}(X)$ and let $\delta \in (0, \infty]$. Then by Observation 4.14,

$$\mathcal{G}^s_\delta(E) = \inf\left\{\sum_{n\in\mathbb{N}} \gamma_s(O_n) : (O_n : n \in \mathbb{N}) \subset \mathfrak{O}_X|_\delta, \bigcup_{n\in\mathbb{N}} O_n \supset E\right\}.$$

Then for every $\varepsilon > 0$, there exists $(O_n : n \in \mathbb{N}) \subset \mathfrak{O}_X|_\delta$ such that $\bigcup_{n\in\mathbb{N}} O_n \supset E$ and

$$\sum_{n\in\mathbb{N}} \gamma_s(O_n) \le \mathcal{G}^s_\delta(E) + \varepsilon.$$

Let $O = \bigcup_{n\in\mathbb{N}} O_n$. Then $O \in \mathfrak{O}_X$ and $O \supset E$. Now that $(O_n : n \in \mathbb{N}) \subset \mathfrak{O}_X|_\delta$ and $\bigcup_{n\in\mathbb{N}} O_n \supset O$, we have

$$\mathcal{G}^s_\delta(O) \le \sum_{n\in\mathbb{N}} \gamma_s(O_n) \le \mathcal{G}^s_\delta(E) + \varepsilon.$$

Since $E \subset O$, we have $\mathcal{G}^s_\delta(E) \le \mathcal{G}^s_\delta(O)$. Thus we have

$$\mathcal{G}^s_\delta(E) \le \mathcal{G}^s_\delta(O) \le \mathcal{G}^s_\delta(E) + \varepsilon.$$

Then letting $\delta \to 0$, we have

$$\mathcal{G}^s(E) \le \mathcal{G}^s(O) \le \mathcal{G}^s(E) + \varepsilon.$$

This proves (a).

2. Let us prove (b). By (a), for every $k \in \mathbb{N}$ there exists $O_k \in \mathfrak{O}_X$ such that $O_k \supset E$ and

$$\mathcal{G}^s(E) \le \mathcal{G}^s(O_k) \le \mathcal{G}^s(E) + \frac{1}{k}.$$

Let $G = \bigcap_{k\in\mathbb{N}} O_k$. Then G is a G_δ-set and $G \supset E$. Since $E \subset G$ and $G \subset O_k$ for every $k \in \mathbb{N}$, we have

$$\mathcal{G}^s(E) \le \mathcal{G}^s(G) \le \mathcal{G}^s(O_k) \le \mathcal{G}^s(E) + \frac{1}{k}.$$

Letting $k \to \infty$, we have

$$\mathcal{G}^s(E) \le \mathcal{G}^s(G) \le \mathcal{G}^s(E)$$

so that we have $\mathcal{G}^s(G) = \mathcal{G}^s(E)$. This completes the proof of (b). ∎

[IV] Hausdorff Dimension of Sets

Consider the Hausdorff outer measure $\mathcal{H}^s$ on a separable metric space (X, ρ). With $E \in \mathfrak{P}(X)$, let us study the behavior of $\mathcal{H}^s(E)$ as the parameter $s \in [0, \infty)$ varies.

Lemma 4.18. *For $s \in [0, \infty)$ and $\delta \in (0, \infty]$, consider the outer measure $\mathcal{H}^s_\delta$ on a separable metric space (X, ρ). Let $0 \le s' < s'' < \infty$. Then for every $E \in \mathfrak{P}(X)$, we have*

$$\mathcal{H}^{s''}_\delta(E) \le \delta^{s''-s'} \mathcal{H}^{s'}_\delta(E).$$

Proof. By Definition 4.4 for $\mathcal{H}^s_\delta$, we have

$$\text{(1)} \qquad \mathcal{H}^{s''}_\delta(E) = \inf \left\{ \sum_{n \in \mathbb{N}} \gamma_{s''}(V_n) : (V_n : n \in \mathbb{N}) \subset \mathfrak{P}(X)|_\delta, \bigcup_{n \in \mathbb{N}} V_n \supset E \right\}.$$

If $V_n = \emptyset$, then we have $\gamma_{s''}(V_n) = 0 = \gamma_{s'}(V_n)$. If $V_n \neq \emptyset$, then we have

$$\gamma_{s''}(V_n) = |V_n|^{s''} = |V_n|^{s''-s'} |V_n|^{s'} \le \delta^{s''-s'} \gamma_{s'}(V_n).$$

Substituting these estimates into (1), we obtain

$$\begin{aligned} \text{(2)} \qquad \mathcal{H}^{s''}_\delta(E) &\le \delta^{s''-s'} \inf \left\{ \sum_{n \in \mathbb{N}} \gamma_{s'}(V_n) : (V_n : n \in \mathbb{N}) \subset \mathfrak{P}(X)|_\delta, \bigcup_{n \in \mathbb{N}} V_n \supset E \right\} \\ &= \delta^{s''-s'} \mathcal{H}^{s'}_\delta(E). \end{aligned}$$

This completes the proof. ∎

Theorem 4.19. *Consider the Hausdorff outer measure $\mathcal{H}^s$ on a separable metric space (X, ρ). Let $E \in \mathfrak{P}(X)$ be arbitrarily chosen and fixed. Consider $\mathcal{H}^s(E)$ for $s \in [0, \infty)$. Let $0 \le s' < s'' < \infty$.*
(a) *We have $\mathcal{H}^{s'}(E) \ge \mathcal{H}^{s''}(E)$.*
(b) *If $\mathcal{H}^{s'}(E) < \infty$, then $\mathcal{H}^{s''}(E) = 0$.*
(c) *If $\mathcal{H}^{s''}(E) > 0$, then $\mathcal{H}^{s'}(E) = \infty$.*
(d) *If there exists $s^* \in [0, \infty)$ such that $\mathcal{H}^{s^*}(E) \in (0, \infty)$, then s^* is unique.*

Proof. 1. Let us prove (a). According to Lemma 4.18, we have for every $\delta \in (0, \infty]$

$$\mathcal{H}^{s''}_\delta(E) \le \delta^{s''-s'} \mathcal{H}^{s'}_\delta(E).$$

In particular for $\delta \in (0, 1]$, $s'' - s' > 0$ implies $\delta^{s''-s'} \le 1$ and thus $\mathcal{H}^{s''}_\delta(E) \le \mathcal{H}^{s'}_\delta(E)$. Then we have

$$\mathcal{H}^{s''}(E) = \lim_{\delta \downarrow 0} \mathcal{H}^{s''}_\delta(E) \le \lim_{\delta \downarrow 0} \mathcal{H}^{s'}_\delta(E) = \mathcal{H}^{s'}(E).$$

This proves (a).

2. Let us prove (b). Now we have according to Lemma 4.18

(1) $$\mathcal{H}^{s''}(E) = \lim_{\delta\downarrow 0} \mathcal{H}_\delta^{s''}(E) \leq \lim_{\delta\downarrow 0} \delta^{s''-s'} \mathcal{H}_\delta^{s'}(E).$$

If $\mathcal{H}^{s'}(E) < \infty$, then since $\mathcal{H}^{s'}(E) = \lim_{\delta\downarrow 0} \mathcal{H}_\delta^{s'}(E)$, we have

(2) $$\lim_{\delta\downarrow 0} \mathcal{H}_\delta^{s'}(E) < \infty.$$

Observe that since $s'' - s' > 0$, we have

(3) $$\lim_{\delta\downarrow 0} \delta^{s''-s'} = 0.$$

Substituting (2) and (3) in (1), we obtain $\mathcal{H}^{s''}(E) = 0$. This proves (b).

3. Let us prove (c). According to Lemma 4.18, we have

$$\mathcal{H}_\delta^{s''}(E) \leq \delta^{s''-s'} \mathcal{H}_\delta^{s'}(E) \quad \text{for every } \delta \in (0, \infty].$$

Let $\delta \in (0, \infty)$. Multiplying the last inequality by $\delta^{s'-s''} \in (0, \infty)$, we have

(4) $$\delta^{s'-s''} \mathcal{H}_\delta^{s''}(E) \leq \mathcal{H}_\delta^{s'}(E) \leq \mathcal{H}^{s'}(E).$$

If $\mathcal{H}^{s''}(E) > 0$, then since $\mathcal{H}^{s''}(E) = \lim_{\delta\downarrow 0} \mathcal{H}_\delta^{s''}(E)$, we have

(5) $$\lim_{\delta\downarrow 0} \mathcal{H}_\delta^{s''}(E) > 0.$$

Observe that since $s' - s'' < 0$, we have

(6) $$\lim_{\delta\downarrow 0} \delta^{s'-s''} = \infty.$$

Substituting (5) and (6) in (4), we obtain $\mathcal{H}^{s'}(E) = \infty$. This proves (c).

4. Let us prove (d). Let $s_1, s_2 \in [0, \infty)$ and $s_1 < s_2$.
If $\mathcal{H}^{s_1}(E) \in (0, \infty)$, then $\mathcal{H}^{s_2}(E) = 0$ by (b) and thus $\mathcal{H}^{s_2}(E) \notin (0, \infty)$.
If $\mathcal{H}^{s_2}(E) \in (0, \infty)$, then $\mathcal{H}^{s_1}(E) = \infty$ by (c) and thus $\mathcal{H}^{s_1}(E) \notin (0, \infty)$. ∎

Definition 4.20. *Consider the Hausdorff outer measure $\mathcal{H}^s$, where $s \in [0, \infty)$, on a separable metric space (X, ρ). Let $E \in \mathfrak{P}(X)$. Let $\mathcal{H}^{(\bullet)}(E)$ be a function on $s \in [0, \infty)$ defined by setting*

$$\mathcal{H}^{(s)}(E) = \mathcal{H}^s(E).$$

Let $\mathcal{R}(\mathcal{H}^{(\bullet)}(E))$ be the range of $\mathcal{H}^{(\bullet)}(E)$. Thus we have $\mathcal{R}(\mathcal{H}^{(\bullet)}(E)) \subset [0, \infty]$.

Observation 4.21. By Definition 4.20, $\mathcal{H}^{(\bullet)}(E)$ is a nonnegative extended real-valued function on $s \in [0, \infty)$. Then according to (a) of Theorem 4.19, $\mathcal{H}^{(\bullet)}(E)$ is a decreasing

function on $s \in [0, \infty)$.
Since $\mathcal{H}^0$ is a counting outer measure, $\mathcal{H}^0(E) = \infty$ if E is an infinite set and $\mathcal{H}^0(E) = k \in \mathbb{Z}_+$ if E is a finite set with k elements.

Theorem 4.22. *Consider the Hausdorff outer measure $\mathcal{H}^s$, where $s \in [0, \infty)$, on a separable metric space (X, ρ). Let $E \in \mathfrak{P}(X)$. Consider the function $\mathcal{H}^{(\bullet)}(E)$ on $s \in [0, \infty)$.*
(a) *$\mathcal{H}^{(\bullet)}(E)$ is a decreasing nonnegative extended real-valued function on $s \in [0, \infty)$.*
(b) *There are four possibilities for the range $\mathcal{R}(\mathcal{H}^{(\bullet)}(E))$ of $\mathcal{H}^{(\bullet)}(E)$:*
1° $\mathcal{R}(\mathcal{H}^{(\bullet)}(E)) = \{\infty\}$.
2° $\mathcal{R}(\mathcal{H}^{(\bullet)}(E)) = \{0\}$.
3° $\mathcal{R}(\mathcal{H}^{(\bullet)}(E)) = \{\infty, 0\}$.
4° $\mathcal{R}(\mathcal{H}^{(\bullet)}(E)) = \{\infty, c, 0\}$ *where* $c \in (0, \infty)$.
(c) *If $\mathcal{H}^{(\bullet)}(E)$ is not constant on $s \in [0, \infty)$, then it has exactly one point of discontinuity.*
(d) *If $\mathcal{H}^{(\bullet)}(E)$ assumes a value $c \in (0, \infty)$ at some $s^* \in [0, \infty)$, then $\mathcal{H}^s(E) = \infty$ for $s \in [0, s^*)$ (with the understanding that $[0, 0) = \emptyset$) and $\mathcal{H}^s(E) = 0$ for $s \in (s^*, \infty)$.*
(e) *$\mathcal{H}^{(\bullet)}(E)$ is identically equal to 0 on $s \in [0, \infty)$ if and only if $E = \emptyset$.*

Proof. 1. (a) is by (a) of Theorem 4.19.

2. (b) is from the fact that $\mathcal{H}^{(\bullet)}(E)$ can assume at most one finite positive value according to (d) of Theorem 4.19.

3. Let us prove (c). Suppose $\mathcal{H}^{(\bullet)}(E)$ is not constant on $s \in [0, \infty)$. Then we have case 3° and case 4°.

In case 3°, since $\mathcal{H}^{(\bullet)}(E)$ is decreasing on $s \in [0, \infty)$ there exists just one point of discontinuity.

In case 4°, since $\mathcal{H}^{(\bullet)}(E)$ is decreasing on $s \in [0, \infty)$ and since there exists at most one point $s^* \in [0, \infty)$ such that $\mathcal{H}^{s^*}(E) \in (0, \infty)$ according to (d) of Theorem 4.19, $\mathcal{H}^{(\bullet)}(E)$ has just one point of discontinuity.

4. Let us prove (d). If $\mathcal{H}^{(\bullet)}(E)$ assumes a finite positive value at some $s^* \in [0, \infty)$, then since $\mathcal{H}^{(\bullet)}(E)$ is a decreasing function and since $\mathcal{H}^{(\bullet)}(E)$ cannot assume a finite positive value at another point in $[0, \infty)$ according to (d) of Theorem 4.19, we have $\mathcal{H}^s(E) = \infty$ on $[0, s^*)$ and $\mathcal{H}^s(E) = 0$ on (s^*, ∞).

5. Let us prove (e). Let $E = \emptyset$. Then since $\mathcal{H}^0$ is an outer measure on X, we have $\mathcal{H}^0(E) = \mathcal{H}^0(\emptyset) = 0$. Then since $\mathcal{H}^{(\bullet)}(E)$ is a decreasing nonnegative extended real-valued function on $s \in [0, \infty)$ according to (a) of Theorem 4.19, we have $\mathcal{H}^s(E) = 0$ for all $s \in [0, \infty)$.

Conversely suppose $\mathcal{H}^s(E) = 0$ for all $s \in [0, \infty)$. We have in particular $\mathcal{H}^0(E) = 0$. Since $\mathcal{H}^0$ is the counting outer measure, $\mathcal{H}^0(E) = 0$ implies that $E = \emptyset$. ∎

Observation 4.23. Summarizing (b) of Theorem 4.22, we have three possible cases for the behavior of the decreasing nonnegative extended real-valued function $\mathcal{H}^{(\bullet)}(E)$ on $s \in [0,\infty)$:
Case (a). $\mathcal{H}^{(\bullet)}(E)$ is identically equal to ∞ on $s \in [0,\infty)$.
Case (b). $\mathcal{H}^{(\bullet)}(E)$ is identically equal to 0 on $s \in [0,\infty)$.
Case (c). $\mathcal{H}^{(\bullet)}(E)$ is not constant on $s \in [0,\infty)$ and has exactly one point of discontinuity $s^\star \in [0,\infty)$.

Definition 4.24. (Hausdorff Dimension of a Set) *Consider the Hausdorff outer measure $\mathcal{H}^s$, where $s \in [0,\infty)$, on a separable metric space (X,ρ). We define the Hausdorff dimension of a set $E \in \mathfrak{P}(X)$ denoted by $\dim_H(E)$ as follows.*
(a) *If $\mathcal{H}^{(\bullet)}(E)$ is identically equal to ∞ on $s \in [0,\infty)$, then we define $\dim_H(E) = \infty$.*
(b) *If $\mathcal{H}^{(\bullet)}(E)$ is identically equal to 0 on $s \in [0,\infty)$, then we define $\dim_H(E) = 0$.*
(c) *If $\mathcal{H}^{(\bullet)}(E)$ has a single point of discontinuity $s^\star \in [0,\infty)$, then we define $\dim_H(E) = s^\star$.*

Theorem 4.25. *Consider the Hausdorff outer measure $\mathcal{H}^s$, where $s \in [0,\infty)$, on a separable metric space (X,ρ) and the Hausdorff dimension $\dim_H(E)$ of a set $E \in \mathfrak{P}(X)$.*
(a) *We have*

$$\begin{aligned}\dim_H(E) &= \sup\{s \in [0,\infty) : \mathcal{H}^s(E) = \infty\}\\ &= \sup\{s \in [0,\infty) : \mathcal{H}^s(E) > 0\}\\ &= \inf\{s \in [0,\infty) : \mathcal{H}^s(E) = 0\}\\ &= \inf\{s \in [0,\infty) : \mathcal{H}^s(E) < \infty\},\end{aligned}$$

with the usual understanding for supremum and infimum of an empty subset.
(b) *If $\mathcal{H}^s(E) \in (0,\infty)$, then $\dim_H(E) = s$.*

Proof. 1. (a) is immediate from Definition 4.24.

2. If $\mathcal{H}^s(E) \in (0,\infty)$, then s is the unique point of discontinuity of $\mathcal{H}^{(\bullet)}(E)$ by (d) of Theorem 4.22 and thus $\dim_H(E) = s$ by (c) of Definition 4.24. ∎

Next we show that if a set increases then its Hausdorff dimension increases.

Theorem 4.26. *Consider the Hausdorff outer measure $\mathcal{H}^s$, where $s \in [0,\infty)$, on a separable metric space (X,ρ) and the Hausdorff dimension $\dim_H(E)$ of a set $E \in \mathfrak{P}(X)$.*
(a) *If $E, F \in \mathfrak{P}(X)$ and $E \subset F$, then $\dim_H(E) \le \dim_H(F)$.*
(b) *For $(E_n : n \in \mathbb{N}) \subset \mathfrak{P}(X)$, we have*

$$\dim_H\Big(\bigcup_{n\in\mathbb{N}} E_n\Big) = \sup\{\dim_H(E_n) : n \in \mathbb{N}\}.$$

Proof. 1. Let us prove (a). Let $E \subset F$. For every $s \in [0, \infty)$, we have $\mathcal{H}^s(E) \le \mathcal{H}^s(F)$ by the monotonicity of the outer measure $\mathcal{H}^s$. Thus we have

$$(1) \qquad \mathcal{H}^{(\bullet)}(E) \le \mathcal{H}^{(\bullet)}(F) \quad \text{on } [0, \infty).$$

Since both $\mathcal{H}^{(\bullet)}(E)$ and $\mathcal{H}^{(\bullet)}(F)$ are nonnegative extended real-valued on $[0, \infty)$, the last inequality implies that we have

$$\{s \in [0, \infty) : \mathcal{H}^s(E) = 0\} \supset \{s \in [0, \infty) : \mathcal{H}^s(F) = 0\},$$

and consequently we have

$$(2) \qquad \inf \{s \in [0, \infty) : \mathcal{H}^s(E) = 0\} \le \inf \{s \in [0, \infty) : \mathcal{H}^s(F) = 0\}.$$

According to Theorem 4.25, the first infimum in (2) is equal to $\dim_H(E)$ and the second is equal to $\dim_H(F)$. Thus we have $\dim_H(E) \le \dim_H(F)$.

2. Let us prove (b). Let $(E_n : n \in \mathbb{N}) \subset \mathfrak{P}(X)$. Then for every $i \in \mathbb{N}$, we have $E_i \subset \bigcup_{n\in\mathbb{N}} E_n$ and then $\dim_H(E_i) \le \dim_H \left(\bigcup_{n\in\mathbb{N}} E_n\right)$ by (a). Thus we have

$$(3) \qquad t := \sup \{\dim_H(E_n) : n \in \mathbb{N}\} \le \dim_H \left(\bigcup_{n\in\mathbb{N}} E_n\right).$$

Let $\varepsilon > 0$ be arbitrarily given. Then for every $i \in \mathbb{N}$, we have by the definition of t in (3)

$$(4) \qquad t + \varepsilon > \dim_H(E_i).$$

Now $\dim_H(E_i) = \inf \{s \in [0, \infty) : \mathcal{H}^s(E_i) = 0\}$ according to Theorem 4.25. Thus (4) implies that $\mathcal{H}^{t+\varepsilon}(E_i) = 0$. Then by the countable subadditivity of the outer measure $\mathcal{H}^{t+\varepsilon}$ we have

$$(5) \qquad \mathcal{H}^{t+\varepsilon}\left(\bigcup_{i\in\mathbb{N}} E_i\right) \le \sum_{i\in\mathbb{N}} \mathcal{H}^{t+\varepsilon}(E_i) = 0.$$

According to Theorem 4.25, $\dim_H \left(\bigcup_{i\in\mathbb{N}} E_i\right) = \inf \{s \in [0, \infty) : \mathcal{H}^s\left(\bigcup_{i\in\mathbb{N}} E_i\right) = 0\}$. Thus (5) implies that

$$(6) \qquad \dim_H \left(\bigcup_{i\in\mathbb{N}} E_i\right) \le t + \varepsilon.$$

Since this holds for every $\varepsilon > 0$, we have

$$\dim_H \left(\bigcup_{i\in\mathbb{N}} E_i\right) = t = \sup \{\dim_H(E_n) : n \in \mathbb{N}\}.$$

This completes the proof. ■

Theorem 4.27. *Let E and F be two disjoint sets in a separable metric space (X, ρ). Assume further that $E, F \in \mathfrak{M}(\mathcal{H}^s)$ for every $s \in [0, \infty)$. Then we have*

$$\dim_H(E \cup F) = \dim_H(E) + \dim_H(F). \tag{1}$$

Proof. Since $E, F \in \mathfrak{M}(\mathcal{H}^s)$ and $E \cap F = \emptyset$, we have

$$\mathcal{H}^s(E \cup F) = \mathcal{H}^s(E) + \mathcal{H}^s(F) \quad \text{for } s \in [0, \infty). \tag{2}$$

For an arbitrary set $A \subset X$, $\dim_H(A)$ is defined by Definition 4.24 in terms of a decreasing nonnegative extended real-valued function $\mathcal{H}^{(\bullet)}(A)$ on $s \in [0, \infty)$, whose behavior is described in Observation 4.23. Then the equality (2) implies the equality (1). ∎

Corollary 4.28. *Let (X, ρ) be a separable metric space and consider $\mathfrak{B}_X$. If $E, F \in \mathfrak{B}_X$ and $E \cap F = \emptyset$, then we have*

$$\dim_H(E \cup F) = \dim_H(E) + \dim_H(F).$$

Proof. According to Theorem 4.6, $\mathcal{H}^s$ is a Borel outer measure on (X, ρ) for every $s \in [0, \infty)$ and then we have $\mathfrak{B}_X \subset \mathfrak{M}(\mathcal{H}^s)$ for every $s \in [0, \infty)$. Thus this Corollary is a particular case of Theorem 4.27. ∎

[V] Hausdorff Outer Measures under Hölder and Lipschitz Mapping

Definition 4.29. (Hölder and Lipschitz Mapping) (a) *A mapping f of a set E in a metric space (X, ρ_X) into a metric space (Y, ρ_Y) is called a Hausdorff mapping of exponent $\alpha > 0$ with coefficient $\beta > 0$ if it satisfies the condition:*

$$\rho_Y\big(f(x'), f(x'')\big) \le \beta \rho_X(x', x'')^\alpha \quad \text{for } x', x'' \in E.$$

(b) *A Hölder mapping of exponent 1 is called a Lipschitz mapping.*

Observation 4.30. Every Hölder mapping f of a set E in a metric space (X, ρ_X) into a metric space (Y, ρ_Y) is uniformly continuous on E.

Proof. Let f be a Hölder mapping of exponent $\alpha > 0$ with coefficient $\beta > 0$. For any $\varepsilon > 0$, let $\delta = \left(\frac{\varepsilon}{\beta}\right)^{1/\alpha}$. Then for any $x', x'' \in E$ such that $\rho_X(x', x'') < \delta$, we have

$$\rho_Y\big(f(x'), f(x'')\big) \le \beta \rho_X(x', x'')^\alpha < \beta \delta^\alpha = \beta \frac{\varepsilon}{\beta} = \varepsilon.$$

This proves the uniform continuity of f on E. ■

Notations. When we treat two metric spaces (X, ρ_X) and (Y, ρ_Y) simultaneously, we write $\mathcal{H}_X^s$ and $\mathcal{H}_{X,\delta}^s$ for a Hausdorff outer measure on (X, ρ_X) and similarly $\mathcal{H}_Y^s$ and $\mathcal{H}_{Y,\delta}^s$ for a Hausdorff outer measure on (Y, ρ_Y).

Theorem 4.31. *Let (X, ρ_X) and (Y, ρ_Y) be two separable metric spaces. Consider the Hausdorff outer measure $\mathcal{H}_X^s$ where $s \in [0, \infty)$ on (X, ρ_X) and the Hausdorff outer measure $\mathcal{H}_Y^s$ where $s \in [0, \infty)$ on (Y, ρ_Y). Let f be a Hölder mapping of exponent $\alpha > 0$ with coefficient $\beta > 0$ of a set $E \in \mathfrak{P}(X)$ into Y. Then we have:*
(a) $\mathcal{H}_Y^{s/\alpha}\big(f(E)\big) \le \beta^{s/\alpha} \mathcal{H}_X^s(E)$, *that is,* $\mathcal{H}_Y^s\big(f(E)\big) \le \beta^s \mathcal{H}_X^{\alpha s}(E)$, *for every* $s \in [0, \infty)$.
(b) $\dim_H\big(f(E)\big) \le \frac{1}{\alpha} \dim_H(E)$.
In particular if f is a Lipschitz mapping with coefficient $\beta > 0$, then we have:
(c) $\mathcal{H}_Y^s\big(f(E)\big) \le \beta^s \mathcal{H}_X^s(E)$ *for every* $s \in [0, \infty)$.
(d) $\dim_H\big(f(E)\big) \le \dim_H(E)$.

Proof. 1. Let us prove (a). For $\delta \in (0, \infty]$, let

(1) $$(V_n : n \in \mathbb{N}) \subset \mathfrak{P}(X)|_\delta \quad \text{and} \quad \bigcup_{n \in \mathbb{N}} V_n \supset E.$$

Since $E \subset \bigcup_{n \in \mathbb{N}} V_n$, we have $E \subset \bigcup_{n \in \mathbb{N}} (E \cap V_n)$ and then $f(E) \subset \bigcup_{n \in \mathbb{N}} f(E \cap V_n)$. Thus $\big(f(E \cap V_n) : n \in \mathbb{N}\big)$ is a covering sequence for the set $f(E)$. In order to find the denomination of this covering sequence, we find the diameter $|f(E \cap V_n)|$ for $n \in \mathbb{N}$. Thus let $x', x'' \in E \cap V_n$. Then we have

(2) $$\rho_Y\big(f(x'), f(x'')\big) \le \beta \rho_X(x', x'')^\alpha \le \beta |V_n|^\alpha \le \beta \delta^\alpha,$$

and then

$$|f(E \cap V_n)| \le \beta |V_n|^\alpha \le \beta \delta^\alpha. \tag{3}$$

Thus we have

$$\big(f(E \cap V_n) : n \in \mathbb{N}\big) \subset \mathfrak{P}(Y)|_{\beta\delta^\alpha} \quad \text{and} \quad \bigcup_{n\in\mathbb{N}} f(E \cap V_n) \supset f(E). \tag{4}$$

Then by Definition 4.4 we have

$$\begin{aligned}(5)\quad \mathcal{H}^{s/\alpha}_{Y,\beta\delta^\alpha}\big(f(E)\big) &= \inf\Big\{\sum_{n\in\mathbb{N}} \gamma_{s/\alpha}\big(f(E\cap V_n)\big) : \\ &\qquad \big(f(E \cap V_n) : n \in \mathbb{N}\big) \subset \mathfrak{P}(Y)|_{\beta\delta^\alpha}, \bigcup_{n\in\mathbb{N}} f(E\cap V_n) \supset f(E)\Big\} \\ &\le \sum_{n\in\mathbb{N}} |f(E\cap V_n)|^{s/\alpha} \le \sum_{n\in\mathbb{N}} \beta^{s/\alpha}|V_n|^s = \beta^{s/\alpha}\sum_{n\in\mathbb{N}} |V_n|^s.\end{aligned}$$

Since this holds for every sequence $(V_n : n \in \mathbb{N})$ satisfying condition (1), we have

$$\begin{aligned}(6)\quad \mathcal{H}^{s/\alpha}_{Y,\beta\delta^\alpha}\big(f(E)\big) &\le \beta^{s/\alpha} \inf\Big\{\sum_{n\in\mathbb{N}} |V_n|^s : (V_n : n \in \mathbb{N}) \subset \mathfrak{P}(X)|_\delta, \bigcup_{n\in\mathbb{N}} V_n \supset E\Big\} \\ &= \beta^{s/\alpha}\mathcal{H}^s_{X,\delta}(E).\end{aligned}$$

Then by Definition 4.5, we have

$$\begin{aligned}(7)\quad \mathcal{H}^{s/\alpha}_Y\big(f(E)\big) &= \lim_{\beta\delta^\alpha\downarrow 0} \mathcal{H}^{s/\alpha}_{Y,\beta\delta^\alpha}\big(f(E)\big) = \lim_{\delta\downarrow 0} \mathcal{H}^{s/\alpha}_{Y,\beta\delta^\alpha}\big(f(E)\big) \\ &\le \beta^{s/\alpha} \lim_{\delta\downarrow 0} \mathcal{H}^s_{X,\delta}(E) = \beta^{s/\alpha}\mathcal{H}^s_X(E).\end{aligned}$$

This proves (a).

2. Let us prove (b). Now $\mathcal{H}^{s/\alpha}_Y\big(f(E)\big), \beta^{s/\alpha}\mathcal{H}^s_X(E) \in [0,\infty]$. By (a), we have $\mathcal{H}^{s/\alpha}_Y\big(f(E)\big) \le \beta^{s/\alpha}\mathcal{H}^s_X(E)$. Thus we have

$$\begin{aligned}\big\{s \in [0,\infty) : \mathcal{H}^s_X(E) = 0\big\} &= \big\{s \in [0,\infty) : \beta^{s/\alpha}\mathcal{H}^s_X(E) = 0\big\} \\ &\subset \big\{s \in [0,\infty) : \mathcal{H}^{s/\alpha}_Y\big(f(E)\big) = 0\big\},\end{aligned}$$

and then

$$\inf\big\{s \in [0,\infty) : \mathcal{H}^s_X(E) = 0\big\} \ge \inf\big\{s \in [0,\infty) : \mathcal{H}^{s/\alpha}_Y\big(f(E)\big) = 0\big\}. \tag{8}$$

Then we have

$$\begin{aligned}\dim_H\big(f(E)\big) &= \inf\left\{\tfrac{s}{\alpha} \in [0,\infty) : \mathcal{H}_Y^{s/\alpha}\big(f(E)\big) = 0\right\} \quad \text{by Theorem 4.25}\\ &= \tfrac{1}{\alpha}\inf\left\{s \in [0,\infty) : \mathcal{H}_Y^{s/\alpha}\big(f(E)\big) = 0\right\}\\ &\le \tfrac{1}{\alpha}\inf\left\{s \in [0,\infty) : \mathcal{H}_X^{s}(E) = 0\right\} \quad \text{by (8)}\\ &= \tfrac{1}{\alpha}\dim_H(E) \quad \text{by Theorem 4.25.}\end{aligned}$$

This proves (b).

3. (c) and (d) are obtained by setting $\alpha = 1$ in (a) and (b) respectively. ∎

Theorem 4.32. *Let (X, ρ_X) and (Y, ρ_Y) be two separable metric spaces. Let $E \in \mathfrak{P}(X)$. Let f and g be two mappings of E into Y such that with some $\alpha > 0$ and $\beta > 0$ we have*

(1) $$\rho_Y\big(f(x'), f(x'')\big) \le \beta\rho_Y\big(g(x'), g(x'')\big)^\alpha \quad \text{for } x', x'' \in E.$$

Then we have:
(a) $\mathcal{H}_Y^{s/\alpha}\big(f(E)\big) \le \beta^{s/\alpha}\mathcal{H}_Y^{s}\big(g(E)\big)$, *for every $s \in [0,\infty)$.*
(b) $\dim_H\big(f(E)\big) \le \frac{1}{\alpha}\dim_H\big(g(E)\big)$.

Proof. 1. Let us prove (a). Let $s \in [0,\infty)$ and $\delta \in (0,\infty]$. By Definition 4.4, we have

(2) $$\mathcal{H}_{Y,\delta}^{s}\big(g(E)\big) = \inf\left\{\sum_{n\in\mathbb{N}} \gamma_s(V_n) : (V_n : n \in \mathbb{N}) \subset \mathfrak{P}(Y)|_\delta, \bigcup_{n\in\mathbb{N}} V_n \supset g(E)\right\}.$$

Then for $\varepsilon > 0$, there exists $(V_n^* : n \in \mathbb{N}) \subset \mathfrak{P}(Y)|_\delta$ such that $\bigcup_{n\in\mathbb{N}} V_n^* \supset g(E)$ and

(3) $$\sum_{n\in\mathbb{N}} \gamma_s(V_n^*) \le \mathcal{H}_{Y,\delta}^{s}\big(g(E)\big) + \varepsilon \le \mathcal{H}_Y^{s}\big(g(E)\big) + \varepsilon.$$

Let $(W_n^* : n \in \mathbb{N})$ be a sequence of sets in Y defined by setting

(4) $$W_n^* = f\big(g^{-1}(V_n^*)\big) \quad \text{for } n \in \mathbb{N}.$$

We claim that

(5) $$(W_n^* : n \in \mathbb{N}) \subset \mathfrak{P}(Y)|_{\beta\delta^\alpha} \quad \text{and} \quad \bigcup_{n\in\mathbb{N}} W_n^* \supset f(E).$$

Now we have $\bigcup_{n\in\mathbb{N}} V_n^* \supset g(E)$ and this implies

$$E \subset g^{-1}\big(g(E)\big) \subset g^{-1}\left(\bigcup_{n\in\mathbb{N}} V_n^*\right) = \bigcup_{n\in\mathbb{N}} g^{-1}(V_n^*)$$

and then

$$f(E) \subset f\left(\bigcup_{n\in\mathbb{N}} g^{-1}(V_n^*)\right) = \bigcup_{n\in\mathbb{N}} f\big(g^{-1}(V_n^*)\big) = \bigcup_{n\in\mathbb{N}} W_n^*.$$

This shows that $(W_n^* : n \in \mathbb{N})$ is a covering sequence for $f(E)$. Next we determine the denomination of this covering sequence. To estimate $|W_n^*|$, let $y', y'' \in W_n^*$. Then there exist $x', x'' \in g^{-1}(V_n^*) \subset X$ such that $y' = f(x')$ and $y'' = f(x'')$. By (1) we have

$$\rho_Y(y', y'') = \rho_Y\big(f(x'), f(x'')\big) \le \beta \rho_Y\big(g(x'), g(x'')\big)^\alpha. \tag{6}$$

Since $x', x'' \in g^{-1}(V_n^*)$, we have $g(x'), g(x'') \in V_n^*$ so that we have

$$\rho_Y\big(g(x'), g(x'')\big) \le |V_n^*| \le \delta$$

and then substituting this in (6) we have

$$\rho_Y(y', y'') \le \beta \rho_Y\big(g(x'), g(x'')\big)^\alpha leq \beta \delta^\alpha$$

and thus

$$|W_n^*| = \sup_{y', y'' \in W_n^*} \rho_Y(y', y'') \le \beta \delta^\alpha.$$

This shows that $W_n^* \in \mathfrak{P}(Y)|_{\beta\delta^\alpha}$ and completes the proof of (5).

We showed above that $g(x'), g(x'') \in V_n^*$. Then we have $\rho_Y\big(g(x'), g(x'')\big) \le |V_n^*|$ and $\rho_Y\big(g(x'), g(x'')\big)^\alpha \le |V_n^*|^\alpha$. Substituting this in (6), we have $\rho_Y(y', y'') \le \beta |V_n^*|^\alpha$. Then we have

$$|W_n^*| = \sup_{y', y'' \in W_n^*} \rho_Y(y', y'') \le \beta |V_n^*|^\alpha. \tag{7}$$

Now by Definition 4.4 we have

$$\begin{aligned}
\mathcal{H}^{s/\alpha}_{Y,\beta\delta^\alpha}\big(f(E)\big) &= \inf\left\{\sum_{n\in\mathbb{N}} \gamma_{s/\alpha}(W_n) : (W_n : n \in \mathbb{N}) \subset \mathfrak{P}(Y)|_{\beta\delta^\alpha}, \bigcup_{n\in\mathbb{N}} W_n \supset f(E)\right\} \\
&\le \sum_{n\in\mathbb{N}} \gamma_{s/\alpha}(W_n^*) = \sum_{n\in\mathbb{N}} \beta^{s/\alpha} |V_n^*|^s \quad \text{by (7)} \\
&= \beta^{s/\alpha} \sum_{n\in\mathbb{N}} \gamma_s(V_n^*) \le \beta^{s/\alpha}\{\mathcal{H}^s_Y\big(g(E)\big) + \varepsilon\} \quad \text{by (3).}
\end{aligned}$$

Since this holds for an arbitrary $\varepsilon > 0$, we have

$$\mathcal{H}^{s/\alpha}_{Y,\beta\delta^\alpha}\big(f(E)\big) \le \beta^{s/\alpha} \mathcal{H}^s_Y\big(g(E)\big).$$

Then letting $\delta \downarrow 0$, we have

$$\mathcal{H}^{s/\alpha}_Y\big(f(E)\big) \le \beta^{s/\alpha} \mathcal{H}^s_Y\big(g(E)\big).$$

This completes the proof of (a).

2. (b) is proved by the same argument as in (b) of Theorem 4.31. ∎

Definition 4.33. (Bi-Lipschitz Mapping and Isometric Mapping) (a) *A mapping f of a set E in a metric space (X, ρ_X) into a metric space (Y, ρ_Y) is called a bi-Lipschitz mapping with coefficients β_1 and β_2 if there exist $0 < \beta_1 \leq \beta_2 < \infty$ such that*

$$\beta_1 \rho_X(x', x'') \leq \rho_Y\big(f(x'), f(x'')\big) \leq \beta_2 \rho_X(x', x'') \quad \text{for } x', x'' \in E.$$

(b) *We say that f is an isometric mapping if*

$$\rho_Y\big(f(x'), f(x'')\big) = \rho_X(x', x'') \quad \text{for } x', x'' \in E.$$

(Thus an isometric mapping is a bi-Lipschitz mapping with coefficients $\beta_1 = \beta_2 = 1$.)

Proposition 4.34. *Let (X, ρ_X) and (Y, ρ_Y) be two metric spaces. If f is a bi-Lipschitz mapping with coefficients β_1 and β_2 of a set $E \in \mathfrak{P}(X)$ into Y, then f is a homeomorphism of E onto $f(E)$ in the relative topologies of E and $f(E)$. Moreover the inverse mapping f^{-1} of f is a bi-Lipschitz mapping with coefficients β_2^{-1} and β_1^{-1} of $f(E) \subset Y$ into X.*

Proof. f is a Lipschitz mapping with coefficient β_2 of $E \in \mathfrak{P}(X)$ into Y. Thus f is a continuous mapping of E into Y according to Observation 4.30. The condition $\beta_1 \rho_X(x', x'') \leq \rho_Y\big(f(x'), f(x'')\big)$ for $x', x'' \in E$ implies that if $f(x') = f(x'')$ then $x' = x''$. Thus f is a one-to-one mapping. Let f^{-1} be the inverse mapping of f mapping $f(E) \subset Y$ one-to-one onto E. For every $y', y'' \in f(E)$, we have unique $x', x'' \in E$ such that $x' = f^{-1}(y')$ and $x'' = f^{-1}(y'')$. The two conditions

$$\beta_1 \rho_X(x', x'') \leq \rho_Y\big(f(x'), f(x'')\big) \quad \text{and} \quad \rho_Y\big(f(x'), f(x'')\big) \leq \beta_2 \rho_X(x', x'')$$

are rewritten respectively as

$$\rho_X\big(f^{-1}(y'), f^{-1}(y'')\big) \leq \beta_1^{-1} \rho_Y(y', y'') \quad \text{and} \quad \beta_2^{-1} \rho_Y(y', y'') \leq \rho_X\big(f^{-1}(y'), f^{-1}(y'')\big).$$

Thus f^{-1} is a bi-Lipschitz mapping with coefficients β_2^{-1} and β_1^{-1} of $f(E)$ into X. Since f and f^{-1} are Lipschitz mappings, they are continuous mappings. This shows that f is a homeomorphism. ∎

Theorem 4.35. *Let (X, ρ_X) and (Y, ρ_Y) be two separable metric spaces. Let f be a bi-Lipschitz mapping with coefficients β_1 and β_2 of a set $E \in \mathfrak{P}(X)$ into Y. Then we have:*
(a) $\beta_1^s \mathcal{H}_X^s(E) \leq \mathcal{H}_Y^s\big(f(E)\big) \leq \beta_2^s \mathcal{H}_X^s(E)$ *for every $s \in [0, \infty)$.*
(b) $\dim_H(f(E)) = \dim_H(E)$.
In particular if f is an isometric mapping, then we have:
(c) $\mathcal{H}_Y^s\big(f(E)\big) = \mathcal{H}_X^s(E)$ *for every $s \in [0, \infty)$.*

Proof. If f is a bi-Lipschitz mapping with coefficients β_1 and β_2, then f is a Lipschitz mapping of E with coefficient β_2 so that by (c) of Theorem 4.31 we have the estimate $\mathcal{H}_Y^s(f(E)) \leq \beta_2^s \mathcal{H}_X^s(E)$ for every $s \in [0, \infty)$. By Proposition 4.34, f^{-1} is a Lipschitz mapping of $f(E)$ with coefficient β_1^{-1} so that by (c) of Theorem 4.31 again we have $\mathcal{H}_X^s(f^{-1}(f(E))) \le \beta_1^{-s}\mathcal{H}_Y^s(f(E))$, that is, $\beta_1^s\mathcal{H}_X^s(E) \le \mathcal{H}_Y^s(f(E))$, for every $s \in [0,\infty)$. This proves (a). Also by (d) of Theorem 4.31, we have $\dim_H(f(E)) \le \dim_H(E)$ as well as $\dim_H\big((f^{-1}(T(E)))\big) \le \dim_H(f(E))$, that is, $\dim_H(E) \le \dim_H(f(E))$. This proves (b). Since an isometry is a bi-Lipschitz mapping with coefficients $\beta_1 = \beta_2 = 1$, (c) follows from (a). ∎

An isometric mapping of X onto X is of particular interest. We show below that such a mapping preserves not only $\mathcal{H}^s$ but also $\mathfrak{M}(\mathcal{H}^s)$.

Theorem 4.36. *Let (X, ρ_X) be a separable metric space. Let f be an isometric mapping of X onto X. Then f is a homeomorphism of X onto X. Moreover the inverse mapping f^{-1} is an isometric mapping of X onto X.*

Proof. This theorem is a particular case of Proposition 4.34. ∎

Theorem 4.37. (Invariance of $\mathcal{H}^s$ and $\mathfrak{M}(\mathcal{H}^s)$ under Isometric Transformation) *Let (X, ρ_X) be a separable metric space. Let f be an isometric mapping of X onto X. Let $E \in \mathfrak{P}(X)$. Then we have:*
(a) $\mathcal{H}^s(f(E)) = \mathcal{H}^s(E)$ *for every* $s \in [0,\infty)$.
(b) $\dim_H(f(E)) = \dim_H(E)$.
(c) $\mathcal{H}^s(f^{-1}(E)) = \mathcal{H}^s(E)$ *for every* $s \in [0,\infty)$.
(d) $f(E) \in \mathfrak{M}(\mathcal{H}^s)$ *if and only if* $E \in \mathfrak{M}(\mathcal{H}^s)$.

Proof. 1. (a) is from (c) of Theorem 4.35. (b) is from (b) of Theorem 4.35.

2. f^{-1} is an isometric mapping of X onto X according to Theorem 4.36. Thus replacing f in (a) with f^{-1} we obtain (c).

3. Let us prove (c). By definition $\mathfrak{M}(\mathcal{H}^s)$ consists of all $E \in \mathfrak{P}(X)$ such that for every $A \in \mathfrak{P}(X)$ we have $\mathcal{H}^s(A) = \mathcal{H}^s(A \cap E) + \mathcal{H}^s(A \cap E^c)$. Suppose $E \in \mathfrak{M}(\mathcal{H}^s)$. Let us show that $f(E) \in \mathfrak{M}(\mathcal{H}^s)$. Now if f is an isometric mapping of X onto X, then f is a one-to-one mapping of X onto X so that the inverse mapping f^{-1} exists and furthermore f^{-1} is an isometric mapping of X onto X according to Theorem 4.36. For an arbitrary $A \in \mathfrak{P}(X)$, we have

$$\begin{aligned}&\mathcal{H}^s(A \cap f(E)) = \mathcal{H}^s\big(f(f^{-1}(A)) \cap f(E)\big)\\ &= \mathcal{H}^s\big(f(f^{-1}(A) \cap E)\big) = \mathcal{H}^s\big(f^{-1}(A) \cap E\big) \quad \text{by (a).}\end{aligned}$$

Since f is a one-to-one mapping of Xx onto X, we have $f(E)^c = f(E^c)$. Then we have

$$\begin{aligned}&\mathcal{H}^s(A \cap f(E)^c) = \mathcal{H}^s\big(f(f^{-1}(A)) \cap f(E^c)\big)\\ &= \mathcal{H}^s\big(f(f^{-1}(A) \cap E^c)\big) = \mathcal{H}^s\big(f^{-1}(A) \cap E^c\big) \quad \text{by (a).}\end{aligned}$$

Then by the fact that $E \in \mathfrak{M}(\mathcal{H}^s)$ we have

$$\mathcal{H}^s\big(A \cap f(E)\big) + \mathcal{H}^s\big(A \cap f(E)^c\big) = \mathcal{H}^s\big(f^{-1}(A) \cap E\big) + \mathcal{H}^s\big(f^{-1}(A) \cap E^c\big)$$
$$= \mathcal{H}^s\big(f^{-1}(A)\big) = \mathcal{H}^s(A) \quad \text{by (c).}$$

This shows that $f(E) \in \mathfrak{M}(\mathcal{H}^s)$ if $E \in \mathfrak{M}(\mathcal{H}^s)$. Conversely if $f(E) \in \mathfrak{M}(\mathcal{H}^s)$, then since $E = f^{-1}\big(f(E)\big)$ and since f^{-1} is an isometric mapping of X onto X, we have $E \in \mathfrak{M}(\mathcal{H}^s)$ by our result above. ∎

§5 Hausdorff Measures on Linear Spaces

[VI] Hausdorff Outer Measure on Separable Normed Linear Spaces

Let (X, ρ) be a separable metric space. For $s \in [0, \infty)$ and $\delta \in (0, \infty]$, the outer measure $\mathcal{H}_\delta^s$ on X is defined to be the outer measure generated by the Hausdorff premeasure γ_s on the covering class $\mathfrak{P}(X)|_\delta$ for X.
Let us consider the special case that the metric space (X, ρ) is in fact a normed linear space $(X, \|\cdot\|_X)$ over the field of scalars $\mathbb{R}$. We show that the outer measure $\mathcal{H}_\delta^s$ on X is generated by the Hausdorff premeasure γ_s on the covering class $\mathfrak{W}_X|_\delta$ for X where $\mathfrak{W}_X$ is the collection of all convex sets in X.
We then treat $\mathcal{H}_\delta^s$ under transformations on the normed linear space $(X, \|\cdot\|_X)$.

Review. (Normed Linear Spaces)
• Let $(X, \|\cdot\|_X)$ be a normed linear space over the field of scalars $\mathbb{R}$. If we define $\rho_X(x', x'') := \|x' - x''\|_X$ for $x', x'' \in X$ then ρ_X is a metric on X. Thus we treat a normed linear space as an example of metric space.
• Let X be a linear space over the field of scalars $\mathbb{R}$. For $x_0 \in X$, $E, F \subset X$, $\lambda \in \mathbb{R}$ and $\alpha, \beta \in \mathbb{R}$, we define

$$E + x_0 = \{x + x_0 : x \in E\}.$$
$$\lambda E = \{\lambda x \in X : x \in E\}.$$
$$\alpha E + \beta F = \{\alpha x' + \beta x'' \in X : x' \in E, x'' \in F\}.$$

Note that $E - F = 1 \cdot E + (-1)F$, a particular case of $\alpha E + \beta F$, not to be confused with $E \setminus F$.

Review. (Convex Sets)
• A set E in a linear space X over the field of scalars $\mathbb{R}$ is said to be convex if the line segment connecting any two points in E is contained in E.
($\emptyset$ and a singleton in X are convex sets by default. X itself is trivially a convex set.)
• Let $\{E_\alpha : \alpha \in A\}$ be an arbitrary collection of convex sets. Then the intersection $\bigcap_{\alpha \in A} E_\alpha$ is a convex set.
• For $E \subset X$, let $h(E)$ be the intersection of all convex sets containing E. Then $h(E)$ is convex set and indeed it is the smallest convex set containing E. We call $h(E)$ the convex hull of E. We have diameter $|h(E)| = |E|$.
• If E is a convex set in a normed linear space X, then its closure $\overline{E}$ is a convex set.
• An open ball in a normed linear space X is a convex set. So is a closed ball.

[VI.1] Construction of Hausdorff Outer Measure on Separable Normed Linear Space

Proposition 5.1. *Let $(X, \|\cdot\|_X)$ be a separable normed linear space over the field of scalers $\mathbb{R}$. The collection $\mathfrak{W}_X$ of all convex sets in X is a fine-covering class for X.*

Proof. The collection $\mathfrak{S}_X^o$ of all open balls in X and $\emptyset$ is a fine-covering class for X according to (a) of Theorem 2.22. Now an open ball is a convex set. Thus $\mathfrak{W}_X \supset \mathfrak{S}_X^o$. Then by Observation 2.20, $\mathfrak{W}_X$ is a fine-covering class for X. ∎

Theorem 5.2. (Construction of Hausdorff Outer Measure $\mathcal{H}^s$ on a Separable Normed Linear Space $(X, \|\cdot\|_X)$) *Let $(X, \|\cdot\|_X)$ be a separable normed linear space over the field of scalars $\mathbb{R}$. Let $\mathfrak{W}_X$ be the collection of all convex sets in X. For $s \in [0, \infty)$ and $\delta \in (0, \infty]$, let us define set functions $\mathcal{W}_\delta^s$ and $\mathcal{W}^s$ on $\mathfrak{P}(X)$ by setting for every $E \in \mathfrak{P}(X)$*

$$(1) \qquad \mathcal{W}_\delta^s(E) = \inf\left\{\sum_{n\in\mathbb{N}} \gamma_s(W_n) : (W_n : n \in \mathbb{N}) \subset \mathfrak{W}_X|_\delta, \bigcup_{n\in\mathbb{N}} W_n \supset E\right\}$$

$$(2) \qquad \mathcal{W}^s(E) = \lim_{\delta\downarrow 0} \mathcal{W}_\delta^s(E).$$

Then $\mathcal{W}_\delta^s$ and $\mathcal{W}^s$ are outer measures on X and furthermore we have

$$(3) \qquad \mathcal{W}_\delta^s = \mathcal{H}_\delta^s \quad \text{for every } \delta \in (0, \infty] \text{ and } s \in [0, \infty).$$

$$(4) \qquad \mathcal{W}^s = \mathcal{H}^s \quad \text{for every } s \in [0, \infty).$$

Proof. 1. By Proposition 5.1, the collection $\mathfrak{W}_X$ is a finite-covering class for X. Then for every $\delta \in (0, \infty]$ the truncation $\mathfrak{W}_X|_\delta$ is a finite-covering class for X according to Proposition 2.25. Then $\mathcal{W}_\delta^s$ as defined by (1) is an outer measure on X generated by the premeasure γ_s on the covering class $\mathfrak{W}_X|_\delta$ according to Theorem 2.4. Now the collection $\mathfrak{W}_X|_\delta$ decreases as $\delta \downarrow 0$ and this implies that for every $E \in \mathfrak{P}(X)$ we have $\mathcal{W}_\delta^s(E)$ increasing as $\delta \downarrow 0$ according to Proposition 2.8. Then $\mathcal{W}^s$ as defined by (2) is an outer measure according to Theorem 2.9.

2. Let us show that $\mathcal{W}_\delta^s = \mathcal{H}_\delta^s$ for every $\delta \in (0, \infty]$. Now by Definition 4.4 we have for every $E \in \mathfrak{P}(X)$

$$(5) \qquad \mathcal{H}_\delta^s(E) = \inf\left\{\sum_{n\in\mathbb{N}} \gamma_s(V_n) : (V_n : n \in \mathbb{N}) \subset \mathfrak{P}(X)|_\delta, \bigcup_{n\in\mathbb{N}} V_n \supset E\right\}.$$

Since $\mathfrak{W}_X \subset \mathfrak{O}(X)$ we have $\mathfrak{W}_X|_\delta \subset \mathfrak{P}(X)|_\delta$ and this implies $\mathcal{H}_\delta^s(E) \leq \mathcal{W}_\delta^s(E)$ according to Proposition 2.8. It remains to show the reverse inequality: $\mathcal{H}_\delta^s(E) \geq \mathcal{W}_\delta^s(E)$.

We claim that for three collections of nonnegative extended real numbers, we have

$$(6) \qquad \begin{aligned} &\left\{\sum_{n\in\mathbb{N}} \gamma_s(V_n) : (V_n : n \in \mathbb{N}) \subset \mathfrak{P}(X)|_\delta, \bigcup_{n\in\mathbb{N}} V_n \supset E\right\} \\ &\subset \left\{\sum_{n\in\mathbb{N}} \gamma_s(W_n) : (W_n : n \in \mathbb{N}) \subset \mathfrak{W}_X|_\delta, \bigcup_{n\in\mathbb{N}} W_n \supset E\right\} \\ &\subset [0, \infty]. \end{aligned}$$

Then (6) implies

$$\text{(7)} \qquad \inf\left\{\sum_{n\in\mathbb{N}}\gamma_s(V_n) : (V_n : n\in\mathbb{N})\subset\mathfrak{P}(X)|_\delta, \bigcup_{n\in\mathbb{N}} V_n\supset E\right\}$$
$$\geq \inf\left\{\sum_{n\in\mathbb{N}}\gamma_s(W_n) : (W_n : n\in\mathbb{N})\subset\mathfrak{W}_X|_\delta, \bigcup_{n\in\mathbb{N}} W_n\supset E\right\},$$

that is, $\mathcal{H}^s_\delta(E)\geq \mathcal{W}^s_\delta(E)$, and we are done.

Let us prove (6). Take an arbitrary sequence $(V_n : n\in\mathbb{N})\subset\mathfrak{P}(X)|_\delta$ such that $\bigcup_{n\in\mathbb{N}} V_n\supset E$. According to Convention 4.3, we can assume without loss of generality that $V_n\neq\emptyset$ for every $n\in\mathbb{N}$. Then we have $\gamma_s(V_n)=|V_n|^s$. For $h(V_n)$, the convex hull of V_n, we have diameter $|h(V_n)|=|V_n|\leq\delta$. Thus the sequence $\big(h(V_n) : n\in\mathbb{N}\big)\subset\mathfrak{P}(X)|_\delta$ and $\bigcup_{n\in\mathbb{N}} h(V_n)\supset\bigcup_{n\in\mathbb{N}} V_n\supset E$. Note that since $h(V_n)$ is a convex set we actually have $\big(h(V_n) : n\in\mathbb{N}\big)\subset\mathfrak{W}_X|_\delta$. Note also that $\gamma_s\big(h(V_n)\big)=|h(V_n)|^s=|V_n|^s=\gamma_s(V_n)$ and

$$\sum_{n\in\mathbb{N}}\gamma_s\big(h(V_n)\big)=\sum_{n\in\mathbb{N}}\gamma_s(V_n).$$

Thus we have shown that corresponding to every sequence $(V_n : n\in\mathbb{N})\subset\mathfrak{P}(X)|_\delta$ such that $\bigcup_{n\in\mathbb{N}} V_n\supset E$, we have a sequence $(W_n : n\in\mathbb{N})\subset\mathfrak{W}_X|_\delta$ such that $\bigcup_{n\in\mathbb{N}} W_n\supset E$ and moreover $\sum_{n\in\mathbb{N}}\gamma_s(W_n)=\sum_{n\in\mathbb{N}}\gamma_s(V_n)$. This proves the first set-inclusion in (6). The second is trivial.

3. Finally we have $\mathcal{W}^s=\lim_{\delta\downarrow 0}\mathcal{W}^s_\delta=\lim_{\delta\downarrow 0}\mathcal{H}^s_\delta=\mathcal{H}^s$. ∎

[VI.2] Hausdorff Outer Measure under Transformation on a Separable Normed Linear Space

Definition 5.3. *Let X be a linear space over the field of scalars $\mathbb{R}$. Let $x_0\in X$ be arbitrarily chosen and fixed. A translation by x_0 in X is a mapping T of X onto X defined by setting $T(x)=x+x_0$ for every $x\in X$. Thus for $E\in\mathfrak{P}(X)$ we have $T(E)=E+x_0$.*

Proposition 5.4. *Let $(X,\|\cdot\|_X)$ be a normed linear space over the field of scalars $\mathbb{R}$. Then every translation on X is an isometric mapping of X onto X.*

Proof. Consider an arbitrary translation T in X given by $T(x)=x+x_0$ for every $x\in X$ where x_0 is an arbitrary element of X. Then for the metric ρ on X derived from the norm $\|\cdot\|_X$ on X, we have for any $x',x''\in X$

$$\rho\big(T(x'),T(x'')\big)=\|T(x')-T(x'')\|_X=\|(x'+x_0)-(x''+x_0)\|_X$$
$$=\|x'-x''\|_X=\rho(x',x'').$$

This shows that T is an isometric mapping of X onto X. ∎

Theorem 5.5. (Invariance of $\mathcal{H}^s$ and $\mathfrak{M}(\mathcal{H}^s)$ under Translation) *Let $(X, \|\cdot\|_X)$ be a separable normed linear space over the field of scalars $\mathbb{R}$. Let T be a translation on X by $x_0 \in X$. Let $E \in \mathfrak{P}(X)$. Then we have:*
(a) *$\mathcal{H}^s(T(E)) = \mathcal{H}^s(E)$, that is, $\mathcal{H}^s(E + x_0) = \mathcal{H}^s(E)$, for every $s \in [0, \infty)$.*
(b) *$\dim_H(T(E)) = \dim_H(E)$, that is, $\dim_H(E + x_0) = \dim_H(E)$.*
(c) *$\mathcal{H}^s(T^{-1}(E)) = \mathcal{H}^s(E)$, that is, $\mathcal{H}^s(E - x_0) = \mathcal{H}^s(E)$, for every $s \in [0, \infty)$.*
(d) *$T(E) \in \mathfrak{M}(\mathcal{H}^s)$ if and only if $E \in \mathfrak{M}(\mathcal{H}^s)$, that is, $E + x_0 \in \mathfrak{M}(\mathcal{H}^s)$ if and only if $E \in \mathfrak{M}(\mathcal{H}^s)$.*

Proof. According to Proposition 5.4, a translation T on X is an isometric mapping of X onto X. Thus our theorem is a particular case of Theorem 4.37.

Definition 5.6. *Let X be a linear space over the field of scalars $\mathbb{R}$. Let $\lambda \in \mathbb{R}$ and $\lambda > 0$. A mapping T of X into X defined by setting $T(x) = \lambda x$ for every $x \in X$ is called a scaling of X by factor λ. Thus for $E \in \mathfrak{P}(X)$ we have $T(E) = \lambda E$.*

Observation 5.7. Let X be a linear space over the field of scalars $\mathbb{R}$. A scaling T of X by a factor $\lambda > 0$ is a one-to-one mapping of X onto X and its inverse mapping T^{-1} is given by $T^{-1}(x) = \frac{1}{\lambda}x$, that is, T^{-1} is a scaling of X by a factor $\frac{1}{\lambda}$.

Theorem 5.8. *Let $(X, \|\cdot\|_X)$ be a separable normed linear space over the field of scalars $\mathbb{R}$. Let T be a scaling of X by a factor $\lambda > 0$. Let $E \in \mathfrak{P}(X)$. Then we have:*
(a) *$\mathcal{H}^s(T(E)) = \lambda^s \mathcal{H}^s(E)$, that is, $\mathcal{H}^s(\lambda E) = \lambda^s \mathcal{H}^s(E)$, for every $s \in [0, \infty)$.*
(b) *$\dim_H(T(E)) = \dim_H(E)$, that is, $\dim_H(\lambda E) = \dim_H(E)$.*
(c) *$\mathcal{H}^s(T^{-1}(E)) = \frac{1}{\lambda^s}\mathcal{H}^s(E)$, that is, $\mathcal{H}^s(\frac{1}{\lambda}E) = \frac{1}{\lambda^s}\mathcal{H}^s(E)$, for every $s \in [0, \infty)$.*
(d) (Invariance of $\mathfrak{M}(\mathcal{H}^s)$ under Scaling): *$T(E) \in \mathfrak{M}(\mathcal{H}^s)$ if and only if $E \in \mathfrak{M}(\mathcal{H}^s)$, that is, $\lambda E \in \mathfrak{M}(\mathcal{H}^s)$ if and only if $E \in \mathfrak{M}(\mathcal{H}^s)$.*

Proof. 1. Let us prove (a). Let $\delta \in (0, \infty]$. By Definition 4.4, we have

$$(1)\qquad \mathcal{H}^s_\delta(E) = \inf\left\{\sum_{n\in\mathbb{N}} \gamma_s(V_n) : (V_n : n \in \mathbb{N}) \subset \mathfrak{P}(\mathbb{R}^n)|_\delta, \bigcup_{n\in\mathbb{N}} V_n \supset E\right\}.$$

Let $\lambda > 0$. Then $V \in \mathfrak{P}(\mathbb{R}^n)|_\delta$ if and only if $\lambda V \in \mathfrak{P}(\mathbb{R}^n)|_{\lambda\delta}$. Thus we have

$$(2)\qquad (V_n : n \in \mathbb{N}) \subset \mathfrak{P}(\mathbb{R}^n)|_\delta, \bigcup_{n\in\mathbb{N}} V_n \supset E$$
$$\Leftrightarrow (\lambda V_n : n \in \mathbb{N}) \subset \mathfrak{P}(\mathbb{R}^n)|_{\lambda\delta}, \bigcup_{n\in\mathbb{N}} \lambda V_n \supset \lambda E.$$

Now by Definition 4.4, we have

$$(3)\qquad \mathcal{H}^s_{\lambda\delta}(\lambda E)=\inf\left\{\sum_{n\in\mathbb{N}}\gamma_s(\lambda V_n):(\lambda V_n:n\in\mathbb{N})\subset\mathfrak{P}(\mathbb{R}^n)|_{\lambda\delta},\bigcup_{n\in\mathbb{N}}\lambda V_n\supset\lambda E\right\}.$$

By Convention 4.3, we have

$$\gamma_s(V_n)=|V_n|^s \quad\text{and}\quad \gamma_s(\lambda V_n)=|\lambda V_n|^s=\lambda^s|V_n|^s=\lambda^s\gamma_s(V_n).$$

Substituting this in (3), we obtain

$$\begin{aligned}(4)\qquad \mathcal{H}^s_{\lambda\delta}(\lambda E)&=\inf\left\{\lambda^s\sum_{n\in\mathbb{N}}\gamma_s(V_n):(V_n:n\in\mathbb{N})\subset\mathfrak{P}(\mathbb{R}^n)|_{\delta},\bigcup_{n\in\mathbb{N}}V_n\supset E\right\}\\&=\lambda^s\mathcal{H}^s_\delta(E)\quad\text{by (1).}\end{aligned}$$

Then we have

$$(5)\qquad \lim_{\delta\downarrow 0}\mathcal{H}^s_{\lambda\delta}(\lambda E)=\lambda^s\lim_{\delta\downarrow 0}\mathcal{H}^s_\delta(E).$$

According to Definition 4.5, we have $\mathcal{H}^s(\lambda E)=\lim_{\delta\downarrow 0}\mathcal{H}^s_{\lambda\delta}(\lambda E)$ and $\mathcal{H}^s(E)=\lim_{\delta\downarrow 0}\mathcal{H}^s_\delta(E)$. Then (5) implies that $\mathcal{H}^s(\lambda E)=\lambda^s\mathcal{H}^s(E)$. This proves (a).

2. (b) follows immediately from (a) and Theorem 4.25.

3. By Observation 5.7, T^{-1} is a scaling of X by factor $\frac{1}{\lambda}$. Thus (a) implies (c).

4. Let us prove (d). Let us show that$T(E)\in\mathfrak{M}(\mathcal{H}^s)$ if and only if $E\in\mathfrak{M}(\mathcal{H}^s)$. By definition $\mathfrak{M}(\mathcal{H}^s)$ consists of all $E\in\mathfrak{P}(X)$ such that

$$\mathcal{H}^s(A)=\mathcal{H}^s(A\cap E)+\mathcal{H}^s(A\cap E^c)\quad\text{for every } A\in\mathfrak{P}(X).$$

Suppose $E\in\mathfrak{M}(\mathcal{H}^s)$. Let us show that $T(E)\in\mathfrak{M}(\mathcal{H}^s)$. Now T is a scaling of X by a factor $\lambda>0$. By Observation 5.7, T is a one-to-one mapping of X onto X and its inverse mapping T^{-1} is a scaling of X by a factor $\frac{1}{\lambda}>0$. Then for an arbitrary $A\in\mathfrak{P}(X)$, we have

$$\mathcal{H}^s(A\cap T(E))=\mathcal{H}^s\big(T(T^{-1}(A)\cap E)\big)=\lambda^s\mathcal{H}^s(T^{-1}(A)\cap E)\quad\text{by (a)}.$$

Since T is a one-to-one mapping of X onto X, we have $T(E)^c=T(E^c)$. Thus we have

$$\begin{aligned}&\mathcal{H}^s(A\cap T(E)^c)=\mathcal{H}^s\big(T(T^{-1}(A))\cap T(E^c)\big)\\&=\mathcal{H}^s\big(T(T^{-1}(A)\cap E^c)\big)=\lambda^s\mathcal{H}^s(T^{-1}(A)\cap E^c)\quad\text{by (a).}\end{aligned}$$

Thus we have

$$\begin{aligned}&\mathcal{H}^s(A\cap T(E))+\mathcal{H}^s(A\cap T(E)^c)\\&=\lambda^s\mathcal{H}^s(T^{-1}(A)\cap E)+\lambda^s\mathcal{H}^s(T^{-1}(A)\cap E^c)\\&=\lambda^s\big\{\mathcal{H}^s(T^{-1}(A)\cap E)+\mathcal{H}^s(T^{-1}(A)\cap E^c)\big\}\\&=\lambda^s\mathcal{H}^s(T^{-1}(A))\quad\text{by the fact that } E\in\mathfrak{M}(\mathcal{H}^s)\\&=\lambda^s\frac{1}{\lambda^s}\mathcal{H}^s(A)=\mathcal{H}^s(A)\quad\text{by (c).}\end{aligned}$$

This shows that $T(E) \in \mathfrak{M}(\mathcal{H}^s)$ if $E \in \mathfrak{M}(\mathcal{H}^s)$.

Conversely if $T(E) \in \mathfrak{M}(\mathcal{H}^s)$, then since $E = T^{-1}(T(E))$ and since T^{-1} is a scaling of X, we have $E \in \mathfrak{M}(\mathcal{H}^s)$ by our argument above. ■

Let $(X, \|\cdot\|_X)$ and $(Y, \|\cdot\|_Y)$ be separable normed linear spaces over the field of scalars $\mathbb{R}$. Let us consider linear mapping of X into Y.

Definition 5.9. *Let $(X, \|\cdot\|_X)$ and $(Y, \|\cdot\|_Y)$ be normed linear spaces over the field of scalars $\mathbb{R}$. A mapping L of X into Y is called a linear mapping if it satisfies the condition:*

$$L(\alpha' x' + \alpha'' x'') = \alpha' L(x') + \alpha'' L(x'') \quad \text{for } x', x'' \in X, \alpha', \alpha'' \in \mathbb{R}.$$

(a) *With the closed ball $\overline{S}(0,1) = \{x \in X : \|x\|_X \leq 1\}$ in X, we define*

$$\|L\|_* = \sup_{x \in \overline{S}(0,1)} \|L(x)\|_Y \in [0, \infty].$$

(b) *We say that L is a bounded linear mapping of X into Y if $\|L\|_* < \infty$.*

Proposition 5.10. *Let $(X, \|\cdot\|_X)$ and $(Y, \|\cdot\|_Y)$ be normed linear spaces over the field of scalars $\mathbb{R}$. Let L be a linear mapping of X into Y. Then we have*

$$\text{(1)} \qquad \|L(x)\|_Y \leq \|L\|_* \|x\|_X \quad \text{for every } x \in X,$$

$$\text{(2)} \qquad \|L(x') - L(x'')\|_Y \leq \|L\|_* \|x' - x''\|_X \quad \text{for every } x', x'' \in X.$$

Moreover if L is a bounded linear mapping then L is a Lipschitz mapping with Lipschitz coefficient equal to $\|L\|_*$.

Proof. 1. Let us prove (1). For $x = 0 \in X$, (1) holds trivially. Let $x_0 \in X$ and $x_0 \neq 0 \in X$. Then $\|x_0\|_X > 0$ and $\|x_0\|_X^{-1} x_0 \in \overline{S}(0,1)$. Thus we have

$$\begin{aligned} \|L(x_0)\|_Y &= \left\| L\left(\|x_0\|_X \frac{x_0}{\|x_0\|_X} \right) \right\|_Y = \|x_0\|_X \left\| L\left(\frac{x_0}{\|x_0\|_X} \right) \right\|_Y \\ &\leq \|x_0\|_X \sup_{x \in \overline{S}(0,1)} \|L(x)\|_Y = \|L\|_* \|x_0\|_X. \end{aligned}$$

2. To prove (2), note that for any $x', x'' \in X$, we have

$$\|L(x') - L(x'')\|_Y = \|L(x' - x'')\|_Y \leq \|L\|_* \|x' - x''\|_X$$

by the linearity of L and by (1).

3. If L is a bounded linear mapping then $\|L\|_* < \infty$ and (2) shows that L is a Lipschitz mapping with Lipschitz coefficient equal to $\|L\|_*$. ■

Proposition 5.11. *Let $(X, \|\cdot\|_X)$ and $(Y, \|\cdot\|_Y)$ be normed linear spaces over the field of scalars $\mathbb{R}$. Let L be a linear mapping of X into Y. Then L is a bounded linear mapping of X if and only if L is continuous on X.*

Proof. 1. According to Proposition 5.10, for any linear mapping L of X into Y we have

$$(1) \qquad \|L(x') - L(x'')\|_Y \leq \|L\|_* \|x' - x''\|_X \quad \text{for every } x', x'' \in X.$$

If L is a bounded linear mapping then $\|L\|_* < \infty$ and (1) implies that L is uniformly continuous on X.

2. Suppose a linear mapping L of X into Y is continuous on X. Then in particular L is continuous at $0 \in X$. Thus for every $\varepsilon > 0$ there exists $\delta > 0$ such that

$$\|L(x) - L(0)\|_Y < \varepsilon \quad \text{for every } x \in X \text{ such that } \|x - 0\|_X < \delta,$$

that is,

$$(2) \qquad \|L(x)\|_Y < \varepsilon \quad \text{for every } x \in X \text{ such that } \|x\|_X < \delta.$$

Let $x \in \overline{S}(0,1)$. We have $\|x\|_X \leq 1$. Consider $\frac{\delta}{2}x \in X$. We have $\|\frac{\delta}{2}x\|_X = \frac{\delta}{2}\|x\|_X \leq \frac{\delta}{2}$ since $\|x\|_X \leq 1$. Then by (2) we have

$$\left\|L\left(\frac{\delta}{2}x\right)\right\|_Y < \varepsilon, \text{ that is, } \|L(x)\|_Y < \frac{2}{\delta}\varepsilon.$$

Since this holds for every $x \in \overline{S}(0,1)$, we have

$$\|L\|_* = \sup_{x \in \overline{S}(0,1)} \|L(x)\|_Y \leq \frac{2}{\delta}\varepsilon < \infty.$$

This shows that L is a bounded linear mapping. ∎

Theorem 5.12. *Let $(X, \|\cdot\|_X)$ and $(Y, \|\cdot\|_Y)$ be normed linear spaces over the field of scalars $\mathbb{R}$. Let L be a bounded linear mapping, or equivalently a continuous linear mapping, of X into Y. Let $E \in \mathfrak{P}(X)$. Then we have:*
(a) $\mathcal{H}^s_Y(L(E)) \leq (\|L\|_*)^s \mathcal{H}^s_X(E)$.
(b) $\dim_H(L(E)) \leq \dim_H(E)$.

Proof. (Let L be a linear mapping of X into Y. We showed in Proposition 5.11 that L is a bounded linear mapping of X into Y if and only if L is continuous on X.)

Now let L be a bounded linear mapping of X into Y. According to Proposition 5.10, L is a Lipschitz mapping with Lipschitz coefficient equal to $\|L\|_*$. Then Theorem 4.31 implies (a) and (b). ∎

Let $(X, \|\cdot\|_X)$ and $(Y, \|\cdot\|_Y)$ be normed linear spaces over the field of scalars $\mathbb{R}$. Let L be a linear mapping of X into Y. We show next that if the linear space X is finite-dimensional then L is a bounded linear mapping, that is, $\|L\|_* < \infty$, and in particular L is continuous on X.

Proposition 5.13. *Consider the normed linear space $(\mathbb{R}^n, \|\cdot\|_n)$, where $\|\cdot\|_n)$ is the n-dimensional Euclidean norm, and $(Y, \|\cdot\|_Y)$, another normed linear spaces over the field of scalars $\mathbb{R}$. Let L be a linear mapping of $\mathbb{R}^n$ into Y. Then we have $\|L\|_* < \infty$, that is, L is a bounded linear mapping, and in particular L is continuous on $\mathbb{R}^n$.*

Proof. Let $(X, \|\cdot\|_X)$ and $(Y, \|\cdot\|_Y)$ be normed linear spaces over the field of scalars $\mathbb{R}$. Let L be a linear mapping of X into Y. Then by Definition 5.9, we have

$$\|L\|_* = \sup_{x \in \overline{S}(0,1)} \|L(x)\|_Y \in [0, \infty].$$

Let us show that when $(X, \|\cdot\|_X) = (\mathbb{R}^n, \|\cdot\|_n)$ then we have $\|L\|_* < \infty$.

Let $\{e_1, \ldots, e_n\} \subset \mathbb{R}^n$ be the standard basis of the linear space $\mathbb{R}^n$, that is,

$$e_1 = (1, 0, \ldots, 0), \ldots, e_n = (0, \ldots, 0, 1).$$

Then every $x \in \mathbb{R}^n$ is given by

$$x = \xi_1 e_1 + \cdots + \xi_n e_n \quad \text{where } \xi_i, \ldots, \xi_n \in \mathbb{R},$$

and

$$\|x\|_n = \sqrt{\xi_1^2 + \cdots + \xi_n^2}.$$

In particular for $x \in \overline{S}(0.1)$ we have

$$\sqrt{\xi_1^2 + \cdots + \xi_n^2} \leq 1 \quad \text{and thus} \quad |\xi_1|, \ldots, |\xi_n| \leq 1.$$

Then since L is a linear mapping we have

$$L(x) = L(\xi_1 e_1 + \cdots + \xi_n e_n) = L(\xi_1 e_1) + \cdots + L(\xi_n e_n)$$

and then

$$\begin{aligned}
\|L(x)\|_Y &= \|L(\xi_1 e_1) + \cdots + L(\xi_n e_n)\|_Y \\
&\leq \|L(\xi_1 e_1)\|_Y + \cdots + \|L(\xi_n e_n)\|_Y \\
&= |\xi_1| \|L(e_1)\|_Y + \cdots + |\xi_n| \|L(e_n)\|_Y \\
&\leq \|L(e_1)\|_Y + \cdots + \|L(e_n)\|_Y.
\end{aligned}$$

Let $M = \max\{\|L(e_1)\|_Y, \ldots, \|L(e_n)\|_Y\} < \infty$. Then we have $\|L(x)\|_Y \leq nM$. Then

$$\|L\|_* = \sup_{x \in \overline{S}(0,1)} \|L(x)\|_Y \leq nM < \infty.$$

This completes the proof. ■

Theorem 5.14. *Consider the normed linear space* $(\mathbb{R}^n, \|\cdot\|_n)$, *where* $\|\cdot\|_n)$ *is the* n-*dimensional Euclidean norm, and* $(Y, \|\cdot\|_Y)$, *another normed linear spaces over the field of scalars* $\mathbb{R}$. *Let* L *be a linear mapping of* $\mathbb{R}^n$ *into* Y. *Let* $E \in \mathfrak{P}(\mathbb{R}^n)$. *Then we have:*
(a) $\mathcal{H}_Y^s(L(E)) \leq (\|L\|_*)^s \mathcal{H}_{\mathbb{R}^n}^s(E)$.
(b) $\dim_H(L(E)) \leq \dim_H(E)$.

Proof. According to Proposition 5.13, every linear mapping L of $\mathbb{R}^n$ into Y is a bounded linear mapping. Then Theorem 5.12 implies (a) and (b). ∎

[VII] Hausdorff Outer Measure on $(\mathbb{R}^N, \rho_e)$

We have been treating Hausdorff outer measure on a separable normed linear space $(X, \|\cdot\|_X)$ over the field of scalars $\mathbb{R}$. Let us now consider the particular case that $(X, \|\cdot\|_X) = (\mathbb{R}^N, \rho_e)$ where ρ_e is the Euclidean norm on $\mathbb{R}^N$.

Definition 5.15. *For an arbitrary* $x = (\xi_1, \ldots, \xi_N) \in \mathbb{R}^N$ *and* $\eta > 0$, *consider a subset* $Q \in \mathfrak{P}(\mathbb{R}^N)$ *defined by setting*

$$Q = [\xi_1, \xi_1 + \eta] \times \cdots \times [\xi_N, \xi_N + \eta].$$

We call Q *a parallel closed cube of size* η *in* $\mathbb{R}^N$. *We write* $\mathfrak{Q}_{\mathbb{R}^N}$ *for the collection of all parallel closed cubes in* $\mathbb{R}^N$ *and* $\emptyset$.

Proposition 5.16. $\mathfrak{Q}_{\mathbb{R}^N}$ *is a fine-covering class for* $\mathbb{R}^N$.

Proof. Let $\eta > 0$ be arbitrarily given. It is easily shown that $\mathbb{R}^N$ is equal to the union of countably many adjacent parallel closed cubes of size η. Then by Definition 2.19 (Fine-covering Class), $\mathfrak{Q}_{\mathbb{R}^N}$ is a fine-covering class for $\mathbb{R}^N$. ∎

Observation 5.17. Consider $(\mathbb{R}^N, \rho_e)$. With $s \in [0, \infty)$ and $\delta \in (0, \infty]$, define a set function $\mathcal{Q}^s_\delta$ on $\mathfrak{P}(\mathbb{R}^N)$ by setting for every $E \in \mathfrak{P}(\mathbb{R}^N)$

$$\mathcal{Q}^s_\delta(E) = \inf \left\{ \sum_{n \in \mathbb{N}} \gamma_s(Q_n) : (Q_n : n \in \mathbb{N}) \subset \mathfrak{Q}_{\mathbb{R}^N}|_\delta, \bigcup_{n \in \mathbb{N}} Q_n \supset E \right\}.$$

By Theorem 2.4, $\mathcal{Q}^s_\delta$ is an outer measure on $\mathbb{R}^N$. As $\delta \downarrow$, we have $\mathfrak{Q}_{\mathbb{R}^N}|_\delta \downarrow$ and thus $\mathcal{Q}^s_\delta(E) \uparrow$ by Proposition 2.8. If we define $\mathcal{Q}^s = \lim_{\delta \downarrow 0} \mathcal{Q}^s_\delta$, then $\mathcal{Q}^s$ is an outer measure on $\mathbb{R}^N$ according to Theorem 2.28. We show below that $\mathcal{Q}^s = \mathcal{H}^s$.

Observation 5.18. Consider $(X, \|\cdot\|_X)$, a separable normed linear space over the field of scalars $\mathbb{R}$, of which our $(\mathbb{R}^N, \rho_e)$ is a particular case. Let $\mathfrak{W}_X$ be the collection of all convex sets in X. (We showed in Proposition 5.1 that $\mathfrak{W}_X$ is a fine-covering class for X.) For $s \in [0, \infty)$ and $\delta \in (0, \infty]$, let us define a set function $\mathcal{W}^s_\delta$ on $\mathfrak{P}(X)$ by setting for every $E \in \mathfrak{P}(X)$

$$\mathcal{W}^s_\delta(E) = \inf \left\{ \sum_{n \in \mathbb{N}} \gamma_s(W_n) : (W_n : n \in \mathbb{N}) \subset \mathfrak{W}_X|_\delta, \bigcup_{n \in \mathbb{N}} W_n \supset E \right\}.$$

We showed in Theorem 5.2 that $\mathcal{W}^s_\delta = \mathcal{H}^s_\delta$ for every $\delta \in (0, \infty]$ and $s \in [0, \infty)$ and then

$$\mathcal{W}^s = \lim_{\delta \downarrow 0} \mathcal{W}^s_\delta = \lim_{\delta \downarrow 0} \mathcal{H}^s_\delta = \mathcal{H}^s.$$

In the case $(\mathbb{R}^N, \rho_e)$, we compare $\mathcal{Q}^s_\delta$ and $\mathcal{W}^s_\delta$.

Proposition 5.19. *For every convex set W in $\mathbb{R}^N$ with diameter $|W| < \infty$, there exists a parallel closed cube Q such that $W \subset Q$ and diameter $|Q| = \sqrt{N}|W|$.*

Proof. Let W be a convex set W in $\mathbb{R}^N$ with diameter $|W| < \infty$. We can construct a parallel closed cube Q of size $|W|$ containing W. Then the diameter $|Q| = \sqrt{N}|W|$. ■

Theorem 5.20. (Construction of Hausdorff Outer Measure $\mathcal{H}^s$ on $(\mathbb{R}^N, \rho_e)$) *Consider $(\mathbb{R}^N, \rho_e)$. Let $\mathfrak{Q}_{\mathbb{R}^N}$ be the collection of all parallel closed cubes in $\mathbb{R}^N$ and $\emptyset$. With $s \in [0, \infty)$ and $\delta \in (0, \infty]$, define a set function $\mathcal{Q}^s_\delta$ on $\mathfrak{P}(\mathbb{R}^N)$ by setting for every $E \in \mathfrak{P}(\mathbb{R}^N)$*

$$\mathcal{Q}^s_\delta(E) = \inf\left\{\sum_{n\in\mathbb{N}} \gamma_s(Q_n) : (Q_n : n \in \mathbb{N}) \subset \mathfrak{Q}_{\mathbb{R}^N}|_\delta, \bigcup_{n\in\mathbb{N}} Q_n \supset E\right\}.$$

Let us define a set function $\mathcal{Q}^s$ on $\mathfrak{P}(\mathbb{R}^N)$ by setting

$$\mathcal{Q}^s(E) = \lim_{\delta\downarrow 0} \mathcal{Q}^s_\delta(E).$$

Then $\mathcal{Q}^s = \mathcal{H}^s$.

Proof. 1. Let us compare $\mathcal{Q}^s_\delta$ and $\mathcal{W}^s_\delta$ given respectively by

$$\mathcal{Q}^s_\delta(E) = \inf\left\{\sum_{n\in\mathbb{N}} |Q_n|^s : (Q_n : n \in \mathbb{N}) \subset \mathfrak{Q}_{\mathbb{R}^N}|_\delta, \bigcup_{n\in\mathbb{N}} Q_n \supset E\right\}. \tag{1}$$

$$\mathcal{W}^s_\delta(E) = \inf\left\{\sum_{n\in\mathbb{N}} |W_n|^s : (W_n : n \in \mathbb{N}) \subset \mathfrak{W}_X|_\delta, \bigcup_{n\in\mathbb{N}} W_n \supset E\right\}. \tag{2}$$

Now a parallel closed Q in $\mathbb{R}^N$ is a convex set in $\mathbb{R}^N$. Thus we have $\mathfrak{Q}_{\mathbb{R}^N} \subset \mathfrak{W}_X$ and then $\mathfrak{Q}_{\mathbb{R}^N}|_\delta \subset \mathfrak{W}_X|_\delta$. According to Proposition 2.8, this implies that

$$\mathcal{W}^s_\delta(E) \le \mathcal{Q}^s_\delta(E). \tag{3}$$

2. Consider three sets of nonnegative extended real numbers given by

$$\Gamma_0(E) = \left\{\sum_{n\in\mathbb{N}} |W_n|^s : (W_n : n \in \mathbb{N}) \subset \mathfrak{W}_X|_\delta, \bigcup_{n\in\mathbb{N}} W_n \supset E\right\}.$$

$$\Gamma_1(E) = \left\{N^{\frac{1}{2}}\sum_{n\in\mathbb{N}} |W_n|^s : (W_n : n \in \mathbb{N}) \subset \mathfrak{W}_X|_\delta, \bigcup_{n\in\mathbb{N}} W_n \supset E\right\}.$$

$$\Gamma_2(E) = \left\{\sum_{n\in\mathbb{N}} |Q_n|^s : (Q_n : n \in \mathbb{N}) \subset \mathfrak{Q}_{\mathbb{R}^N}|_{\sqrt{N}\delta}, \bigcup_{n\in\mathbb{N}} Q_n \supset E\right\}.$$

Let us show that

$$\Gamma_0(E) = \Gamma_1(E) \subset \Gamma_2(E). \tag{4}$$

Now it is obvious that $\Gamma_0(E) = \Gamma_1(E)$. It remains to show that $\Gamma_1(E) \subset \Gamma_2(E)$. To show this, we show that corresponding to every element in Γ_1 there exists an element in Γ_2. Now let $(W_n : n \in \mathbb{N}) \subset \mathfrak{W}_X|_\delta$ be such that $\bigcup_{n\in\mathbb{N}} W_n \supset E$. By Proposition 5.19, there exists $Q_n \in \mathfrak{Q}_{\mathbb{R}^N}$ such that $W_n \subset Q_n$ and $|Q_n| = \sqrt{N}|W_n| \leq \sqrt{N}\delta$ and thus $Q_n \in \mathfrak{Q}_{\mathbb{R}^N}|_{\sqrt{N}\delta}$. Now $|Q_n|^s = (\sqrt{N})^s|W_n|^s$ and $\sum_{n\in\mathbb{N}} |Q_n|^s = (\sqrt{N})^s \sum_{n\in\mathbb{N}} |W_n|^s$.

Now (4) implies that $\inf \Gamma_2(E) \leq \inf \Gamma_0(E)$. Observe that

$$\inf \Gamma_0(E) = \inf \left\{ \sum_{n\in\mathbb{N}} |W_n|^s : (W_n : n \in \mathbb{N}) \subset \mathfrak{W}_X|_\delta, \bigcup_{n\in\mathbb{N}} W_n \supset E \right\} = \mathcal{W}^s_\delta(E)$$

and

$$\inf \Gamma_2(E) = \inf \left\{ \sum_{n\in\mathbb{N}} |Q_n|^s : (Q_n : n \in \mathbb{N}) \subset \mathfrak{Q}_{\mathbb{R}^N}|_{\sqrt{N}\delta}, \bigcup_{n\in\mathbb{N}} Q_n \supset E \right\} = \mathcal{Q}^s_{\sqrt{N}\delta}(E).$$

Thus we have $\mathcal{Q}^s_{\sqrt{N}\delta}(E) \leq \mathcal{W}^s_\delta(E)$. Since this holds for every $\delta \in (0, \infty]$, we have

(5) $$\mathcal{Q}^s_\delta(E) \leq \mathcal{W}^s_{\frac{1}{\sqrt{N}}\delta}(E).$$

By (3) and (5), we have

(6) $$\mathcal{W}^s_\delta(E) \leq \mathcal{Q}^s_\delta(E) \leq \mathcal{W}^s_{\frac{1}{\sqrt{N}}\delta}(E).$$

Letting $\delta \downarrow 0$, we have $\mathcal{W}^s(E) \leq \mathcal{Q}^s(E) \leq \mathcal{W}^s(E)$ so that $\mathcal{Q}^s(E) = \mathcal{W}^s(E)$. But $\mathcal{W}^s(E) = \mathcal{H}^s(E)$. Thus $\mathcal{Q}^s(E) = \mathcal{H}^s(E)$. ∎

Our next objective is a comparison of the Hausdorff outer measure $\mathcal{H}^N$ and the Lebesgue outer measure μ_L^N on $(\mathbb{R}^N, \rho_e)$. To facilitate the comparison, we prepare an alternate construction of μ_L^N as defined by Definition 3.46.

Proposition 5.21. (Alternate Construction of Lebesgue Outer Measure μ_L^N on $(\mathbb{R}^N, \rho_e)$) *Consider $(\mathbb{R}^N, \rho_e)$ and the fine-covering class $\mathfrak{Q}_{\mathbb{R}^N}$ for $\mathbb{R}^N$. Let v be a premeasure on $\mathfrak{Q}_{\mathbb{R}^N}$ defined by setting $v(Q) =$ volume of Q for $Q \in \mathfrak{Q}_{\mathbb{R}^N}$ and $v(\emptyset) = 0$. Let μ be the outer measure on $\mathbb{R}^N$ generated by the premeasure v on the fine-covering class $\mathfrak{Q}_{\mathbb{R}^N}$ for $\mathbb{R}^N$. Then we have $\mu = \mu_L^N$.*

Proof. Let $E \in \mathfrak{P}(\mathbb{R}^N)$. Then since μ is the outer measure on $\mathbb{R}^N$ generated by the premeasure v on the fine-covering class $\mathfrak{Q}_{\mathbb{R}^N}$ for $\mathbb{R}^N$, we have

$$\mu(E) = \inf \left\{ \sum_{n\in\mathbb{N}} v(Q_n) : (Q_n : n \in \mathbb{N}) \subset \mathfrak{Q}_{\mathbb{R}^N}, \bigcup_{n\in\mathbb{N}} Q_n \supset E \right\}.$$

On the other hand, by Definition 3.46 we have

$$\mu_L^N(E) = \inf \left\{ \sum_{n\in\mathbb{N}} v(R_n) : (R_n : n \in \mathbb{N}) \subset \mathfrak{V}_{oc}, \bigcup_{n\in\mathbb{N}} R_n \supset E \right\},$$

where $\mathfrak{V}_{oc}$ is the collection of all open-closed rectangles $\prod_{i=1}^{N}(\alpha_i, \beta_i] \subset \mathbb{R}^N$ and $\emptyset$.

Consider two collections of nonnegative extended real numbers given by

$$\Gamma_1(E) = \Big\{ \sum_{n \in \mathbb{N}} v(Q_n) : (Q_n : n \in \mathbb{N}) \subset \mathfrak{Q}_{\mathbb{R}^N}, \bigcup_{n \in \mathbb{N}} Q_n \supset E \Big\}.$$

$$\Gamma_2(E) = \Big\{ \sum_{n \in \mathbb{N}} v(R_n) : (R_n : n \in \mathbb{N}) \subset \mathfrak{V}_{oc}, \bigcup_{n \in \mathbb{N}} R_n \supset E \Big\}.$$

To show that $\Gamma_1(E) \subset \Gamma_2(E)$, we show that corresponding to an arbitrary $\gamma \in \Gamma_1(E)$ there exists a member of $\Gamma_2(E)$. Let $\gamma = \sum_{n \in \mathbb{N}} v(Q_n)$ be an arbitrary member of $\Gamma_1(E)$. Now corresponding to $Q_n \in \mathfrak{Q}_{\mathbb{R}^N}$ there exists $R_n \in \mathfrak{V}_{oc}$ such that $v(R_n) = v(Q_n)$. Then $\sum_{n \in \mathbb{N}} v(R_n) = \sum_{n \in \mathbb{N}} v(Q_n) = \gamma$ and thus we have a corresponding element in $\Gamma_2(E)$.

Similarly, to show that $\Gamma_2(E) \subset \Gamma_1(E)$, we show that corresponding to an arbitrary $\gamma \in \Gamma_2(E)$ there exists a member of $\Gamma_1(E)$. Let $\gamma = \sum_{n \in \mathbb{N}} v(R_n)$ be an arbitrary member of $\Gamma_2(E)$. Now corresponding to $R_n \in \mathfrak{V}_{oc}$ there exists $Q_n \in \mathfrak{Q}_{\mathbb{R}^N}$ such that $v(Q_n) = v(R_n)$. Then $\sum_{n \in \mathbb{N}} v(Q_n) = \sum_{n \in \mathbb{N}} v(R_n) = \gamma$ and thus we have a corresponding element in $\Gamma_1(E)$.

Thus we have shown that $\Gamma_1(E) \subset \Gamma_2(E)$ and $\Gamma_2(E) \subset \Gamma_1(E)$. Therefore we have $\Gamma_1(E) \subset \Gamma_2(E)$. This implies that $\inf \Gamma_1(E) = \inf \Gamma_2(E)$. But $\inf \Gamma_1(E) = \mu(E)$ and $\inf \Gamma_2(E) = \mu_L^N(E)$. Thus we have $\mu(E) = \mu_L^N(E)$. This completes the proof. ■

Theorem 5.22. **(Comparison of Hausdorff Outer Measure $\mathcal{H}^N$ and Lebesgue Outer Measure μ_L^N on $(\mathbb{R}^N, \rho_e)$)** *Consider $\mathcal{H}^N$ and μ_L^N on $(\mathbb{R}^N, \rho_e)$. We have*

$$\mathcal{H}^N = N^{\frac{N}{2}} \mu_L^N,$$

that is, $\mathcal{H}^N(E) = N^{\frac{N}{2}} \mu_L^N(E)$ for every $E \in \mathfrak{P}(\mathbb{R}^N)$.

Proof. Let $E \in \mathfrak{P}(\mathbb{R}^N)$. By Proposition 5.21, we have

$$(1) \qquad \mu_L^N(E) = \inf \Big\{ \sum_{n \in \mathbb{N}} v(Q_n) : (Q_n : n \in \mathbb{N}) \subset \mathfrak{Q}_{\mathbb{R}^N}, \bigcup_{n \in \mathbb{N}} Q_n \supset E \Big\}.$$

According to Theorem 5.20, we have

$$(2) \qquad \mathcal{H}^N(E) = \mathcal{Q}^N(E) = \lim_{\delta \downarrow 0} \mathcal{Q}_\delta^N(E),$$

where

$$(3) \qquad \mathcal{Q}_\delta^N(E) = \inf \Big\{ \sum_{n \in \mathbb{N}} \gamma_N(Q_n) : (Q_n : n \in \mathbb{N}) \subset \mathfrak{Q}_{\mathbb{R}^N}|_\delta, \bigcup_{n \in \mathbb{N}} Q_n \supset E \Big\}.$$

By Convention 4.3, we have $\gamma_N(Q_n) = |Q_n|^N$. Let $\eta > 0$ be the size of the cube Q_n. Then $|Q_n| = \sqrt{N}\eta$ and $|Q_n|^N = (\sqrt{N}\eta)^N = N^{\frac{N}{2}}\eta^N = N^{\frac{N}{2}} v(Q_n)$. Thus we have $\gamma_N(Q_n) = N^{\frac{N}{2}} v(Q_n)$. Substituting this in (3), we obtain

$$(4) \qquad \mathcal{Q}_\delta^N(E) = N^{\frac{N}{2}} \cdot \inf \Big\{ \sum_{n \in \mathbb{N}} v(Q_n) : (Q_n : n \in \mathbb{N}) \subset \mathfrak{Q}_{\mathbb{R}^N}|_\delta, \bigcup_{n \in \mathbb{N}} Q_n \supset E \Big\}.$$

We verify next that

$$(5) \qquad \inf\left\{\sum_{n\in\mathbb{N}} v(Q_n) : (Q_n : n\in\mathbb{N}) \subset \mathfrak{Q}_{\mathbb{R}^N}|_\delta, \bigcup_{n\in\mathbb{N}} Q_n \supset E\right\}$$

$$= \inf\left\{\sum_{n\in\mathbb{N}} v(Q_n) : (Q_n : n\in\mathbb{N}) \subset \mathfrak{Q}_{\mathbb{R}^N}, \bigcup_{n\in\mathbb{N}} Q_n \supset E\right\} = \mu_L^N(E).$$

Observe that the second equality in (5) is valid by (1). To prove the first equality in (5), let us consider two sets of nonnegative extended real numbers given by

$$\Gamma_1(E) = \left\{\sum_{n\in\mathbb{N}} v(Q_n) : (Q_n : n\in\mathbb{N}) \subset \mathfrak{Q}_{\mathbb{R}^N}, \bigcup_{n\in\mathbb{N}} Q_n \supset E\right\}.$$

$$\Gamma_2(E) = \left\{\sum_{n\in\mathbb{N}} v(Q_n) : (Q_n : n\in\mathbb{N}) \subset \mathfrak{Q}_{\mathbb{R}^N}|_\delta, \bigcup_{n\in\mathbb{N}} Q_n \supset E\right\}.$$

Now $\mathfrak{Q}_{\mathbb{R}^N}|_\delta \subset \mathfrak{Q}_{\mathbb{R}^N}$ implies that $\Gamma_2(E) \subset \Gamma_1(E)$. To show $\Gamma_1(E) \subset \Gamma_2(E)$, we show that corresponding to an arbitrary $\gamma \in \Gamma_1(E)$ there exists an element in $\Gamma_2(E)$. Let $\gamma \in \Gamma_1(E)$ be given by $\gamma = \sum_{n\in\mathbb{N}} v(Q_n)$ where $(Q_n : n \in \mathbb{N}) \subset \mathfrak{Q}_{\mathbb{R}^N}$ and $\bigcup_{n\in\mathbb{N}} Q_n \supset E$. Let us decompose Q_n into finitely many, say $p_n \in \mathbb{N}$, members of $\mathfrak{Q}_{\mathbb{R}^N}|_\delta$, say $Q_{n,1}, \dots, Q_{n,p_n}$, such that $v(Q_n) = \sum_{i=1}^{p_n} v(Q_{n,i})$. Let

$$(C_n : n \in \mathbb{N}) = (Q_{1,1}, \dots, Q_{1,p_1}; Q_{2,1}, \dots, Q_{1,p_2}; \dots) \subset \mathfrak{Q}_{\mathbb{R}^N}|_\delta.$$

Then we have $\sum_{n\in\mathbb{N}} v(C_n) = \sum_{n\in\mathbb{N}} v(Q_n)$. This shows that corresponding to an arbitrary element of $\Gamma_1(E)$ there exists an element of $\Gamma_2(E)$. This shows that $\Gamma_1(E) \subset \Gamma_2(E)$. Now we have $\Gamma_1(E) = \Gamma_2(E)$. This implies that $\inf \Gamma_1(E) = \inf \Gamma_2(E)$ and proves the first equality in (5). Substituting (5) in (4), we have

$$(6) \qquad \mathcal{Q}_\delta^N(E) = N^{\frac{N}{2}} \mu_L^N(E).$$

Then we have

$$(7) \qquad \lim_{\delta\downarrow 0} \mathcal{Q}_\delta^N(E) = N^{\frac{N}{2}} \mu_L^N(E).$$

Substituting (7) into (2), we have

$$\mathcal{H}^N(E) = N^{\frac{N}{2}} \mu_L^N(E).$$

This completes the proof. ∎

Observation 5.23. Theorem 5.22 implies that

$$\mathcal{H}^1 = \mu_L^1, \quad \mathcal{H}^2 = 2\mu_L^2, \quad \mathcal{H}^3 = 3^{\frac{3}{2}}\mu_L^3, \quad \mathcal{H}^4 = 16\mu_L^4, \quad \mathcal{H}^5 = 5^{\frac{5}{2}}\mu_L^5, \quad \dots.$$

Theorem 5.24. *Consider the Hausdorff outer measure $\mathcal{H}^s$ on $(\mathbb{R}^N, \rho_e)$.*
(a) $\mathcal{H}^s(\mathbb{R}^n) = \infty$ *for* $s \in [0, n]$.
(b) $\mathcal{H}^s(\mathbb{R}^n) = 0$ *for* $s \in (n, \infty)$.
(c) $\dim_H(\mathbb{R}^n) = n$.

Proof. 1. Let us prove (a). According to Theorem 5.22, we have $\mathcal{H}^n = n^{\frac{n}{2}}\mu_L^n$ and then we have $\mathcal{H}^n(\mathbb{R}^n) = n^{\frac{n}{2}}\mu_L^n(\mathbb{R}^n) = \infty$. Then since $\mathcal{H}^\bullet(\mathbb{R}^n)$ is a decreasing function we have $\mathcal{H}^s(\mathbb{R}^n) = \infty$ for $s \in [0, n]$.

2. Let us prove (b). Since $\mathcal{H}^s$ is a Borel outer measure we have $\mathfrak{B}_{\mathbb{R}^n} \subset \mathfrak{M}(\mathcal{H}^s)$, the σ-algebra of $\mathcal{H}^s$-measurable set. Now $\mathbb{R}^n$ can be represented as the union of countably many disjoint left-open and right-closed unit cubes $\{Q_i : i \in \mathbb{N}\}$ in $\mathbb{R}^n$. Now we have $\{Q_i : i \in \mathbb{N}\} \subset \mathfrak{B}_{\mathbb{R}^n} \subset \mathfrak{M}(\mathcal{H}^s)$. Then since $\mathcal{H}^s$ is a measure on $\mathfrak{M}(\mathcal{H}^s)$, we have $\mathcal{H}^s(\mathbb{R}^n) = \sum_{i \in \mathbb{N}} \mathcal{H}^s(Q_i)$. On the other hand we have $\mathcal{H}^n(Q_i) = n^{\frac{n}{2}}\mu_L^n(Q_i) = n^{\frac{n}{2}} < \infty$. This implies according to (b) of Theorem 4.19 that $\mathcal{H}^s(Q_i) = 0$ for $s > n$. Thus we have $\mathcal{H}^s(\mathbb{R}^n) = 0$ for $s \in (n, \infty)$.

3. (a) and (b) imply that $\dim_H(\mathbb{R}^n) = n$ by (c) of Definition 4.24. ∎

Corollary 5.25. *Let* $E \in \mathfrak{P}(\mathbb{R}^n)$.
(a) $\mathcal{H}^s(E) = 0$ *for every* $s \in (n, \infty)$.
(b) $\dim_H(E) \leq n$.

Proof. 1. Let us prove (a). Now $E \subset \mathbb{R}^n$ implies $\mathcal{H}^s(E) \leq \mathcal{H}^s(\mathbb{R}^n)$. But $\mathcal{H}^s(\mathbb{R}^n) = 0$ for $s \in (n, \infty)$ by (b) of Theorem 5.24. Thus $\mathcal{H}^s(E) = 0$ for every $s \in (n, \infty)$.

2. Since $E \subset \mathbb{R}^n$, we have $\dim_H(E) \leq \dim_H(\mathbb{R}^n)$ by Theorem 4.26. But $\dim_H(\mathbb{R}^n) = n$ by (c) of Theorem 5.24. Therefore we have $\dim_H(E) \leq n$. ∎

Theorem 5.26. *Consider the Hausdorff outer measure $\mathcal{H}^n$ on $(\mathbb{R}^N, \rho_e)$.*
(a) *Let Q be a parallel closed cube of size $\eta > 0$ in $\mathbb{R}^n$. Then we have*

$$\mathcal{H}^n(Q) = n^{\frac{n}{2}}\eta^n \quad \textit{and} \quad \dim_H(Q) = n.$$

(b) *Let $E \in \mathfrak{P}(\mathbb{R}^n)$ be a bounded set in $\mathbb{R}^n$. Then we have $\mathcal{H}^n(E) < \infty$.*
(c) *Let O be a non-empty open set in $\mathbb{R}^n$. Then we have $\mathcal{H}^n(O) > 0$ and $\dim_H(O) = n$.*
(d) *Let $E \in \mathfrak{P}(\mathbb{R}^n)$ be such that interior $E^\circ \neq \emptyset$. Then $\mathcal{H}^n(E) > 0$ and $\dim_H(E) = n$.*

Proof. 1. Let us prove (a). According to Theorem 5.22, we have $\mathcal{H}^n = n^{\frac{n}{2}}\mu_L^n$. Thus we have $\mathcal{H}^n(Q) = n^{\frac{n}{2}}\mu_L^n(Q) = n^{\frac{n}{2}}\eta^n$.

According to (a) of Theorem 4.25, we have $\dim_H(Q) = \sup\{s \in [0, \infty) : \mathcal{H}^s(Q) > 0\}$. Thus the result above, $\mathcal{H}^n(Q) = n^{\frac{n}{2}}\eta^n > 0$, implies $\dim_H(Q) \geq n$. On the other hand, $Q \subset \mathbb{R}^n$ implies $\dim_H(Q) \leq \dim_H(\mathbb{R}^n) = n$ by (c) of Theorem 5.24. Thus we have $\dim_H(Q) = n$.

2. Let E be a bounded set in $\mathbb{R}^n$. Then $E \subset \overline{B}(0, \frac{\eta}{2})$ for sufficiently large $\eta > 0$. Now

$$\overline{B}\Big(0, \frac{\eta}{2}\Big) \subset \Big[-\frac{\eta}{2}, \frac{\eta}{2}\Big]_1 \times \cdots \times \Big[-\frac{\eta}{2}, \frac{\eta}{2}\Big]_n := Q.$$

Then Q is a parallel closed cube of size η and thus $\mathcal{H}^n(Q) = n^{\frac{n}{2}}\eta^n$ by (a). Then the set-inclusion $E \subset \overline{B}(0, \frac{\eta}{2}) \subset Q$ implies $\mathcal{H}^n(E) \leq \mathcal{H}^n(Q) < \infty$.

3. Let us prove (c). Let O be a non-empty open set in $\mathbb{R}^n$. Then for sufficiently small $\eta > 0$, there exists a parallel closed cube Q of size η contained in O. Then we have $\mathcal{H}^n(O) \geq \mathcal{H}^n(Q) = n^{\frac{n}{2}}\eta^n > 0$ by (a). This implies that $\dim_H(O) = n$ by the same argument as in 1) above.

4. (d) is proved by the same argument as for (c). ∎

Below we have more examples of finding Hausdorff dimensions of sets.

Proposition 5.27. *If G is a bounded non-empty open set in $\mathbb{R}^n$, then* $\dim_H(G) = n$.

Proof. According to Theorem 5.22, we have $\mathcal{H}^n = n^{\frac{n}{2}}\mu_L^n$. Now for a bounded non-empty open set G in $\mathbb{R}^n$, we have $\mu_L^n(G) \in (0,\infty)$ and then $\mathcal{H}^n(G) = n^{\frac{n}{2}}\mu_L^n(G) \in (0,\infty)$. By (b) of Theorem 4.25, this implies that $\dim_H(G) = n$. ∎

Proposition 5.28. *We have* $\dim_H(\mathbb{R}^n) = n$.

Proof. In $\mathbb{R}^n$ there exists a countable collection $\{G_k : k \in \mathbb{N}\}$ of bounded non-empty open sets such that $\mathbb{R}^n = \bigcup_{k\in\mathbb{N}} G_k$. According to Proposition 5.27, we have $\dim_H(G_k) = n$. Then by Theorem 4.26 we have

$$\dim_H(\mathbb{R}^n) = \dim_H\left(\bigcup_{k\in\mathbb{N}} G_k\right) = \sup\left\{\dim_H(G_k) : k \in \mathbb{N}\right\} = n.$$

This completes the proof. ∎

Proposition 5.29. *For every non-empty open set O in $\mathbb{R}^n$, we have* $\dim_H(O) = n$.

Proof. The set inclusion $O \subset \mathbb{R}^n$ implies $\dim_H(O) \leq \dim_H(\mathbb{R}^n) = n$ by Proposition 5.28. Now O contains a bounded non-empty open set G so that $\dim_H(O) \geq \dim_H(G) = n$ by Proposition 5.27. Thus we have $n \leq \dim_H(O) \leq n$ so that $\dim_H(O) = n$. ∎

Proposition 5.30. *Let $E \subset \mathbb{R}^n$ be a σ-finite set with respect to the outer measure $\mathcal{H}^n$. Then we have* $\dim_H(E) = n$.

Proof. Let $E \subset \mathbb{R}^n$ be a σ-finite set with respect to the outer measure $\mathcal{H}^n$, that is, there exists a countable collection $\{E_k : k \in \mathbb{N}\}$ of subsets of $\mathbb{R}^n$ such that $0 < \mathcal{H}^n(E_k) < \infty$ for every $k \in \mathbb{N}$ and $E = \bigcup_{k\in\mathbb{N}} E_k$. Now $\mathcal{H}^n(E_k) \in (0,\infty)$ implies $\dim_H(E_k) = n$ by (b) of Theorem 4.25. Then by Theorem 4.26 we have

$$\dim_H(E) = \dim_H\left(\bigcup_{k\in\mathbb{N}} E_k\right) = \sup\left\{\dim_H(E_k) : k \in \mathbb{N}\right\} = n.$$

This completes the proof. ∎

[VIII] Hausdorff Outer Measure of Fractional Dimension

We show first that if $\dim_H(E) \in [0,1)$ for $E \in \mathfrak{P}(\mathbb{R}^n)$ then E is a totally disconnected set.

Definition 5.31. *A set E in a topological space X is said to be totally disconnected if for every pair of distinct points x_1 and x_2 in E there exists a pair of disjoint open sets O_1 and O_2 in X such that $x_1 \in O_1$, $x_2 \in O_2$, and $E \subset O_1 \cup O_2$.*
(Note that by default $\emptyset$ and a singleton in X are totally disconnected sets.)

Theorem 5.32. *Every subset of a totally disconnected set in a topological space is a totally disconnected set.*

Proof. This is an immediate consequence of Definition 5.31. ■

Example. Let us call $x \in \mathbb{R}^n$ a rational point in $\mathbb{R}^n$ if the n coordinates of x are all rational numbers. Similarly we call $x \in \mathbb{R}^n$ an irrational point in $\mathbb{R}^n$ if the n coordinates of x are all irrational numbers. Let E be the set of all rational points in $\mathbb{R}^n$ and let F the set of all irrational points in $\mathbb{R}^n$. Then E is a totally disconnected set in $\mathbb{R}^n$ and so is F.

Proof. Let us show that F is a totally disconnected set in $\mathbb{R}^n$. (The same argument will show that E is a totally disconnected set in $\mathbb{R}^n$.)

Let $x_1, x_2 \in E$ and $x_1 \neq x_2$. Let x_1 and x_2 be represented as $x_1 = (a_1, \dots, a_n)$ and $x_2 = (b_1, \dots, b_n)$ respectively where $a_1, \dots, a_n, b_1, \dots, b_n$ are all irrational numbers. Since $x_1 \neq x_2$ we have $a_i \neq b_i$ for at least one $i = 1, \dots, n$. We may assume without loss of generality that $i = 1$ and $a_1 < b_1$. Let γ be a rational number such that $a_1 < \gamma < b_1$. Let

$$O_1 = (-\infty, \gamma) \times \mathbb{R}_2 \times \cdots \times \mathbb{R}_n \quad \text{and} \quad O_2 = (\gamma, \infty) \times \mathbb{R}_2 \times \cdots \times \mathbb{R}_n.$$

Then O_1 and O_2 are two disjoint open sets in $\mathbb{R}^n$ and moreover $x_1 \in O_1$, $x_2 \in O_2$ and $F \subset O_1 \cup O_2$. This shows that F is a totally disconnected set in $\mathbb{R}^n$. ■

Theorem 5.33. *Let $E \in \mathfrak{P}(\mathbb{R}^n)$. If $\dim_H(E) < 1$, then E is a totally disconnected set.*

Proof. 1. Let $x_1 \in E$ be arbitrarily chosen and fixed. Define a mapping f of $\mathbb{R}^n$ into $\mathbb{R}$ by setting $f(x) = |x - x_1|$ for $x \in \mathbb{R}^n$. Then for any $x', x'' \in \mathbb{R}^n$ we have

$$|f(x') - f(x'')| = \Big| |x' - x_1| - |x'' - x_1| \Big| \leq |x' - x''|.$$

This shows that f is a Lipschitz mapping. Then according to (d) of Theorem 4.31 we have $\dim_H(f(E)) \leq \dim_H(E) < 1$. This implies $\mathcal{H}^1(f(E)) = 0$. Now by Theorem 5.22 and Observation 5.23, we have $\mathcal{H}^1 = \mu_L^1$ and thus $\mu_L^1(f(E)) = \mathcal{H}^1(f(E)) = 0$. This implies that the set $f(E) \subset \mathbb{R}$ cannot contain any open interval in $\mathbb{R}$.

2. Let $x_2 \in E$, $x_2 \neq x_1$ be arbitrarily chosen. We have $f(x_2) = |x_2 - x_1| > 0$. Consider the open interval $(0, f(x_2)) \subset \mathbb{R}$. Since the set $f(E)$ cannot contain any open interval in $\mathbb{R}$, there exists $\xi \in (0, f(x_2))$ such that $\xi \notin f(E)$. Now since f is a Lipschitz mapping of $\mathbb{R}^n$ into $\mathbb{R}$, f is a continuous mapping. This implies that $O_1 := f^{-1}\big((-\infty, \xi)\big)$ and $O_2 := f^{-1}\big((\xi, \infty)\big)$ are open sets in $\mathbb{R}^n$. Clearly O_1 and O_2 are disjoint. Then since $f(x_1) = |x_1 - x_1| = 0 < \xi$, we have $x_1 \in O_1$. Since $f(x_2) > \xi$, we have $x_2 \in O_2$. For any $x \in E$, since $\xi \notin f(E)$, we have $f(x) \neq \xi$ and thus $x \in O_1 \cup O_2$. This shows that $E \subset O_1 \cup O_2$. Thus we have shown that for any pair of distinct points x_1 and x_2 in E there exists a pair of disjoint open sets O_1 and O_2 in $\mathbb{R}^n$ such that $x_1 \in O_1$, $x_2 \in O_2$ and $E \subset O_1 \cup O_2$. ∎

Our next goal is to show that for every $\gamma \in [0, 1)$ there exists $E \in \mathfrak{P}(\mathbb{R}^n)$ such that $\dim_H(E) = \gamma$. We start by constructing some $E \in \mathfrak{P}(\mathbb{R})$ with $\dim_H \in [0, 1)$ although a set with fractional Hausdorff dimension need not be a subset of $\mathbb{R}$.

Definition 5.34. (Cantor Ternary Set) *Consider $J = [0, 1]$.*
At Step 1, we decompose J into three subintervals of length $\frac{1}{3}$, remove the middle open subinterval and let J_0 and J_1 be the remaining two closed subintervals of length $\frac{1}{3}$.
At Step 2, we apply the process of removal above to J_0 resulting in two closed subintervals $J_{0,0}$ and $J_{0,1}$ of length $\left(\frac{1}{3}\right)^2$ and apply the process of removal to J_1 resulting in two closed subintervals $J_{1,0}$ and $J_{1,1}$ of length $\left(\frac{1}{3}\right)^2$ and thus we have 2^2 closed subintervals of length $\left(\frac{1}{3}\right)^2$.
Thus iterating, at Step k, we have 2^k closed subintervals of length $\left(\frac{1}{3}\right)^k$.
We iterate the process of removal indefinitely and call the resulting set a Cantor ternary set. Thus a Cantor ternary set is obtained by removing $2^0 + 2^1 + 2^2 + \cdots = \aleph_0$ disjoint open subintervals from J.
Let us call $\frac{1}{3}$ the ratio of removal in a Cantor ternary set.

Next we are interested in generalizing a Cantor ternary set by considering a ratio of removal other than $\frac{1}{3}$. We want to remove an open interval V from the center of $J = [0, 1]$ in such a way that the remaining two closed intervals each has a positive length $\lambda > 0$. This implies that the positive length of the open interval V must be $1 - 2\lambda > 0$. This implies that $1 > 2\lambda$ and therefore $\lambda \in \left(0, \frac{1}{2}\right)$.

Definition 5.35. (Cantor Set C_λ with Ratio $\lambda \in \left(0, \frac{1}{2}\right)$) *Let $J = [0, 1]$ and $\lambda \in \left(0, \frac{1}{2}\right)$.*
Let $\mathfrak{J}_0 = \{J\}$, a collection of $1 = 2^0$ closed intervals of length $1 = \lambda^0$.
At Step 1, we remove an open interval in the center of J so that each of the remaining two closed intervals J_0 and J_1 has length λ. Let $\mathfrak{J}_1$ be the collection of the two closed intervals of length λ.
At Step 2, we remove an open interval in the center of each one of J_0 and J_1 so that each of the remaining 2^2 closed intervals $J_{0,0}, J_{0,1}, J_{1,0}, J_{1,1}$ has length λ^2. Let $\mathfrak{J}_2$ be the collection of these 2^2 closed intervals of length λ^2.
We iterate this process of removal.

At Step $k \in \mathbb{N}$, we remove an open interval in the center of 2^{k-1} closed intervals of length λ^{k-1} so that we have 2^k closed intervals of length λ^k. Let $\mathfrak{J}_k$ be the collection of these 2^k closed intervals of length λ^k.
We iterate the process of removal indefinitely and call the resulting set a Cantor set C_λ with ratio $\lambda \in \left(0, \frac{1}{2}\right)$.

Proposition 5.36. *Let $J = [0,1]$ and $\lambda \in \left(0, \frac{1}{2}\right)$. Let $\mathfrak{J}_0 = \{J\}$.*
At Step 1, we let $\mathfrak{J}_1 = \{J_0, J_1\}$ where J_0 and J_1 are two closed intervals of length λ obtained by removing an open interval V of length $1 - 2\lambda$ from the center of J.
At Step 2, we let $\mathfrak{J}_2 = \{J_{0,0}, J_{0,1}, J_{1,0}, J_{1,1}\}$ where $J_{0,0}$ and $J_{0,1}$ are two closed intervals each with length λ^2 obtained by removing an open interval V_0 of length $\lambda(1-2\lambda)$ from the center of J_0 and similarly $J_{1,0}$ and $J_{1,1}$ intervals each with length λ^2 obtained by removing an open interval V_1 of length $\lambda(1-2\lambda)$ from the center of J_1.
Thus iterating, at Step k, we let $\mathfrak{J}_k = \{J_{i_1 i_2 \dots i_k} : i_j = 0, 1 \text{ for } j = 1, \dots, k\}$ where $J_{i_1 i_2 \dots i_{k-1} 0}$ and $J_{i_1 i_2 \dots i_{k-1} 1}$ are two closed intervals of length λ^k obtained by removing an open interval $V_{i_1 i_2 \dots i_{k-1}}$ of length $(1-2\lambda)^{k-1}$ from the center of $J_{i_1 i_2 \dots i_{k-1}}$. Thus $\mathfrak{J}_k$ is a collection of 2^k disjoint closed intervals of length λ^k. Let E_k be the union of members of the class $\mathfrak{J}_k$, that is,

$$E_k = \bigcup_{i_1=0}^{1} \cdots \bigcup_{i_k=0}^{1} J_{i_1 i_2 \dots i_k}.$$

Then we have

$$C_\lambda = \bigcap_{k \in \mathbb{Z}_+} E_k.$$

Proof. This Proposition is essentially a restatement of Definition 5.35. ∎

Theorem 5.37. *The Cantor set C_λ with $\lambda \in \left(0, \frac{1}{2}\right)$ has the following properties:*
(a) *C_λ is a null set in the measure space $(\mathbb{R}, \mathfrak{B}_\mathbb{R}, \mu_L)$.*
(b) *$G := [0,1] \setminus C_\lambda$ is a union of countably many disjoint open intervals in $[0,1]$. Moreover we have $\mu_L(G) = 1$ and G is dense in $[0,1]$.*
(c) *C_λ is an uncountable set. Indeed the cardinality of C_λ is equal to $\mathfrak{c}$, the continuum.*
(d) *C_λ is a compact set in $\mathbb{R}$.*
(e) *C_λ is a perfect set in $\mathbb{R}$, that is, C_λ is identical with the set of all its limit points.*
(f) *C_λ is nowhere dense in $\mathbb{R}$, that is, the interior of its closure, $\left(\overline{C_\lambda}\right)^\circ$, is an empty set.*

Proof. 1. The set E_k in Proposition 5.36 is the union of 2^k disjoint closed intervals of length λ^k in $\mathbb{R}$ and thus $E_k \in \mathfrak{B}_\mathbb{R}$ and $\mu_L(E_k) = 2^k \lambda^k = (2\lambda)^k$. Since $\left(E_k : k \in \mathbb{Z}_+\right)$ is a decreasing sequence, we have $C_\lambda = \bigcap_{k \in \mathbb{Z}_+} E_k = \lim_{k \to \infty} E_k$ and this implies that $C_\lambda \in \mathfrak{B}_\mathbb{R}$. Then since $E_0 = [0,1]$ and $\mu_L([0,1]) = 1 < \infty$, we have

$$\mu_L(C_\lambda) = \lim_{k \to \infty} \mu_L(E_k) = \lim_{k \to \infty} (2\lambda)^k = 0$$

since $2\lambda < 1$. This shows that C_λ is a null set in the measure space $(\mathbb{R}, \mathfrak{B}_{\mathbb{R}}, \mu_L)$.

2. C_λ is obtained by removing an open interval in $[0,1]$ countably many times. Thus $G := [0,1] \setminus C_\lambda$ is a union of countably many disjoint open intervals in $[0,1]$.

Now $G \cap C_\lambda = \emptyset$ and $G \cup C_\lambda = [0,1]$. This implies $\mu_L(G) + \mu_L(C_\lambda) = \mu_L([0,1]) = 1$. Then since $\mu_L(C_\lambda) = 0$ by (a), we have $\mu_L(G) = 1$.

In general, if E is a null set in $(\mathbb{R}, \mathfrak{B}_{\mathbb{R}}, \mu_L)$ then E^c is a dense subset of $\mathbb{R}$, that is, for every open interval I in $\mathbb{R}$ we have $E^c \cap I \neq \emptyset$. Then since C_λ is a null set in $(\mathbb{R}, \mathfrak{B}_{\mathbb{R}}, \mu_L)$, C_λ^c is a dense subset of $\mathbb{R}$ and this implies that $G := [0,1] \setminus C_\lambda$ is a dense subset of $[0,1]$.

3. C_λ is the result of indefinitely iterated process of deleting open intervals from $[0,1]$. In the first step an open interval is deleted from $[0,1]$, leaving 2 disjoint closed intervals. In the second step an open interval is deleted from each of the 2 disjoint closed intervals, leaving 2^2 disjoint closed intervals. In the third step an open interval is deleted from each of the 2^2 disjoint closed intervals, leaving 2^3 disjoint closed intervals and so on indefinitely. The two endpoints of each of the 2^k disjoint closed intervals in the k-th step of deletion are never deleted in subsequent steps of deletion and are thus elements of C_λ. Thus the cardinality of C_λ is at least equal to $2^{\aleph_0} = \mathfrak{c}$, the continuum. Since $C_\lambda \subset [0,1]$ and the cardinality of $[0,1]$ is equal to $\mathfrak{c}$, the cardinality of C_λ is equal to $\mathfrak{c}$.

4. The set E_k in Proposition 5.36 is a union of 2^k closed intervals in $\mathbb{R}$ and is thus a closed set in $\mathbb{R}$. Then $C_\lambda = \bigcap_{k \in \mathbb{Z}_+} E_k$ is a closed set in $\mathbb{R}$. Moreover $C_\lambda \subset [0,1]$ is a bounded set in $\mathbb{R}$. Therefore C_λ is a compact set in $\mathbb{R}$.

5. Let F be the set of all limit points of C_λ. Since C_λ is a closed set we have $F \subset C_\lambda$. It remains to show that $C_\lambda \subset F$. Thus we are to show that for every $x_0 \in C_\lambda$ and for every $\delta > 0$ there exists $x \in C_\lambda$ such that $x \neq x_0$ and $x \in (x_0 - \delta, x_0 + \delta)$. Now by Proposition 5.36, we have $x_0 \in C_\lambda = \bigcap_{k \in \mathbb{Z}_+} E_k$ where E_k is the union of 2^k disjoint closed intervals of length λ^k. Thus for sufficiently large $k \in \mathbb{N}$ we have a closed interval $J_{i_1 i_2 \dots i_k} \subset (x_0 - \delta, x_0 + \delta)$. Now the two endpoints of $J_{i_1 i_2 \dots i_k}$ are in C_λ. Thus there exists a point $x \in C_\lambda$ such that $x \neq x_0$ and $x \in (x_0 - \delta, x_0 + \delta)$.

6. Let us show that $(\overline{C_\lambda})^\circ = \emptyset$. Since C_λ is a closed set we have $\overline{C_\lambda} = C_\lambda$ and then $(\overline{C_\lambda})^\circ = C_\lambda^\circ$. To show $C_\lambda^\circ = \emptyset$, assume the contrary. Then C_λ° is a non-empty open set in $\mathbb{R}$ so that $\mu_L(C_\lambda^\circ) > 0$. This contradict the fact that $\mu_L(C_\lambda) = 0$. Therefore we have $C_\lambda^\circ = \emptyset$ and then $(\overline{C_\lambda})^\circ = \emptyset$. ∎

Proposition 5.38. *Consider the collection $\mathfrak{J}_k$ of 2^k disjoint closed intervals of length λ^k in $\mathbb{R}$ for $k \in \mathbb{Z}_+$ as defined in Proposition 5.36 and let $\mathfrak{J} = \bigcup_{k \in \mathbb{Z}_+} \mathfrak{J}_k$. Then we have:*

(a) *If x is an endpoint of a member of $\mathfrak{J}$, then $x \in C_\lambda$.*

(b) *If $x \in C_\lambda$ and x is not the left endpoint of any member of $\mathfrak{J}$, then there exists a sequence $(x_m : m \in \mathbb{N})$ such that x_m is the left endpoint of a member of $\mathfrak{J}$, $x_m < x$, and $x_m \uparrow x$. Similarly if $x \in C_\lambda$ and x is not the right endpoint of any member of $\mathfrak{J}$, then there exists a sequence $(x_m : m \in \mathbb{N})$ such that x_m is the right endpoint of a member of $\mathfrak{J}$, $x_m > x$, and $x_m \downarrow x$.*

(c) *If $x \notin C_\lambda$ and there exists a point $\xi \in C_\lambda$ such that $x < \xi$, then there exists $\alpha \in C_\lambda$ such that $x < \alpha$, α is the left endpoint of a member of $\mathfrak{J}$ and $[x, \alpha) \cap C_\lambda = \emptyset$. Similarly if $y \notin C_\lambda$*

and there exists a point $\eta \in C_\lambda$ such that $\eta < y$, then there exists $\beta \in C_\lambda$ such that $\beta < y$, β is the right endpoint of a member of $\mathfrak{J}$ and $(\beta, y] \cap C_\lambda = \emptyset$.

Proof. 1. Since C_λ is the result of removing open intervals in the middle of member closed intervals in $\mathfrak{J}$, every endpoint of a member closed interval in $\mathfrak{J}$ remains.

2. According to Proposition 5.36, $C_\lambda = \bigcap_{k \in \mathbb{N}} E_k$ and E_k is the union of 2^k disjoint closed intervals of length λ^k and we write $\mathfrak{J}_k$ for the collection of these closed intervals and write $\mathfrak{J} = \bigcup_{k \in \mathbb{N}} \mathfrak{J}_k$.

Suppose $x \in C_\lambda$ but x is not the left endpoint of any member interval of $\mathfrak{J}$. Then $x \in E_k$ for every $k \in \mathbb{N}$ and thus $x \in J_k$ for some $J_k \in \mathfrak{J}_k$ for every $k \in \mathbb{N}$. Let ξ_k be the left endpoint of J_k. For an arbitrary $\varepsilon > 0$, we have $\lambda^k < \varepsilon$ for sufficiently large $k \in \mathbb{N}$, that is, the length of J_k is less than ε for sufficiently large $k \in \mathbb{N}$. Thus $\xi_k < x$ and $x - \xi_k < \varepsilon$ for sufficiently large $k \in \mathbb{N}$. Then we have $\lim_{k \to \infty} \xi_k = x$. For every $m \in \mathbb{N}$, let $x_m = \max\{\xi_1, \dots, \xi_m\}$. Then $(x_m : m \in \mathbb{N})$ is a sequence of left endpoints of members of $\mathfrak{J}$ such that $x_m < x$ and $x_m \uparrow x$.

Similarly for the case that $x \in C_\lambda$ and x is not the right endpoint of any member of $\mathfrak{J}$.

3. Suppose $x \notin C_\lambda$ and there exists a point $\xi \in C_\lambda$ such that $x < \xi$. Then $[x, \infty) \cap C_\lambda$ is nonempty and is a compact set. Let α be the nearest point in the compact set $[x, \infty) \cap C_\lambda$ from x. Then $[x, \alpha) \cap C_\lambda = \emptyset$ so that α must be a left endpoint of a member of $\mathfrak{J}$ by (b). Similarly for the case that $y \notin C_\lambda$ and there exists a point $\eta \in C_\lambda$ such that $\eta < y$. ■

Definition 5.39. *For $\lambda \in \left(0, \frac{1}{2}\right)$, define a function s^* of λ by setting*

$$s^*(\lambda) = \frac{\log 2}{\log \lambda^{-1}} = -\frac{\log 2}{\log \lambda} \in (0, 1).$$

Lemma 5.40. *Consider the function $s^*(\lambda)$ for $\lambda \in \left(0, \frac{1}{2}\right)$ as in Definition 5.39. Then we have*

$$2^x = \lambda^{-x s^*(\lambda)} \quad \textit{for every } x \in \mathbb{R}.$$

Proof. For $x \in \mathbb{R}$, we have

$$\begin{aligned} 2^x &= e^{\log 2^x} = e^{x \log 2} = \exp\left\{x \frac{\log 2}{\log \lambda^{-1}} \log \lambda^{-1}\right\} \\ &= \exp\left\{x s^*(\lambda) \log \lambda^{-1}\right\} = \exp\left\{\log(\lambda^{-1})^{x s^*(\lambda)}\right\} \\ &= \lambda^{-x s^*(\lambda)}. \end{aligned}$$ ■

Notations. For $k \in \mathbb{Z}_+$, $\mathfrak{J}_k$ is a collection of 2^k disjoint closed intervals of length λ^k in $\mathbb{R}$

as defined in Proposition 5.36. For our subsequent discussions, it is convenient to represent $\mathfrak{J}_k$ as $\mathfrak{J}_k = \{J_{k,i} : i = 1, \dots, 2^k\}$.

Proposition 5.41. *Consider $\mathfrak{J}_k$ for $k \in \mathbb{Z}_+$ and $s^*(\lambda)$ for $\lambda \in (0, \frac{1}{2})$ as in Definition 5.39.*
(a) *Let $I \in \mathfrak{J}_{k_0}$ for some $k_0 \in \mathbb{Z}_+$. Then for every $k \geq k_0$, we have*

$$|I|^{s^*(\lambda)} = \sum_{\{i \in \mathbb{N} : J_{k,i} \subset I\}} |J_{k,i}|^{s^*(\lambda)}. \tag{1}$$

(b) *For every $k \in \mathbb{Z}_+$, we have*

$$\sum_{i=1}^{2^k} |J_{k,i}|^{s^*(\lambda)} = 1. \tag{2}$$

(c) *Let G be an open interval G in $\mathbb{R}$. Then for every $k \in \mathbb{Z}_+$, we have*

$$4|G|^{s^*(\lambda)} \geq \sum_{\{i \in \mathbb{N} : J_{k,i} \cap G \neq \emptyset\}} |J_{k,i}|^{s^*(\lambda)}, \tag{3}$$

and in particular we have

$$4|G|^{s^*(\lambda)} \geq \sum_{\{i \in \mathbb{N} : J_{k,i} \subset G\}} |J_{k,i}|^{s^*(\lambda)}. \tag{4}$$

Proof. 1. Let us prove (a). If $I \in \mathfrak{J}_{k_0}$ for some $k_0 \in \mathbb{Z}_+$, then the length of I is equal to λ^{k_0} and thus $|I| = \lambda^{k_0}$. Then for every $k \geq k_0$ the interval I contains 2^{k-k_0} members of $\mathfrak{J}_k = \{J_{k,i} : i = 1, \dots, 2^k\}$. Since $|J_{k,i}| = \lambda^k$ for $i = 1, \dots, 2^k$, we have

$$\sum_{\{i \in \mathbb{N} : J_{k,i} \subset I\}} |J_{k,i}|^{s^*(\lambda)} = 2^{k-k_0} \lambda^{ks^*(\lambda)} = \lambda^{(k_0-k)s^*(\lambda)} \lambda^{ks^*(\lambda)} \quad \text{by Lemma 5.40}$$

$$= \lambda^{k_0 s^*(\lambda)} = |I|^{s^*(\lambda)}.$$

2. Let us prove (b). Now $J = [0, 1] \in \mathfrak{J}_0$ and $J_{k,i} \subset J$ for $i = 1, \dots, 2^k$ for every $k \in \mathbb{Z}_+$. Then by (a) we have

$$|J|^{s^*(\lambda)} = \sum_{i=1}^{2^k} |J_{k,i}|^{s^*(\lambda)}. \tag{5}$$

On the other hand, we have

$$|J|^{s^*(\lambda)} = 1^{s^*(\lambda)} = 1. \tag{6}$$

Then (5) and (6) prove (2).

3. Let us prove (c). If an open interval G in $\mathbb{R}$ does not intersect any member of the class $\mathfrak{J}_k = \{J_{k,i} : i = 1, \dots, 2^k\}$ for some $k \in \mathbb{Z}_+$, then the inequality (3) holds trivially.

Suppose G intersects some members of $\mathfrak{J}_k$ for some $k \in \mathbb{Z}_+$. Let k_0 be the smallest $k \in \mathbb{Z}_+$ such that G intersects some members of $\mathfrak{J}_k$. Let $J_{k_0,1}, \dots, J_{k_0,p}$, enumerated from left to right in the real line $\mathbb{R}$, be all the members of $\mathfrak{J}_{k_0}$ that intersect G. Then $p \leq 4$ for otherwise G would intersect some members of $\mathfrak{J}_{k_0-1}$. Consider the maximal case, that is, $p = 4$. Now G intersects $J_{k_0,1}, \dots, J_{k_0,4}$. Since G is an interval we must have $J_{k_0,2}, J_{k_0,3} \subset G$.

Now $G \supset J_{k_0,2}$ implies by means of (1) that

$$|G|^{s^*(\lambda)} \geq |J_{k_0,2}|^{s^*(\lambda)} = \sum_{\{i \in \mathbb{N}: J_{k,i} \subset J_{k_0,2}\}} |J_{k,i}|^{s^*(\lambda)}. \tag{7}$$

Similarly $G \supset J_{k_0,3}$ implies

$$|G|^{s^*(\lambda)} \geq |J_{k_0,3}|^{s^*(\lambda)} = \sum_{\{i \in \mathbb{N}: J_{k,i} \subset J_{k_0,3}\}} |J_{k,i}|^{s^*(\lambda)}. \tag{8}$$

Now we have $G \supset J_{k_0,3}$ and $|J_{k_0,3}| = |J_{k_0,4}|$. This implies that for some $x_0 \in \mathbb{R}$ the translate $G + x_0 \supset J_{k_0,4}$. Then by the same argument as for (7) we obtain

$$|G + x_0|^{s^*(\lambda)} \geq |J_{k_0,4}|^{s^*(\lambda)} = \sum_{\{i \in \mathbb{N}: J_{k,i} \subset J_{k_0,4}\}} |J_{k,i}|^{s^*(\lambda)}.$$

But $|G + x_0| = |G|$. Thus we have

$$|G|^{s^*(\lambda)} \geq |J_{k_0,4}|^{s^*(\lambda)} = \sum_{\{i \in \mathbb{N}: J_{k,i} \subset J_{k_0,4}\}} |J_{k,i}|^{s^*(\lambda)}. \tag{9}$$

By similar argument we obtain

$$|G|^{s^*(\lambda)} \geq |J_{k_0,1}|^{s^*(\lambda)} = \sum_{\{i \in \mathbb{N}: J_{k,i} \subset J_{k_0,1}\}} |J_{k,i}|^{s^*(\lambda)}. \tag{10}$$

Adding (7), (8), (9) and (10) side by side, we have

$$\begin{aligned} 4|G|^{s^*(\lambda)} &\geq \sum_{\{i \in \mathbb{N}: J_{k,i} \subset J_{k_0,1}\}} |J_{k,i}|^{s^*(\lambda)} + \sum_{\{i \in \mathbb{N}: J_{k,i} \subset J_{k_0,2}\}} |J_{k,i}|^{s^*(\lambda)} \\ &\quad + \sum_{\{i \in \mathbb{N}: J_{k,i} \subset J_{k_0,3}\}} |J_{k,i}|^{s^*(\lambda)} + \sum_{\{i \in \mathbb{N}: J_{k,i} \subset J_{k_0,4}\}} |J_{k,i}|^{s^*(\lambda)} \\ &\geq \sum_{\{i \in \mathbb{N}: J_{k,i} \cap G \neq \emptyset\}} |J_{k,i}|^{s^*(\lambda)}, \end{aligned}$$

where the last inequality is by the fact that $J_{k,i} \cap G \neq \emptyset$ implies that $J_{k,i}$ is contained in $J_{k_0,i}$ for some $i = 1, \dots, 4$. This proves (3). ∎

Theorem 5.42. *Consider the Cantor set C_λ with $\lambda \in \left(0, \frac{1}{2}\right)$ and the function s^* of $\lambda \in \left(0, \frac{1}{2}\right)$ defined by*

$$s^*(\lambda) = \frac{\log 2}{\log \lambda^{-1}} = -\frac{\log 2}{\log \lambda} \in (0, 1).$$

Then we have

$$\text{(1)} \qquad \frac{1}{4} \leq \mathcal{H}^{s^*(\lambda)}(C_\lambda) \leq 1,$$

and therefore we have

$$\text{(2)} \qquad \dim_H(C_\lambda) = s^*(\lambda).$$

Proof. 1. Consider $\mathfrak{J}_k = \left\{J_{k,i} : i = 1, \dots, 2^k\right\}$ for $k \in \mathbb{Z}_+$ as defined in Notations preceding Proposition 5.41. By Proposition 5.36 we have

$$\text{(3)} \qquad C_\lambda \subset \bigcup_{i=1}^{2^k} J_{k,i},$$

and by Proposition 5.41 we have

$$\text{(4)} \qquad \sum_{i=1}^{2^k} |J_{k,i}|^{s^*(\lambda)} = 1.$$

For an arbitrary $\delta > 0$, let $k \in \mathbb{Z}_+$ be so large that $|J_{k,i}| = \lambda^k \leq \delta$. Let $\mathfrak{V}$ be a δ covering class for the set $\bigcup_{i=1}^{2^k} J_{k,i}$ consisting of $\emptyset$ and every closed interval I with $|I| \leq \delta$. (See Definition 2.1.) By Definition 4.4 and by means of Theorem 4.15, we have

$$\begin{aligned}\mathcal{H}_\delta^{s^*(\lambda)}\left(\bigcup_{i=1}^{2^k} J_{k,i}\right) &= \inf\left\{\sum_{n\in\mathbb{N}} |V_n|^{s^*(\lambda)} : (V_n : n \in \mathbb{N}) \subset \mathfrak{V}, \bigcup_{n\in\mathbb{N}} V_n \supset \bigcup_{i=1}^{2^k} J_{k,i}\right\} \\ &\leq \sum_{i=1}^{2^k} |J_{k,i}|^{s^*(\lambda)} = 1.\end{aligned}$$

This and (3) imply that $\mathcal{H}_\delta^{s^*(\lambda)}(C_\lambda) \leq \mathcal{H}_\delta^{s^*(\lambda)}\left(\bigcup_{i=1}^{2^k} J_{k,i}\right) \leq 1$. Then we have

$$\text{(5)} \qquad \mathcal{H}^{s^*(\lambda)}(C_\lambda) = \lim_{\delta\downarrow 0} \mathcal{H}_\delta^{s^*(\lambda)}(C_\lambda) \leq 1.$$

2. Let $\delta > 0$ be arbitrarily given. Let G be an open interval such that

$$G \supset \bigcup_{i=1}^{2^k} J_{k,i} \supset C_\lambda.$$

Let $(G_j : j \in \mathbb{N})$ be a sequence of open intervals such that
1° $\bigcup_{j\in\mathbb{N}} G_j = G$.
2° $|G_j| \le \delta$ for every $j \in \mathbb{N}$.
By (4) and (2) of Proposition 5.41, we have

$$4|G|^{s^*(\lambda)} \ge \sum_{\{i\in\mathbb{N}: J_{k,i}\subset G\}} |J_{k,i}|^{s^*(\lambda)} = 1. \tag{6}$$

Now $|G| = |\bigcup_{j\in\mathbb{N}} G_j| \le \sum_{j\in\mathbb{N}} |G_j|$ and thus we have

$$|G|^{s^*(\lambda)} \le \left\{\sum_{j\in\mathbb{N}} |G_j|\right\}^{s^*(\lambda)}.$$

Let $\alpha_j \ge 0$ for $j \in \mathbb{N}$. According to Hölder's inequality for $p \in [0,1]$, we have

$$\left\{\sum_{j\in\mathbb{N}} \alpha_j\right\}^p \le \sum_{j\in\mathbb{N}} \alpha_j^p.$$

Applying this to the right side of (7), we have

$$|G|^{s^*(\lambda)} \le \sum_{j\in\mathbb{N}} |G_j|^{s^*(\lambda)}.$$

Substituting this in the left side of (6) and then dividing both sides by 4, we obtain

$$\sum_{j\in\mathbb{N}} |G_j|^{s^*(\lambda)} \ge \frac{1}{4}. \tag{7}$$

Now consider a δ-covering class $\mathfrak{V}$ for the set C_λ consisting of $\emptyset$ and open intervals with diameter not exceeding δ. Then we have

$$\mathcal{H}_\delta^{s^*(\lambda)}(C_\lambda) = \inf\left\{\sum_{j\in\mathbb{N}} |G_j|^{s^*(\lambda)} : (G_j : j \in \mathbb{N}) \subset \mathfrak{V}\right\} \ge \frac{1}{4}.$$

This implies

$$\mathcal{H}^{s^*(\lambda)}(C_\lambda) = \lim_{\delta\downarrow 0} \mathcal{H}_\delta^{s^*(\lambda)}(C_\lambda) \ge \frac{1}{4}. \tag{8}$$

By (5) and (8) we have (1). Then (1) implies that $\mathcal{H}^{s^*(\lambda)}(C_\lambda) \in (0,\infty)$. This implies according to (b) of Theorem 4.25 that $\dim_H(C_\lambda) = s^*(\lambda)$. This proves (2). ∎

Theorem 5.43. *For every* $s_0 \in (0,1)$ *there exists* $E \subset \mathbb{R}$ *such that* $\dim_H(E) = s_0$. *Indeed for the Cantor set* C_λ *with* $\lambda = \exp\{-\frac{1}{s_0}\log 2\}$, *we have* $\dim_H(C_\lambda) = s_0$.

Proof. By Definition 5.39, we have

$$(1) \qquad s^*(\lambda) = \frac{\log 2}{\log \lambda^{-1}} = -\frac{\log 2}{\log \lambda} \in (0,1) \quad \text{for } \lambda \in \left(0, \tfrac{1}{2}\right).$$

According to Theorem 5.42, for the Cantor set C_λ with $\lambda \in \left(0, \frac{1}{2}\right)$, we have

$$(2) \qquad \dim_H(C_\lambda) = s^*(\lambda).$$

Now the function s^* of $\lambda \in \left(0, \frac{1}{2}\right)$ is a strictly increasing continuous function mapping $\left(0, \frac{1}{2}\right)$ one-to-one onto $(0,1)$. Its inverse function $(s^*)^{-1}$ then is a strictly increasing continuous function mapping $(0,1)$ one-to-one onto $\left(0, \frac{1}{2}\right)$. For $s_0 \in (0,1)$, solving $s_0 = -\left(\log 2/\log \lambda\right)$ for λ, we obtain

$$(3) \qquad \lambda = (s^*)^{-1}(s_0) = \exp\left\{-\frac{1}{s_0}\log 2\right\}.$$

Now consider C_λ with λ given by (3). According to Theorem 5.42 we have

$$\dim_H(C_\lambda) = s^*(\lambda) = s^*\left((s^*)^{-1}(s_0)\right) = s_0.$$

This completes the proof. ■

Theorem 5.44. (Existence of a Set with an Arbitrary Hausdorff Dimension) *For every* $s_0 \in (0,1)$ *and* $n_0 \in \mathbb{N}$, *there exists a set* $E \subset \mathbb{R}^{n_0+1}$ *such that* $\dim_H(E) = n_0 + s_0$.

Proof. According to Corollary 4.28, if $E_1, E_2 \in \mathfrak{B}_{\mathbb{R}^{n_0+1}}$ and $E_1 \cap E_2 = \emptyset$, then we have

$$(1) \qquad \dim_H(E_1 \cup E_2)) = \dim_H(E_1) + \dim_H(E_2).$$

In Theorem 5.43, we showed that for every $s_0 \in (0,1)$ there exists a Cantor set $C_\lambda \in \mathfrak{B}_{\mathbb{R}}$ such that $\dim_H(C_\lambda) = s_0$. In Proposition 5.29, we showed that if $O \in \mathfrak{B}_{\mathbb{R}^{n_0}}$ is a non-empty open set in $\mathbb{R}^{n_0}$, then $\dim_H(O) = n_0$. Now C_λ and O are two disjoint subsets of $\mathbb{R}^{n_0+1}$ and $C_\lambda, O \in \mathfrak{B}_{\mathbb{R}^{n_0+1}}$. Thus by (1) we have

$$(2) \qquad \dim_H(C_\lambda \cup O)) = \dim_H(C_\lambda) + \dim_H(O) = s_0 + n_0.$$

Let $E = C_\lambda \cup O$. Then we have

$$(3) \qquad \dim_H(E) = \dim_H(C_\lambda \cup O)) = n_0 + s_0.$$

This completes the proof. ■

In order to determine the Hausdorff dimension of the Cantor set C_λ, we showed in Theorem 5.42 that $\frac{1}{4} \leq \mathcal{H}^{s^*(\lambda)}(C_\lambda) \leq 1$. We show next that actually $\mathcal{H}^{s^*(\lambda)}(C_\lambda) = 1$.

Theorem 5.45. *Consider the Cantor set C_λ where $\lambda \in \left(0, \frac{1}{2}\right)$ and the function $s^*(\lambda) = \log 2 / \log \lambda^{-1} \in (0, 1)$. Then we have*

$$\mathcal{H}^{s^*(\lambda)}(C_\lambda) = 1.$$

Proof. 1. In Proposition 5.36 we introduced $\mathfrak{J}_k = \{J_{k,i}; i = 1, \ldots, 2^k\}$ for $k \in \mathbb{Z}_+$, let $E_k = \bigcup_{i=1}^{2^k} J_{k,i}$ and then showed that $C_\lambda = \bigcap_{k \in \mathbb{Z}_+} E_k$. Now $C_\lambda \in \mathfrak{B}_\mathbb{R} \subset \mathfrak{M}(\mathcal{H}^{s^*(\lambda)})$ and the outer measure $\mathcal{H}^{s^*(\lambda)}$ on $\mathbb{R}$ is a measure on the σ-algebra $\mathfrak{M}(\mathcal{H}^{s^*(\lambda)})$ of subsets of $\mathbb{R}$. Then since $(E_k : k \in \mathbb{Z}_+)$ is a decreasing sequence of sets with finite measure, we have

$$\mathcal{H}^{s^*(\lambda)}(C_\lambda) = \lim_{k \to \infty} \mathcal{H}^{s^*(\lambda)}(E_k). \tag{1}$$

2. Let us find $\mathcal{H}^{s^*(\lambda)}(E_k)$. According to Theorem 5.5, $\mathcal{H}^{s^*(\lambda)}$ is translation invariant. Thus if we find $\mathcal{H}^{s^*(\lambda)}(J_{k,i})$ then $\mathcal{H}^{s^*(\lambda)}(E_k) = 2^k \mathcal{H}^{s^*(\lambda)}(J_{k,i})$. Let us find $\mathcal{H}^{s^*(\lambda)}(J_{k,i})$.

According to Theorem 5.2, $\mathcal{H}^{s^*(\lambda)} = \mathcal{W}^{s^*(\lambda)}$. The outer measure $\mathcal{W}^{s^*(\lambda)}$ on a separable normed linear space X is based on the covering class $\mathcal{W}_X$ which consists of all convex sets in X. In our case where $X = \mathbb{R}$ the convex sets are the intervals in $\mathbb{R}$ and $\emptyset$. Let $\mathcal{W}_\mathbb{R}$ be the collection of all intervals in $\mathbb{R}$ and $\emptyset$. Then

$$\mathcal{W}_\delta^{s^*(\lambda)}(J_{k,i}) = \inf \left\{ \sum_{n \in \mathbb{N}} |I_n|^{s^*(\lambda)} : (I_n : n \in \mathbb{N}) \subset \mathcal{W}_\mathbb{R}|_\delta, \bigcup_{n \in \mathbb{N}} I_n \supset J_{k,i} \right\}. \tag{2}$$

Since $s^*(\lambda) \in [0, 1]$ we have for $\alpha_1, \ldots, \alpha_m \geq 0$ the inequality

$$(\alpha_1 + \cdots + \alpha_m)^{s^*(\lambda)} \leq \alpha_1^{s^*(\lambda)} + \cdots + \alpha_m^{s^*(\lambda)}.$$

This implies that the infimum in (2) is equal to $|J_{k,i}|^{s^*(\lambda)}$ and thus

$$\mathcal{W}_\delta^{s^*(\lambda)}(J_{k,i}) = |J_{k,i}|^{s^*(\lambda)}$$

and then

$$\mathcal{W}^{s^*(\lambda)}(J_{k,i}) = \lim_{\delta \downarrow 0} \mathcal{W}_\delta^{s^*(\lambda)}(J_{k,i}) = |J_{k,i}|^{s^*(\lambda)}.$$

Thus

$$\mathcal{H}^{s^*(\lambda)}(J_{k,i}) = |J_{k,i}|^{s^*(\lambda)}.$$

Then by (2) of Proposition 5.41 we have

$$\mathcal{H}^{s^*(\lambda)}(E_k) = \sum_{i=1}^{2^k} |J_{k,i}|^{s^*(\lambda)} = 1. \tag{3}$$

Then by (1) and (3), we have $\mathcal{H}^{s^*(\lambda)}(C_\lambda) = 1$. This completes the proof. ∎

§6 Covering Theorems in a Metric Space

Terminologies and Notations. Let (X, ρ) be a metric space.
A subset of X, $B(x,r) := \{y \in X : \rho(x,y) < r\}$, is called an open ball in X with center $x \in X$ and radius $r \in (0,\infty)$.
A subset of X, $\overline{B}(x,r) := \{y \in X : \rho(x,y) \leq r\}$, is called a closed ball in X with center $x \in X$ and radius $r \in (0,\infty)$.
A subset $B \subset X$ is called a ball if it is either an open ball or a closed ball.
For a ball B we write $c(B)$ for the center of B and $r(B)$ for the radius of B.
For a ball B we write $\overline{B}$ for the closure of B, which is a closed ball.
For a ball B we write B° for the interior of B, which is an open ball.

[I] 5r-covering Theorem in a Metric Space

Definition 6.1. *Let B be a ball in a metric space (X,ρ) and let $\alpha \in (0,\infty)$. We write $\alpha \bullet B$ for the ball that is concentric with B but has a radius α times that of B. Thus if $B = B(x,r)$ then $\alpha \bullet B = B(x,\alpha r)$ and if $B = \overline{B}(x,r)$ then $\alpha \bullet B = \overline{B}(x,\alpha r)$.*

Observation 6.2. If X is a linear space over the field of scalars $\mathbb{R}$ then a scalar multiplication by $\lambda \in \mathbb{R}$ of a set $E \subset X$ is defined by setting $\lambda E = \{\lambda x \in X; x \in E\}$. In particular for a ball B in X we have $\lambda B = \{\lambda x \in X; x \in B\}$. Now $\rho(x,y) := |x-y|$ for $x, y \in X$ is a metric on the linear space X and on the metric space (X,ρ) and for $\alpha \in (0,\infty)$ the set $\alpha \bullet B$ is defined by Definition 6.1. Then the two sets αB and $\alpha \bullet B$ are translates of each other. Indeed we have

$$\alpha B = \alpha \bullet B + (\alpha - 1)c(B).$$

Proof. First consider the case that B is a closed ball, that is, $B = \{x \in X : |x - x_0| \leq r\}$ where $x_0 \in X$ and $r \in (0,\infty)$. By Definition 6.1, we have $\alpha \bullet B = \{x \in X : |x - x_0| \leq \alpha r\}$. Then we have

$$\begin{aligned}
\alpha B &= \{\alpha x \in X : x \in B\} = \{x \in X : \alpha^{-1} x \in B\} \\
&= \{x \in X : \left|\alpha^{-1} x - x_0\right| \leq r\} = \{x \in X : |x - \alpha x_0| \leq \alpha r\} \\
&= \{x \in X : |x - x_0| \leq \alpha r\} + (\alpha - 1)x_0 \\
&= \alpha \bullet B + (\alpha - 1)x_0.
\end{aligned}$$

This completes the proof for the case that B is a closed ball.
The case that B is an open ball is treated similarly. ■

5r-covering theorems are based on the following simple fact in a metric space.

Lemma 6.3. *Let B_1 and B_2 be two balls in a metric space (X,ρ). If $B_1 \cap B_2 \neq \emptyset$ and $r(B_2) \geq \frac{1}{2} r(B_1)$, then $B_1 \subset 5 \bullet B_2$.*

Proof. Let $x_1 = c(B_1)$, $x_2 = c(B_2)$, $r_1 = r(B_1)$, and $r_2 = r(B_2)$. Then we have $r_2 \geq \frac{1}{2}r_1$. The ball B_2 is either an open ball or a closed ball.

1. Consider first the case that B_2 is a closed ball. Then we have $B_2 = \overline{B}(x_2, r_2)$. Let $c \in B_1 \cap B_2$. Then for any $x \in B_1$ we have

$$\begin{aligned} \rho(x, x_2) &\leq \rho(x, x_1) + \rho(x_1, c) + \rho(c, x_2) \leq r_1 + r_1 + r_2 \\ &\leq 2r_2 + 2r_2 + r_2 = 5r_2. \end{aligned} \tag{1}$$

This shows that $x \in \overline{B}(x_2, 5r_2) = 5 \bullet B_2$. Since this holds for every $x \in B_1$, we have $B_1 \subset 5 \bullet B_2$.

2. Next consider the case that B_2 is an open ball. Then we have $B_2 = B(x_2, r_2)$. Let $c \in B_1 \cap B_2$. Then we have $\rho(c, x_2) < r_2$ rather than $\rho(c, x_2) \leq r_2$. Then for any $x \in B_1$ we have

$$\begin{aligned} \rho(x, x_2) &\leq \rho(x, x_1) + \rho(x_1, c) + \rho(c, x_2) < r_1 + r_1 + r_2 \\ &\leq 2r_2 + 2r_2 + r_2 = 5r_2. \end{aligned} \tag{2}$$

This shows that $x \in B(x_2, 5r_2) = 5 \bullet B_2$. Since this holds for every $x \in B_1$, we have $B_1 \subset 5 \bullet B_2$. ∎

Review. (Zorn's Lemma)
A comparison relation $a \preceq b$, defined for some pairs of elements of a set S, is called a *partial order* on S, and $(S, \preceq)$ is said to be a *partially ordered set* if the relation $\preceq$ satisfies the following conditions:
1° for every $a \in S$, $a \preceq a$;
2° if $a \preceq b$ and $b \preceq a$, then $a = b$;
3° if $a \preceq b$ and $b \preceq c$, then $a \preceq c$.
(The word "partial" indicates that not all pairs of elements of S are comparable.)
A *maximal element* in a partially ordered set S is an element $m \in S$ such that if $m \preceq a$ is true then $a = m$.
A *chain* in a partially ordered set S is a subset that is itself linearly ordered.
An *upper bound of a chain* is an element beyond every element in the chain.
Zorn's Lemma. *If every chain in a non-empty partially ordered set has an upper bound, then the partially ordered set has a maximal element.*

Observation 6.4. (Disjoint Collection) By definition a disjoint collection is a collection of sets such that every two distinct members are disjoint. Thus a collection consisting of only one set is a disjoint collection by default.

Proposition 6.5. *Let (X, ρ) be a metric space. Let $\mathfrak{F}$ be an arbitrary collection of balls in X with uniformly bounded radii. Then there exists a disjoint subcollection $\mathfrak{G}$ of $\mathfrak{F}$ such that every $B \in \mathfrak{F}$ intersects some $B' \in \mathfrak{G}$ having $r(B') \geq \frac{1}{2}r(B)$.*

Proof. 1. Since the radii of the balls in $\mathcal{F}$ are uniformly bounded, there exists $\alpha \in (0, \infty)$ such that $r(B) \leq \alpha$ for every $B \in \mathcal{F}$.

Let $\mathcal{H}$ be a disjoint subcollection of $\mathcal{F}$ with the property that if $B \in \mathcal{F}$ intersects $B' \in \mathcal{H}$ then $r(B') \geq \frac{1}{2}r(B)$.
(Let us show that such subcollection $\mathcal{H}$ of $\mathcal{F}$ exists. Take an arbitrary $B' \in \mathcal{F}$ having $r(B') \geq \frac{\alpha}{2}$. By Observation 6.4, $\{B'\}$ is a disjoint subcollection of $\mathcal{F}$. If $B \in \mathcal{F}$ intersects B' then B intersects a member B' of $\{B'\}$ with $r(B') \geq \frac{\alpha}{2} \geq \frac{1}{2}r(B)$.)

2. Let $\mathfrak{H}$ be the set of all disjoint subcollection $\mathcal{H}$ of $\mathcal{F}$ with the property that if $B \in \mathcal{F}$ intersects $B' \in \mathcal{H}$ then $r(B') \geq \frac{1}{2}r(B)$.

Let us introduce a partial ordering $\preceq$ in $\mathfrak{H}$ by declaring $\mathcal{H} \preceq \mathcal{H}'$ if $\mathcal{H} \subset \mathcal{H}'$.

Let $\mathfrak{C}$ be a chain in the partially ordered set $\mathfrak{H}$. (A chain in a partially ordered set is a linearly ordered subset.) Let $\mathcal{H}_0 = \bigcup_{\mathcal{H} \in \mathfrak{C}} \mathcal{H}$. From the fact that $\mathcal{H}$ is a disjoint subcollection of $\mathcal{F}$ and $\mathfrak{C}$ is linearly ordered, it follows that $\mathcal{H}_0$ is a disjoint subcollection of $\mathcal{F}$. From the fact that if $B \in \mathcal{F}$ intersects $B' \in \mathcal{H}$ then $r(B') \geq \frac{1}{2}r(B)$, it follows that if $B \in \mathcal{F}$ intersects $B' \in \mathcal{H}_0$ then $r(B') \geq \frac{1}{2}r(B)$. Thus $\mathcal{H}_0 \in \mathfrak{H}$ and then $\mathcal{H}_0$ is an upper bound for the chain $\mathfrak{C}$. This shows that every chain in $\mathfrak{H}$ has an upper bound. Then by Zorn's Lemma, $\mathfrak{H}$ has a maximal element. Let $\mathcal{G}$ be a maximal element in $\mathfrak{H}$. Being an element of $\mathfrak{H}$, $\mathcal{G}$ is a disjoint subcollection of $\mathcal{F}$.

3. Let us show that every $B \in \mathcal{F}$ intersects some $B' \in \mathcal{G}$.

Let $\mathcal{E}$ be the collection of all members of $\mathcal{F}$ that do not intersect any members of $\mathcal{G}$. Let $B_0 \in \mathcal{E}$ be such that $r(B_0) > \frac{1}{2}r(B)$ for every $B \in \mathcal{E}$. (Such B_0 exists since the radii of all members of $\mathcal{E} \subset \mathcal{F}$ are uniformly bounded.)

Now consider the collection $\mathcal{G}' := \mathcal{G} \cup \{B_0\}$. This is a disjoint collection since $B_0 \in \mathcal{E}$. If $B \in \mathcal{F}$ intersects a member of $\mathcal{G}'$ then either B intersects a member of $\mathcal{G}$ or B does not intersect any member of $\mathcal{G}$ and instead intersects B_0. If B intersects a member of $\mathcal{G} \in \mathfrak{H}$ then B intersects $B' \in \mathcal{G}$ with $r(B') \geq \frac{1}{2}r(B)$. If B does not intersect any member of $\mathcal{G}$ and instead intersects B_0 then $B \in \mathcal{E}$ and B intersects B_0 with $r(B_0) > \frac{1}{2}r(B)$. This shows that $\mathcal{G}' \in \mathfrak{H}$, contradicting the maximality $\mathcal{G}$ in $\mathfrak{H}$. Thus every $B \in \mathcal{F}$ must intersect some $B' \in \mathcal{G}$. ■

Theorem 6.6. (5r-covering Theorem in a Metric Space) *Let (X, ρ) be a metric space. Let $\mathcal{F}$ be an arbitrary collection of balls in X with uniformly bounded radii. Then there exists a disjoint subcollection $\mathcal{G}$ of $\mathcal{F}$ such that*

$$\bigcup_{B \in \mathcal{F}} B \subset \bigcup_{B' \in \mathcal{G}} 5 \bullet B'.$$

Proof. By Proposition 6.5, there exists a disjoint subcollection $\mathcal{G}$ of $\mathcal{F}$ such that every $B \in \mathcal{F}$ intersects some $B' \in \mathcal{G}$ having $r(B') \geq \frac{1}{2}r(B)$. Then by Lemma 6.3, we have $B \subset 5 \bullet B'$. This implies that $\bigcup_{B \in \mathcal{F}} B \subset \bigcup_{B' \in \mathcal{G}} 5 \bullet B'$. ■

Remark 6.7. In Theorem 6.6 the condition that the radii of the balls in the collection $\mathcal{F}$ are uniformly bounded is indispensable. Without this condition the assertion of the Theorem would be false. This is shown in the following example.

Example. Consider $(\mathbb{R}, \rho_e)$ where ρ_e is the Euclidean metric on $\mathbb{R}$, that is, $\rho_e(x,y) = |x-y|$ for $x, y \in \mathbb{R}$. Let $\mathcal{F} = \{B(0,k) : k \in \mathbb{N}\}$. The radii of the balls in $\mathcal{F}$ are not uniformly bounded. Now a disjoint subcollection $\mathcal{G}$ of $\mathcal{F}$ must be of the form $\mathcal{G} = \{B(0,k_0)\}$ where $k_0 \in \mathbb{N}$. But $\bigcup_{k\in\mathbb{N}} B(0,k) \not\subset 5 \bullet B(0,k_0)$.

In the 5r-covering theorems above, the subcollection $\mathcal{G}$ is a disjoint collection but it is not necessarily a countable collection. We show below that if the metric space (X, ρ) is a separable metric space and every member of $\mathcal{F}$ is an open ball, then $\mathcal{G}$ can be chosen to be a countable set.

Proposition 6.8. *Let (X, ρ) be a separable metric space. Let $E \subset X$ and let $\{G_\alpha : \alpha \in A\}$ be an open cover of E, that is, $\{G_\alpha : \alpha \in A\}$ is a collection of open sets in X such that $E \subset \bigcup_{\alpha\in A} G_\alpha$. Then there exists a countable subcover of E, that is, there exists a countable subcollection $\{G_{\alpha_i} : i \in \mathbb{N}\}$ of $\{G_\alpha : \alpha \in A\}$ such that $E \subset \bigcup_{i\in\mathbb{N}} G_{\alpha_i}$.*

Proof. Since X is a separable metric space there exists a countable dense subset D of X. Let $\mathcal{Q}$ be the collection of all open balls with centers in D and with positive rational numbers as radii. Then $\mathcal{Q}$ is a countable collection.

Let $E \subset X$ and let $\{G_\alpha : \alpha \in A\}$ be an open cover of E. Let us show first that for every $x \in E$ there exist $\alpha \in A$ and an open ball in $\mathcal{Q}$, which we denote by B_x, such that

$$x \in B_x \subset G_\alpha. \tag{1}$$

Now since $x \in E \subset \bigcup_{\alpha\in A} G_\alpha$, we have $x \in G_\alpha$ for some $\alpha \in A$. Since G_α is an open set containing x, there exists $\varepsilon > 0$ such that $B(x,\varepsilon) \subset G_\alpha$. Let r be a positive rational number such that $0 < r < \varepsilon$. Then we have

$$x \in B(x,r) \subset B(x,\varepsilon) \subset G_\alpha.$$

Since D is a dense subset of X, there exists $y \in D$ such that $y \in B\big(x, \frac{r}{2}\big)$. Consider the open ball $B\big(y, \frac{r}{2}\big) \in \mathcal{Q}$. For every $z \in B\big(y, \frac{r}{2}\big)$, we have $\rho(z,x) \le \rho(z,y) + \rho(y,x) < \frac{r}{2} + \frac{r}{2} = r$ so that $z \in B(x,r)$. Thus $B\big(y, \frac{r}{2}\big) \subset B(x,r)$. Moreover $y \in B\big(x, \frac{r}{2}\big)$ implies $x \in B\big(y, \frac{r}{2}\big)$. Therefore we have

$$x \in B\big(y, \tfrac{r}{2}\big) \subset B(x,r) \subset B(x,\varepsilon) \subset G_\alpha. \tag{2}$$

Our $B\big(y, \frac{r}{2}\big)$ in (2) serves as B_x in (1). This proves (1).

Now (1) implies

$$E \subset \bigcup_{x\in E} B_x \subset \bigcup_{\alpha\in A} G_\alpha. \tag{3}$$

Since $B_x \in \mathcal{Q}$ and $\mathcal{Q}$ is a countable collection, $\{B_x : x \in E\}$ is a countable collection. Let us label $\{B_x : x \in E\}$ as $\{B_i : i \in \mathbb{N}\}$. According to (1), for each $i \in \mathbb{N}$, B_i is contained in some G_α. Pick an arbitrary G_α containing B_i and label it as G_{α_i}. Then $\{G_{\alpha_i} : i \in \mathbb{N}\}$ is a countable subcollection of $\{G_\alpha : \alpha \in A\}$ and we have

$$E \subset \bigcup_{x \in E} B_x = \bigcup_{i \in \mathbb{N}} B_i \subset \bigcup_{i \in \mathbb{N}} G_{\alpha_i}.$$

This completes the proof. ∎

Proposition 6.9. *Let (X, ρ) be a separable metric space. Let $\{B_\alpha : \alpha \in A\}$ be an arbitrary collection of open balls in X. Then there exists a countable subcollection $\{B_{\alpha_i} : i \in \mathbb{N}\}$ of $\{B_\alpha : \alpha \in A\}$ such that*

$$\bigcup_{i \in \mathbb{N}} B_{\alpha_i} = \bigcup_{\alpha \in A} B_\alpha.$$

Proof. Let $\{B_\alpha : \alpha \in A\}$ be an arbitrary collection of open balls in X. Let $E = \bigcup_{\alpha \in A} B_\alpha$. Then $\{B_\alpha : \alpha \in A\}$ is an open cover of E. Then by Proposition 6.8, there exists a countable subcover of E, that is, there exists a countable subcollection $\{B_{\alpha_i} : i \in \mathbb{N}\}$ of $\{B_\alpha : \alpha \in A\}$ such that $E \subset \bigcup_{i \in \mathbb{N}} B_{\alpha_i}$. But $E = \bigcup_{\alpha \in A} B_\alpha \supset \bigcup_{i \in \mathbb{N}} B_{\alpha_i}$. Thus $E = \bigcup_{i \in \mathbb{N}} B_{\alpha_i}$, that is, $\bigcup_{\alpha \in A} B_\alpha = \bigcup_{i \in \mathbb{N}} B_{\alpha_i}$. ∎

Theorem 6.10. (5r-covering Theorem in a Separable Metric Space) *Let (X, ρ) be a separable metric space and let $\mathfrak{F}$ be an arbitrary collection of open balls in X with uniformly bounded radius. Then there exists a countable disjoint subcollection $\mathfrak{G}_0$ of $\mathfrak{F}$ such that*

$$\bigcup_{B \in \mathfrak{F}} B \subset \bigcup_{B \in \mathfrak{G}_0} 5 \bullet B.$$

Proof. Now $\mathfrak{F}$ is a collection of open balls in a separable metric space (X, ρ). Thus by Proposition 6.9 there exists a countable subcollection $\mathfrak{F}_0$ of $\mathfrak{F}$ such that

$$\bigcup_{B \in \mathfrak{F}_0} B = \bigcup_{B \in \mathfrak{F}} B. \tag{1}$$

By Theorem 6.6 there exists a disjoint subcollection $\mathfrak{G}_0$ of $\mathfrak{F}_0$ such that

$$\bigcup_{B \in \mathfrak{F}_0} B \subset \bigcup_{B \in \mathfrak{G}_0} 5 \bullet B. \tag{2}$$

Being a subcollection of the countable collection $\mathfrak{F}_0$, $\mathfrak{G}_0$ is a countable collection. Then by (1) and (2), we have

$$\bigcup_{B \in \mathfrak{F}} B \subset \bigcup_{B \in \mathfrak{G}_0} 5 \bullet B.$$

This completes the proof. ∎

[II] Bounded-doubling Borel-regular Borel Outer Measure

Definition 6.11. *Let μ be a Borel-regular Borel outer measure on a metric space (X,ρ). We call μ a bounded-doubling outer measure if it satisfies the following conditions:*
$1°$ $\mu(B) \in (0,\infty)$ *for every ball* $B \subset X$.
$2°$ *There exists* $\gamma \geq 1$ *such that* $\mu(2 \bullet B) \leq \gamma\mu(B)$ *for every ball* $B \subset X$.

Next we show some examples of bounded-doubling Borel-regular Borel outer measures.

Theorem 6.12. *The Lebesgue outer measure μ_L^n on $(\mathbb{R}^n, \rho_e)$ is a bounded-doubling Borel-regular Borel outer measure.*

Proof. Let B be an arbitrary open ball in $\mathbb{R}^n$ given by $B = \{x \in \mathbb{R}^n : |x - x_0| < r\}$ where $x_0 \in \mathbb{R}^n$ and $r > 0$. By Observation 6.2 we have $2B = 2 \bullet B + (2-1)x_0$, that is, $2 \bullet B = 2B - x_0$. Then by the translation invariance of μ_L^n we have

$$\mu_L^n(2 \bullet B) = \mu_L^n(2B - x_0) = \mu_L^n(2B) = 2^n \mu_L^n(B).$$

Let $\gamma = 2^n$. Then we have $\mu_L^n(2 \bullet B) = \gamma\mu_L^n(B)$. This shows that μ_L^n is a bounded-doubling Borel-regular Borel outer measure. ∎

Theorem 6.13. *The Hausdorff outer measure $\mathcal{H}^s$ on $(\mathbb{R}^n, \rho_e)$ is a bounded-doubling Borel-regular Borel outer measure.*

Proof. Let B be an arbitrary open ball in $\mathbb{R}^n$ given by $B = \{x \in \mathbb{R}^n : |x - x_0| < r\}$ where $x_0 \in \mathbb{R}^n$ and $r > 0$. By Observation 6.2 we have $2B = 2 \bullet B + (2-1)x_0$, that is, $2 \bullet B = 2B - x_0$. Then by Theorem 5.5 and (a) of Theorem 5.8 we have

$$\mathcal{H}^s(2 \bullet B) = \mathcal{H}^s(2B - x_0) = \mathcal{H}^s(2B) = 2^s \mathcal{H}^s(B).$$

Let $\gamma = 2^s$. Then we have $\mathcal{H}^s(2 \bullet B) = \gamma\mathcal{H}^s(B)$. This shows that $\mathcal{H}^s$ is a bounded-doubling Borel-regular Borel outer measure. ∎

Next we list some properties of a bounded-doubling Borel-regular Borel outer measure.

Proposition 6.14. *A bounded-doubling Borel-regular Borel outer measure μ on a metric space (X,ρ) is a σ-finite outer measure.*

Proof. With $x_0 \in X$ arbitrarily chosen, consider the ball $B_k(x_0) = \{x \in X : \rho(x_0, x) < k\}$ where $k \in \mathbb{N}$. Let $x \in X$. Since $\rho(x_0, x) < \infty$, we have $\rho(x_0, x) < k$ for sufficiently large $k \in \mathbb{N}$ and thus $x \in B_k(x_0)$. This implies that $X = \bigcup_{k \in \mathbb{N}} B_k(x_0)$. Now by $1°$ of Definition 6.11 we have $\mu(B_k(x_0)) < \infty$. Thus the outer measure μ is σ-finite. ∎

Proposition 6.15. *Let μ be a bounded-doubling Borel-regular Borel outer measure μ on a metric space (X,ρ). Then for every $\alpha \in [1,\infty)$ there exists a constant $C(\alpha) \geq 1$ such that*

$$\mu(\alpha \bullet B) \leq C(\alpha)\mu(B) \quad \textit{for every ball } B \subset X.$$

Proof. Let μ be a bounded-doubling Borel-regular Borel outer measure μ on a metric space (X, ρ). Then by 2° of Definition 6.11, there exists a constant $\gamma \geq 1$ such that for every ball $B \subset X$ we have

$$\mu(2 \bullet B) \leq \gamma\mu(B). \tag{1}$$

Let $k \in \mathbb{N}$. Then by iterated application of (1) we obtain

$$\mu(2^k \bullet B) \leq \gamma\mu(2^{k-1} \bullet B) \leq \gamma\mu(2^{k-2} \bullet B) \leq \cdots \leq \gamma^k\mu(B). \tag{2}$$

Given $\alpha \in [1, \infty)$, let $k \in \mathbb{N}$ be so large that $\alpha \leq 2^k$. Then by the monotonicity of the outer measure μ and by (2) we have

$$\mu(\alpha \bullet B) \leq \mu(2^k \bullet B) \leq \gamma^k\mu(B). \tag{3}$$

Let $C(\alpha) = \gamma^k$. Then we have

$$\mu(\alpha \bullet B) \leq C(\alpha)\mu(B).$$

This completes the proof. ∎

Lemma 6.16. *Let μ be a Borel outer measure on a metric space (X, ρ). If $\mu(B) \in (0, \infty)$ for every ball $B \subset X$, then any set $E \subset X$ with $\mu(E) < \infty$ can contain at most countably many disjoint balls.*

Proof. Let $E \subset X$ and $\mu(E) < \infty$. Suppose there exists an uncountable collection of disjoint balls $\mathcal{B} = \{B_\gamma : \gamma \in \Gamma\}$ such that $\bigcup_{\gamma\in\Gamma} B_\gamma \subset E$. Let us show that this leads to a contradiction. Now for each $\gamma \in \Gamma$, we have $\mu(B_\gamma) \in (0, \infty)$ so that there exists some $k \in \mathbb{N}$ such that $\mu(B_\gamma) \geq \frac{1}{k}$. For each $k \in \mathbb{N}$, let $\mathcal{B}_k = \{B_\gamma \in \mathcal{B} : \mu(B_\gamma) \geq \frac{1}{k}\}$. Then we have $\mathcal{B} = \bigcup_{k\in\mathbb{N}} \mathcal{B}_k$. Now if $\mathcal{B}_k$ were a finite collection for every $k \in \mathbb{N}$, then $\mathcal{B}$ would be a countable collection, contradicting the assumption that $\mathcal{B}$ is an uncountable collection. Thus there exists $k_0 \in \mathbb{N}$ such that $\mathcal{B}_{k_0}$ is an infinite collection. Select $\{B_n : n \in \mathbb{N}\} \subset \mathcal{B}_{k_0}$. Now the outer measure μ on X is a measure on $\mathfrak{B}_X$. Since $\mathcal{B} \subset \mathfrak{B}_X$ and since $\mathcal{B}$ is a disjoint collection, we have $\sum_{n\in\mathbb{N}} \mu(B_n) = \mu\left(\bigcup_{n\in\mathbb{N}} B_n\right) \leq \mu(E) < \infty$. On the other hand we have $\sum_{n\in\mathbb{N}} \mu(B_n) \geq \sum_{n\in\mathbb{N}} \frac{1}{k_0} = \infty$. This is a contradiction. ∎

Proposition 6.17. *Let μ be a bounded-doubling Borel-regular Borel outer measure μ on a metric space (X, ρ). Let $E \subset X$ with $\mu(E) < \infty$. Then E can contain at most countably many disjoint balls.*

Proof. If μ is a bounded-doubling Borel-regular Borel outer measure μ on a metric space (X, ρ), then by 1° of Definition 6.11 we have $\mu(B) \in (0, \infty)$ for every ball $B \subset X$. According to Lemma 6.16 this implies that any set $E \subset X$ with $\mu(E) < \infty$ can contain at most countably many disjoint balls. ∎

[III] Vitali Covering Theorems

Definition 6.18. (Vitali Cover) *Let (X, ρ) be a metric space and let $E \subset X$. A collection $\mathfrak{F}$ of subsets of X is called a Vitali cover of E if it satisfies the following conditions:*
1° every $F \in \mathfrak{F}$ is a closed set in X with positive diameter, that is, $|F| > 0$;
2° for every $x \in E$ and every $\varepsilon > 0$ there exists $F \in \mathfrak{F}$ with $|F| < \varepsilon$ such that $x \in F$.
(Observe that condition 2° implies $E \subset \bigcup_{F \in \mathfrak{F}} F$.)

Proposition 6.19. *Let $\mathfrak{F}$ be a Vitali cover of a set E in a metric space (X, ρ). Let O be an arbitrary open set in X such that $O \supset E$. Let $\mathfrak{F}|_O$ be the subcollection of $\mathfrak{F}$ consisting of those members of $\mathfrak{F}$ that are contained in O. Then $\mathfrak{F}|_O$ exists and is a Vitali cover of E.*

Proof. 1. Let us prove the existence of $\mathfrak{F}|_O$. Thus we are to show that there exists $F \in \mathfrak{F}$ such that $F \subset O$. Now O is an open set in X such that $O \supset E$. Let $x_0 \in E \subset O$. Let $\varepsilon_0 > 0$ be so small that $B(x_0, \varepsilon_0) \subset O$. By 2° of Definition 6.18, there exists $F_0 \in \mathfrak{F}$ with $|F_0| < \varepsilon_0$ such that $x_0 \in F_0$. This implies that $F_0 \subset B(x_0, \varepsilon_0) \subset O$ and proves the existence of $\mathfrak{F}|_O$.

2. Let us prove that $\mathfrak{F}|_O$ is a Vitali cover of E. Thus we are to show that $\mathfrak{F}|_O$ satisfies conditions 1° and 2° of Definition 6.18.

Now $\mathfrak{F}|_O$ is a subcollection of the Vitali cover $\mathfrak{F}$ of E. Thus if $F \in \mathfrak{F}|_O$ then $F \in \mathfrak{F}$ and hence F is a closed set in X with $|F| > 0$. This shows that $\mathfrak{F}|_O$ satisfies conditions 1° of Definition 6.18.

Let $x \in E$ and $\varepsilon > 0$ be arbitrarily given. Since O is an open set in X and $x \in E \subset O$, we have $B(x, \varepsilon_0) \subset O$ for sufficiently small $\varepsilon_0 > 0$. Take $\varepsilon_0 < \varepsilon$. Since $\mathfrak{F}$ is a Vitali cover of E, there exists $F \in \mathfrak{F}$ such that $|F| < \varepsilon_0$ and $x \in F$. Since $x \in F$, we have $F \subset B(x, \varepsilon_0) \subset O$. Then $F \in \mathfrak{F}|_O$. Thus we have $F \in \mathfrak{F}|_O$ such that $|F| < \varepsilon_0 < \varepsilon$ and $x \in F$. This shows that $\mathfrak{F}|_O$ satisfies conditions 2° of Definition 6.18. ∎

Theorem 6.20. (Vitali Covering Theorem for a Bounded-doubling Borel-regular Borel Outer Measure) *Let μ be a bounded-doubling Borel-regular Borel outer measure on a metric space (X, ρ). Let $E \subset X$. Let $\beta \in (0, \infty)$. Let $\mathfrak{F}$ be a collection of subsets of X satisfying the following conditions:*
1° every member of $\mathfrak{F}$ is a closed ball B in X with center in E and diameter $|B| \leq \beta$;
2° for every $x \in E$ there exists a member of $\mathfrak{F}$ having x as its center;
3° for every $x \in E$ we have $\inf\{r > 0 : \overline{B}(x, r) \in \mathfrak{F}\} = 0$.
Then there exists a countable disjoint subcollection $\{B_i : i \in \mathbb{N}\} \subset \mathfrak{F}$ such that

$$\mu\left(E \setminus \bigcup_{i \in \mathbb{N}} B_i\right) = 0.$$

Proof. 1. Observe that our $\mathfrak{F}$ satisfies the defining conditions of a Vitali cover in Definition 6.18 and thus $\mathfrak{F}$ is an example of Vitali cover.

2. By 1°, radius $r(B)$ of a closed ball B in $\mathcal{F}$ is uniformly bounded by $\frac{\beta}{2}$. Then by Proposition 6.5, there exists a disjoint subcollection $\mathcal{G}$ of $\mathcal{F}$ such that every $B \in \mathcal{F}$ intersects some $B' \in \mathcal{G}$ having $r(B') \geq \frac{1}{2}r(B)$ and then $B \subset 5 \bullet B'$ by Lemma 6.3. Then we have

$$E \subset \bigcup_{B\in\mathcal{F}} B \subset \bigcup_{B\in\mathcal{G}} 5 \bullet B. \tag{1}$$

3. Consider the case that E is a bounded set in X. Then with an arbitrary $x_0 \in E$ and sufficiently large $r > 0$, we have $E \subset B(x_0, r)$. Since every closed ball $B \in \mathcal{F}$ has a center in E and radius $r(B) \leq \frac{\beta}{2}$, we have $B \subset B(x_0, r + \frac{\beta}{2})$. Thus we have

$$\bigcup_{B\in\mathcal{G}} B \subset \bigcup_{B\in\mathcal{F}} B \subset B\big(x_0, r + \tfrac{\beta}{2}\big). \tag{2}$$

Now since μ is a bounded-doubling Borel-regular Borel outer measure, it is a σ-finite outer measure by Proposition 6.14 and then we have $\mu\big(B(x_0, r+\frac{\beta}{2})\big) < \infty$. Now that $\mathcal{G}$ is a disjoint collection of closed balls contained in $B(x_0, r + \frac{\beta}{2})$, the finiteness $\mu\big(B(x_0, r + \frac{\beta}{2})\big) < \infty$ implies that $\mathcal{G}$ is a countable collection, that is, $\mathcal{G} = \{B_i : i \in \mathbb{N}\}$, by Proposition 6.17. Then by (1) we have

$$E \subset \bigcup_{i\in\mathbb{N}} 5 \bullet B_i. \tag{3}$$

Next since $\mathcal{G}$ is a disjoint subcollection of $\mathcal{F} \subset \mathfrak{B}_X$ and since μ is a measure on the σ-algebra $\mathfrak{B}_X$, we have

$$\sum_{i\in\mathbb{N}} \mu(B_i) = \mu\left(\bigcup_{i\in\mathbb{N}} B_i\right) \leq \mu\Big(B\big(x_0, r + \tfrac{\beta}{2}\big)\Big) < \infty. \tag{4}$$

Since μ is a bounded-doubling Borel-regular Borel outer measure, according to Proposition 6.15 there exists a constant $C(5) \geq 1$ such that

$$\mu(5 \bullet B) \leq C(5)\mu(B) \quad \text{for every closed ball } B \text{ in } X. \tag{5}$$

Then by (5) and (4) we have

$$\sum_{i\in\mathbb{N}} \mu(5 \bullet B_i) \leq C(5) \sum_{i\in\mathbb{N}} \mu(B_i) < \infty. \tag{6}$$

This implies

$$\lim_{k\to\infty} \sum_{i>k} \mu(5 \bullet B_i) = 0. \tag{7}$$

Let us show that for every $k \in \mathbb{N}$ we have

$$E \setminus \bigcup_{i=1}^{k} B_i \subset \bigcup_{i>k} 5 \bullet B_i. \tag{8}$$

Let $x \in E \setminus \bigcup_{i=1}^k B_i$. Since $\{B_1, \dots, B_k\}$ is a disjoint collection of closed balls in X, for sufficiently small $r > 0$ the closed ball $\overline{B}(x, r)$ does not intersect any member of $\{B_1, \dots, B_k\}$. By condition 3° we can choose $r > 0$ so that $\overline{B}(x, r) \in \mathcal{F}$. But every closed ball $B \in \mathcal{F}$ intersects a closed ball in $\mathcal{G} = \{B_i : i \in \mathbb{N}\}$ with $r(B_i) \geq \frac{1}{2} r(B)$. Thus our $\overline{B}(x, r) \in \mathcal{F}$ intersects some B_{i_0} with $r(B_{i_0}) \geq \frac{r}{2}$. Then we have $i_0 > k$ and $\overline{B}(x, r) \subset 5 \bullet B_{i_0}$. Thus $x \in \bigcup_{i>k} 5 \bullet B_i$. This proves (8). Then we have

$$\begin{aligned}
\mu\left(E \setminus \bigcup_{i \in \mathbb{N}} B_i\right) &= \mu\left(\lim_{k\to\infty}\left(E \setminus \bigcup_{i=1}^k B_i\right)\right) = \lim_{k\to\infty} \mu\left(E \setminus \bigcup_{i=1}^k B_i\right) \\
&\leq \lim_{k\to\infty} \mu\left(\bigcup_{i>k} 5 \bullet B_i\right) \quad \text{by (8)} \\
&\leq \lim_{k\to\infty} \sum_{i>k} \mu(5 \bullet B_i) = 0 \quad \text{by (7)}.
\end{aligned}$$

This completes the proof for the case that E is a bounded set in X.

4. Let us remove the assumption that E is a bounded set. Consider the measure space $(X, \mathfrak{M}(\mu), \mu)$. Since μ is a Borel outer measure, we have $\mathfrak{B}_X \subset \mathfrak{M}(\mu)$. Let $x_0 \in X$ be arbitrarily fixed and consider the set $S_r = \{x \in X : \rho(x, x_0) = r\}$ for $r \in [0, \infty)$. The collection $\mathfrak{E} = \{S_r : r \in [0, \infty)\}$ is a disjoint collection in $\mathfrak{B}_X$ and moreover $\bigcup_{r \in [0,\infty)} S_r = X$. Since μ is a σ-finite outer measure as we showed in Proposition 6.14, according to Theorem 1.28 we have $\mu(S_r) > 0$ for only countably many values of $r \in [0, \infty)$. Thus we can select a strictly increasing sequence $(r_i : i \in \mathbb{N})$ in $(0, \infty)$ with $\lim_{i\to\infty} r_i = \infty$ such that $\mu(S_{r_i}) = 0$ for all $i \in \mathbb{N}$. Let

$$S_i = \{x \in X : \rho(x, x_0) = r_i\} \quad \text{for } i \in \mathbb{N},$$

so that we have $\mu(S_i) = 0$ for $i \in \mathbb{N}$ and then let

$$\begin{aligned}
V_1 &= \{x \in X : \rho(x, x_0) \in [0, r_1)\} \\
V_i &= \{x \in X : \rho(x, x_0) \in (r_{i-1}, r_i)\} \quad \text{for } i \geq 2.
\end{aligned}$$

Note that $\{V_i : i \in \mathbb{N}\}$ is a collection of bounded open sets, $\{S_i, V_i : i \in \mathbb{N}\}$ is a disjoint collection in $\mathfrak{B}_X$ and moreover $\left(\bigcup_{i \in \mathbb{N}} S_i\right) \cup \left(\bigcup_{i \in \mathbb{N}} V_i\right) = X$. Let $E_i = E \cap V_i$ for $i \in \mathbb{N}$.

Consider the Vitali cover $\mathcal{F}$ of E. For each $i \in \mathbb{N}$, let $\mathcal{F}_i$ be the subcollection of $\mathcal{F}$ consisting of all closed balls in $\mathcal{F}$ with centers $x \in E_i$ and with radius $r > 0$ so small that $\overline{B}(x, r) \subset V_i$. Since E_i is a bounded set, by our result above there exists a countable disjoint subcollection of $\mathcal{F}_i$, denoted by $\{B_{i,j} : j \in \mathbb{N}\}$, such that

$$\mu\left(E_i \setminus \bigcup_{j \in \mathbb{N}} B_{i,j}\right) = 0.$$

Consider the collection $\{B_{i,j} : i \in \mathbb{N}, j \in \mathbb{N}\}$. Since $\{B_{i,j} : j \in \mathbb{N}\}$ is a disjoint collection contained in V_i and since $\{V_i : i \in \mathbb{N}\}$ is a disjoint collection, our $\{B_{i,j} : i \in \mathbb{N}, j \in \mathbb{N}\}$ is a disjoint collection. Now

$$E = E \cap X = E \cap \bigcup_{i \in \mathbb{N}} (V_i \cup S_i) = \bigcup_{i \in \mathbb{N}} (E \cap V_i) \cup \bigcup_{i \in \mathbb{N}} (E \cap S_i) \subset \bigcup_{n \in \mathbb{N}} E_i \cup \bigcup_{n \in \mathbb{N}} S_i$$

and then

$$E \setminus \bigcup_{i \in \mathbb{N}} \bigcup_{j \in \mathbb{N}} B_{i,j} \subset \left(\bigcup_{i \in \mathbb{N}} E_i \setminus \bigcup_{i \in \mathbb{N}} \bigcup_{j \in \mathbb{N}} B_{i,j} \right) \cup \bigcup_{i \in \mathbb{N}} S_i = \bigcup_{i \in \mathbb{N}} \left(E_i \setminus \bigcup_{j \in \mathbb{N}} B_{i,j} \right) \cup \bigcup_{i \in \mathbb{N}} S_i$$

where the last equality is from the fact that E_i and $\bigcup_{j \in \mathbb{N}} B_{k,j}$ are disjoint if $i \neq k$. Thus

$$\mu \left(E \setminus \bigcup_{i \in \mathbb{N}} \bigcup_{j \in \mathbb{N}} B_{i,j} \right) \leq \sum_{i \in \mathbb{N}} \mu \left(E_i \setminus \bigcup_{j \in \mathbb{N}} B_{i,j} \right) + \sum_{i \in \mathbb{N}} \mu(S_i) = 0.$$

This completes the proof. ∎

[IV] Besicovitch Covering Theorem

The next Lemma shows that if the 3rd side of a triangle is longer than the first and the second side then the angle between the 1st and the 2nd side is at least $\frac{\pi}{3}$.

Lemma 6.21. *Suppose $a, b \in \mathbb{R}^2$, $0 < |a| < |a - b|$ and $0 < |b| < |a - b|$. Then we have:*
(a) *$b \notin \overline{B}(a, |a|)$ and $a \notin \overline{B}(b, |b|)$.*
(b) *Let γ be the angle between the two vectors a and b. Then $\gamma \geq \frac{\pi}{3}$.*
(c) *We have*

$$\left| \frac{a}{|a|} - \frac{b}{|b|} \right| \geq 1.$$

Proof. 1. Let $a, b \in \mathbb{R}^2$. Assume that $0 < |a| < |a - b|$ and $0 < |b| < |a - b|$. Now $b \in \overline{B}(a, |a|)$ if and only if $|a - b| \leq |a|$. Thus $|a| < |a - b|$ implies that $b \notin \overline{B}(a, |a|)$. Similarly $|b| < |a - b|$ implies that $a \notin \overline{B}(b, |b|)$.

2. Let the vectors 0, a and b be represented by points o, A and B in the plane $\mathbb{R}^2$. Then the line segments $\overline{OA}$, $\overline{OB}$ and $\overline{BA}$ represent the vectors a, b and $a - b$ respectively. Consider the triangle $\triangle OAB$ and let α, β and γ be the angles sustained by $\overline{OA}$, $\overline{OB}$ and $\overline{BA}$ respectively. Now $0 < |a| < |a - b|$ implies $|\overline{OA}| < |\overline{BA}|$ and similarly $0 < |b| < |a - b|$ implies $|\overline{OB}| < |\overline{BA}|$. Then the angle γ between the vectors a and b is the angle sustained by $\overline{BA}$. Let α and β be the angles sustained by $\overline{OA}$ and $\overline{OB}$ respectively. Then $|\overline{OA}| < |\overline{BA}|$ implies $\alpha \leq \gamma$ and similarly $|\overline{OB}| < |\overline{BA}|$ implies $\beta \leq \gamma$. Then since $\alpha + \beta + \gamma = \pi$, we have $\gamma \geq \frac{\pi}{3}$.

3. Consider the unit vectors $\frac{a}{|a|}$ and $\frac{b}{|b|}$. The angle between these two unit vectors is the angle between a and b and is thus at least $\frac{\pi}{3}$. Then the length of the third side of the triangle with $\frac{a}{|a|}$ and $\frac{b}{|b|}$ as the first and second sides is at least 1. Thus $\left| \frac{a}{|a|} - \frac{b}{|b|} \right| \geq 1$. ∎

Lemma 6.22. *Let $a_1, \dots, a_k$ be distinct points in $\mathbb{R}^n$. Let $\overline{B}(a_1, r_1), \dots, \overline{B}(a_k, r_k)$ be closed balls in $\mathbb{R}^n$ such that*

$$\bigcap_{i=1}^{k} \overline{B}(a_i, r_i) \neq \emptyset, \tag{1}$$

and

$$a_i \notin \overline{B}(a_j, r_j) \quad \textit{for } i \neq j. \tag{2}$$

Then there exists $N(n) \in \mathbb{N}$ such that $k \leq N(n)$.

Proof. By (1) there exists $x \in \bigcap_{i=1}^{k} \overline{B}(a_i, r_i)$. Let us consider the case $0 \in \bigcap_{i=1}^{k} \overline{B}(a_i, r_i)$. The general case can be reduced to this case by a change of coordinate system.

1. Now $0 \in \overline{B}(a_i, r_i)$ for $i = 1, \dots, k$. According to (2), the center of a closed ball in the collection of k closed balls is not contained in another closed ball in the collection. Thus 0 cannot be the center of a closed ball in the collection. Then we have

$$a_i \neq 0 \quad \text{for } i = 1, \dots, k.$$

Since $\overline{B}(a_i, r_i)$ does not contain a_j for $j \neq i$, we have $|a_i - a_j| > r_i$. On the other hand since $0 \in \overline{B}(a_i, r_i)$ we have $|0 - a_i| < r_i$, that is, $|a_i| \leq r_i$. Thus we have

$$|a_i| \leq r_i < |a_i - a_j| \quad \text{for } i \neq j$$

and similarly

$$|a_j| \leq r_j < |a_i - a_j| \quad \text{for } i \neq j.$$

Applying Lemma 6.21 with $a = a_i$ and $b = a_j$ for $i \neq j$, we have

$$\left| \frac{a_i}{|a_i|} - \frac{a_j}{|a_j|} \right| \geq 1 \quad \text{for } i \neq j. \tag{3}$$

2. Let $x_i \in \mathbb{R}^n$ represent the unit vector $\frac{a_i}{|a_i|}$ for $i = 1, \dots, k$. Consider the unit spherical surface $S(0,1) = \{x \in \mathbb{R}^n : |x| = 1\}$. Then we have $x_i \in S(0,1)$ for $i = 1, \dots, k$. Now $\left| \frac{a_i}{|a_i|} - \frac{a_j}{|a_j|} \right| \geq 1$ for $j \neq 1$ implies that the distance $|x_1 - x_j| \geq 1$ for $j \neq 1$. Thus we have

$$x_j \notin S(0,1) \cap B(x_1, 1) \quad \text{for } j \neq 1.$$

Let α be the surface area of $S(0,1)$ and let α_1 be the surface area of $S(0,1) \cap B(x_1, 1)$. Then there can be at most integral part $[\frac{\alpha}{\alpha_1}]$ many x_i on $S(0,1)$. Let $N(n) = [\frac{\alpha}{\alpha_1}]$. Then we have $k \leq N(n)$. ∎

Definition 6.23. (Besicovich Covering) *Let $E \subset \mathbb{R}^n$. For every $x \in E$ construct a closed ball $\overline{B}(x, r(x))$, where $r(x) > 0$, in such a way that* $\sup\{r(x); x \in E\} < \infty$. *We call the collection $\mathcal{B} = \{\overline{B}(x, r(x)) : x \in E\}$ a Besicovich covering of E.*

Theorem 6.24. (Besicovitch Covering Theorem) *Let E be a bounded subset of $\mathbb{R}^n$. Let $\mathcal{B}$ be a Besicovich covering of E.*
(a) *There exists a finite or countable subcollection $\{B_i\}$ of $\mathcal{B}$ such that*

$$E \subset \bigcup_i B_i.$$

Moreover there exists $P(n) \in \mathbb{N}$ such that every $x \in \mathbb{R}^n$ is contained in at most $P(n)$ closed balls in the collection $\{B_i\}$.
(b) *There exist subcollections $\mathcal{B}_1, \ldots, \mathcal{B}_{Q(n)}$ of $\mathcal{B}$, each a finite or countable disjoint collection, such that*

$$E \subset \bigcup_{i=1}^{Q(n)} \bigcup_{B \in \mathcal{B}_i} B$$

where $Q(n)$ is a positive integer depending on the dimension n of $\mathbb{R}^n$ only.

Proof. Consider the Besicovich covering $\mathcal{B} = \{\overline{B}(x, r(x)) : x \in E\}$ where we have $\sup\{r(x) : x \in E\} < \infty$. We are to show that there exists a finite or countable subcollection $\{B_i\}$ of $\mathcal{B}$ such that $E \subset \bigcup_i B_i$, or equivalently $E \setminus \bigcup_i B_i = \emptyset$. We achieve this by deleting gradually smaller closed balls.

1. Let $M_1 = \sup\{r(x) : x \in E\} < \infty$. Choose $x_1 \in E$ such that $r(x_1) \geq \frac{M_1}{2}$ and then choose inductively

$$(1) \qquad \begin{aligned} &x_{i+1} \in E \setminus \textstyle\bigcup_{j=1}^{i} \overline{B}(x_j, r(x_j)), \\ &r(x_{i+1}) \geq \tfrac{M_1}{2}. \end{aligned}$$

We continue the selection process as long as $E \setminus \bigcup_{j=1}^{i} \overline{B}(x_j, r(x_j)) \neq \emptyset$ and there exists $x \in E \setminus \bigcup_{j=1}^{i} \overline{B}(x_j, r(x_j))$ such that $r(x) \geq \frac{M_1}{2}$. The selection process stops when $E \setminus \bigcup_{j=1}^{i} \overline{B}(x_j, r(x_j)) = \emptyset$ in which case we have $E \subset \bigcup_{j=1}^{i} \overline{B}(x_j, r(x_j))$ or when $E \setminus \bigcup_{j=1}^{i} \overline{B}(x_j, r(x_j)) \neq \emptyset$ but $x \in E \setminus \bigcup_{j=1}^{i} \overline{B}(x_j, r(x_j))$ such that $r(x) \geq \frac{M_1}{2}$ does not exist. Let us show that the selection process stops after finitely many steps. Assume the contrary, that is, we can select $\{x_i : i \in \mathbb{N}\}$. By (1), if $i > j$ then $x_i \notin \overline{B}(x_j, r(x_j))$ so that $\rho(x_i, x_j) > r(x_j) \geq \frac{M_1}{2}$. Thus for $i \neq j$, we have $\rho(x_i, x_j) > r(x_j) \geq \frac{M_1}{2}$. Now $\{x_i; i \in \mathbb{N}\} \subset E \subset \overline{E}$. Since E is a bounded set in $\mathbb{R}^n$, $\overline{E}$ is a bounded closed set in $\mathbb{R}^n$ and hence $\overline{E}$ is a compact set in $\mathbb{R}^n$. Then $(x_i : i \in \mathbb{N})$ is a sequence in a compact set $\overline{E}$. Thus there exists a subsequence of $(x_i : i \in \mathbb{N})$ converging to a point in $\overline{E}$. But this is impossible since $\rho(x_i, x_j) > r(x_j) \geq \frac{M_1}{2}$ for $i \neq j$. This shows that the selection process terminates after finitely many steps, say $k_1 \in \mathbb{N}$ steps. Then we have $x_1, \ldots, x_{k_1}$.

Next let

$$M_2 = \sup\left\{ r(x) : x \in E \setminus \bigcup_{i=1}^{k_1} \overline{B}(x_i, r(x_i)) \right\}.$$

Note that $M_2 \leq \frac{M_1}{2}$. Choose

$$x_{k_1+1} \in E \setminus \bigcup_{j=1}^{k_1} \overline{B}(x_j, r(x_j)),$$
$$r(x_{k_1+1}) \geq \tfrac{M_2}{2}$$

and then inductively

$$x_{i+1} \in E \setminus \bigcup_{j=1}^{i} \overline{B}(x_j, r(x_j)),$$
$$r(x_{i+1}) \geq \tfrac{M_2}{2}.$$

Thus proceeding we have a strictly increasing sequence of nonnegative integers

$$0 = k_0 < k_1 < k_2 < k_3 < \cdots$$

and a strictly decreasing sequence of positive real numbers $(M_i : i \in \mathbb{N})$

$$\text{(2)} \qquad \begin{aligned} &M_1 > M_2 > M_3 > \cdots, \\ &M_{i+1} \leq \tfrac{M_i}{2} \text{ for } i \in \mathbb{N} \end{aligned}$$

and a sequence of closed balls $(B_i : i \in \mathbb{N})$ in $\mathcal{B}$ where $B_i = \overline{B}(x_i, r(x_i))$ and $x_i \in E$ for $i \in \mathbb{N}$.

Let us classify the index set $\mathbb{N}$ for the closed balls $\{B_i : i \in \mathbb{N}\}$ into classes $\{I_j : j \in \mathbb{N}\}$ where

$$I_1 = \{1, \ldots, k_1\}, I_2 = \{k_1 + 1, \ldots, k_2\}, I_3 = \{k_2 + 1, \ldots, k_3\}, \ldots$$

and in general

$$I_j = \{k_{j-1} + 1, \ldots, k_j\} \quad \text{for } j \in \mathbb{N}.$$

By the selection of $\{x_i : i \in \mathbb{N}\} \subset E$ as defined above, we have

$$\text{(3)} \qquad \frac{M_j}{2} \leq r(x_i) \leq M_j \quad \text{for } i \in I_j,$$

$$\text{(4)} \qquad x_{j+1} \in E \setminus \bigcup_{i=1}^{j} B_i \quad \text{for } j \in \mathbb{N}.$$

Moreover we claim that for every $i \in I_k$, we have

$$\text{(5)} \qquad x_i \in E \setminus \bigcup_{m \neq k} \bigcup_{j \in I_m} B_j$$

that is, if $i \in I_k$ then x_i is not contained in any closed ball B_j with $j \in \mathbb{N} \setminus \{I_k\}$. Now let us prove (5). Let $m \neq k, j \in I_m$ and $i \in I_k$. If $m < k$, then $x_i \notin B_j$ by (4). If $m < k$, then $r(x_j) < r(x_i)$, $x_j \notin B_i$ by (4), and so $x_i \notin B_j$. This proves (5).

We claim further that if $i', i'' \in I_j$ and $i' \neq i''$ then we have

$$\text{(6)} \qquad \overline{B}(x_{i'}, \tfrac{1}{4}r(x_{i'})) \cap \overline{B}(x_{i''}, \tfrac{1}{4}r(x_{i''})) = \emptyset.$$

Let us prove (6). Let us assume without loss of generality that $i' < i''$. Then by (4) we have $x_{i''} \in E \setminus B_i$ where $B_i = \overline{B}(x_{i'}, r(x_{i'}))$. Then $x_{i''} \notin \overline{B}(x_{i'}, r(x_{i'}))$ and hence $\rho(x_{i'}, x_{i''}) > r(x_{i'}) \geq \frac{M_j}{2}$ by (3). Now $\frac{1}{4}r(x_{i''}) \leq \frac{M_j}{4}$ and $\frac{1}{4}r(x_{i'}) \leq \frac{M_j}{4}$ by (3). Thus $\overline{B}(x_{i'}, \frac{1}{4}r(x_{i'})) \cap \overline{B}(x_{i''}, \frac{1}{4}r(x_{i''})) = \emptyset$. This proves (6).

Let us show

$$E \setminus \bigcup_{i \in \mathbb{N}} B_i = \emptyset, \text{ that is, } E \subset \bigcup_{i \in \mathbb{N}} B_i. \tag{7}$$

By (2) we have $M_{j+1} \leq \frac{M_j}{2}$ for $j \in \mathbb{N}$. Thus we have $\lim_{j \to \infty} M_j = 0$. Then by (3) we have $\lim_{i \to \infty} r(x_i) = 0$. Suppose $E \setminus \bigcup_{i \in \mathbb{N}} B_i \neq \emptyset$. Let $M = \sup \{r(x) : x \in E \setminus \bigcup_{i \in \mathbb{N}} B_i\}$. Then $M \leq M_j$ for every $j \in \mathbb{N}$. Since $\lim_{j \to \infty} M_j = 0$, we have $M = 0$. This is impossible since $r(x) > 0$ for $x \in E$.

Let $N(n)$ be the positive integer defined in Lemma 6.22. Let us show that every $x \in \mathbb{R}^n$ is contained in at most $16N(n)$ closed balls in the collection $\{B_i : i \in \mathbb{N}\}$ where $B_i = \overline{B}(x_i, r(x_i))$ and $x_i \in E$ for $i \in \mathbb{N}$. Let $x \in \mathbb{R}^n$. Suppose x is contained in $p \in \mathbb{N}$ closed balls in the collection $\{B_i : i \in \mathbb{N}\}$, say $B_{i_1}, \ldots, B_{i_p}$. Then $x \in \bigcap_{m=1}^{p} B_{i_m}$. Let us show that $p \leq 16N(n)$. Now since $\bigcup_{j \in \mathbb{N}} I_j = \mathbb{N}$, each i_m belongs to I_j for some $j \in \mathbb{N}$. Then we ask how many I_j can contain an i_m and how many i_m can each I_j contain.

Since $x \in \bigcap_{m=1}^{p} B_{i_m}$, we have $\bigcap_{m=1}^{p} B_{i_m} \neq \emptyset$. Consider the centers $\{x_i : i \in \mathbb{N}\}$ of the closed balls $\{B_i : i \in \mathbb{N}\}$. According to (5), if $i \in I_k$ then x_i is disjoint from B_j for all $j \in I_m$ where $m \neq k$. Thus if $i_m \in I_k$ and $i_{m'} \in I_{k'}$ where $k \neq k'$, then x_{i_m} is disjoint from $B_{i_{m'}}$ and $x_{i_{m'}}$ is disjoint from B_{i_m}. Consider a collection of closed balls in $\mathbb{R}^n$ such that the intersection of all the closed balls in the collection is non-empty and such that the center of every closed ball in the collection is not contained in another closed ball in the collection. According to Lemma 6.22, the cardinality of such a collection of closed balls is bounded above by $N(n)$. Thus we have

$$\text{card}\,\{j \in \mathbb{N} : I_j \cap \{i_1, \ldots, i_p\} \neq \emptyset\} \leq N(n),$$

that is, at most $N(n)$ I_j can contain some i_m.

Let us show next that for each $j \in \mathbb{N}$, the class I_j can contain at most 16^n of $\{i_1, \ldots, i_p\}$, that is,

$$\text{card}\,\big\{I_j \cap \{i_1, \ldots, i_p\}\big\} \leq 16^n.$$

Let $j \in \mathbb{N}$ be fixed and let us write

$$I_j \cap \{i_1, \ldots, i_p\} = \{j_1, \ldots, j_q\}.$$

Consider $\big\{\overline{B}(x_{j_m}, \frac{1}{4}r(x_{j_m})) : m = 1, \ldots, q\big\}$. According to (6), this is a disjoint collection. Since $x \in \overline{B}(x_{i_m}, r(x_{i_m}))$ for $m = 1, \ldots, p$, we have $x \in \overline{B}(x_{j_m}, r(x_{j_m}))$ for $m = 1, \ldots, q$. We have $\frac{M_j}{2} \leq r(x_{j_m}) \leq M_j$ for $m = 1, \ldots, q$. Let $y \in \overline{B}(x_{j_m}, r(x_{j_m}))$.

Then we have

$$\begin{aligned}\rho(y,x) &\leq \rho(y,x_{j_m}) + \rho(x_{j_m},x)\\ &\leq \frac{1}{4}r(x_{j_m}) + r(x_{j_m})\\ &< 2r(x_{j_m}) \leq 2M_j\end{aligned}$$

where the second inequality is from the fact that $x \in \overline{B}(x_{j_m}, r(x_{j_m}))$. Thus $y \in \overline{B}(x, 2M_j)$ and hence $\overline{B}(x_{j_m}, \frac{1}{4}r(x_{j_m})) \subset \overline{B}(x, 2M_j)$ for $m = 1, \dots, q$. Also, since $\frac{M_j}{2} \leq r(x_{j_m})$, we have $\overline{B}(x_{j_m}, \frac{1}{8}M_j) \subset \overline{B}(x_{j_m}, \frac{1}{4}r(x_{j_m}))$. Thus we have

$$\bigcup_{m=1}^{q} \overline{B}(x_{j_m}, \tfrac{1}{8}M_j) \subset \bigcup_{m=1}^{q} \overline{B}(x_{j_m}, \tfrac{1}{4}r(x_{j_m})) \subset \overline{B}(x, 2M_j)$$

and hence

$$\mu_L^n\left(\bigcup_{m=1}^{q} \overline{B}(x_{j_m}, \tfrac{1}{8}M_j)\right) \leq \mu_L^n\left(\bigcup_{m=1}^{q} \overline{B}(x_{j_m}, \tfrac{1}{4}r(x_{j_m}))\right) \leq \mu_L^n\big(\overline{B}(x, 2M_j)\big).$$

Now the disjointness of the collection $\{\overline{B}(x_{j_m}, \frac{1}{4}r(x_{j_m})) : j = 1, \dots, q\}$ implies the disjointness of the collection $\{\overline{B}(x_{j_m}, \frac{1}{8}M_j) : j = 1, \dots, q\}$. Thus we have

$$\sum_{m=1}^{q} \mu_L^n\Big(\overline{B}(x_{j_m}, \tfrac{1}{8}M_j)\Big) \leq \sum_{m=1}^{q} \mu_L^n\Big(\overline{B}(x_{j_m}, \tfrac{1}{4}r(x_{j_m}))\Big) \leq \mu_L^n\big(\overline{B}(x, 2M_j)\big).$$

Now

$$\begin{aligned}&\mu_L^n\Big(\overline{B}(x_{j_m}, \tfrac{1}{8}M_j)\Big) = \left(\tfrac{M_j}{8}\right)^n \mu_L^n((0,1)),\\ &\mu_L^n\big(\overline{B}(x, 2M_j)\big) = (2M_j)^n \mu_L^n((0,1)).\end{aligned}$$

Thus

$$q\left(\frac{M_j}{8}\right)^n \mu_L^n((0,1)) \leq (2M_j)^n \mu_L^n((0,1))$$

and hence

$$q \leq (8 \times 2)^n = 16^n.$$

Let $P(n) = 16^n N(n)$. Then every $x \in \mathbb{R}^n$ is contained in at most $P(n)$ of the collection $\{B_i : i \in \mathbb{N}\}$.

2. Consider the sequence $(B_i : i \in \mathbb{N}) \subset \mathfrak{B}$ where $B_i = \overline{B}(x_i, r(x_i))$ selected above. Recall the classification of the index set $\mathbb{N}$ of the sequence $(B_i : i \in \mathbb{N})$ into classes $\{I_\ell : \ell \in \mathbb{N}\}$. Recall that if $i \in I_\ell$ then $r(x_i) \in [\frac{M_\ell}{2}, M_\ell]$ and $M_{\ell+1} \leq \frac{M_\ell}{2}$ for $\ell \in \mathbb{N}$ and $\lim_{\ell\to\infty} M_\ell = 0$. Consider $B_i = \overline{B}(x_i, r(x_i))$ for $i \in I_1$. Now $I_1 = \{1, \dots, k_1\}$, a finite set. Renumber $\{B_i : i \in I_1\}$ if necessary so that $r(x_1) \geq \cdots \geq r(x_{k_1})$. Then renumber $\{B_i : i \in I_2\}$ if necessary so that $r(x_{k_1+1}) \geq \cdots \geq r(x_{k_2})$ and so on. Thus continuing we renumber $(B_i : i \in \mathbb{N})$ in such a way that $(r(x_i) : i \in \mathbb{N})$is a decreasing sequence of positive real numbers such that $\lim_{i \to \infty} r(x_i) = 0$. For brevity let us write $r_i = r(x_i)$ for $i \in \mathbb{N}$. Then we have $(B_i : i \in \mathbb{N})$ where $B_i = \overline{B}(x_i, r_i)$ and $(r_i : i \in \mathbb{N})$ is a decreasing sequence of positive numbers with $\lim_{i \to \infty} r_i = 0$.

Let $B_{1,1} = B_1$. If $B_{1,1}, \ldots, B_{1,j}$ have been chosen, let $B_{1,j+1}$ be the closed ball with the smallest index from among those members of $\{B_i : i \in \mathbb{N}\}$ that are disjoint from $B_{1,1}, \ldots, B_{1,j}$. Thus continuing we obtain a finite or countable disjoint subcollection $\mathfrak{B}_1$ of $\{B_i : i \in \mathbb{N}\}$. If E is not covered by the union of the members of $\mathfrak{B}_1$, let $B_{2,1}$ be the closed ball with the smallest index among $\{B_i : i \in \mathbb{N}\} \setminus \mathfrak{B}_1$. If $B_{2,1}, \ldots, B_{2,j}$ have been chosen, let $B_{2,j+1}$ be the closed ball with the smallest index from among those members of $\{B_i : i \in \mathbb{N}\} \setminus \mathfrak{B}_1$ that are disjoint from $B_{2,1}, \ldots, B_{2,j}$. Thus continuing we obtain a finite or countable disjoint subcollection $\mathfrak{B}_2$ of $\{B_i : i \in \mathbb{N}\}$. Note that $\mathfrak{B}_1$ and $\mathfrak{B}_2$ are disjoint. If E is not covered by the union of members of $\mathfrak{B}_1 \cup \mathfrak{B}_2$ then let $B_{3,1}$ be the closed ball with the smallest index among $\{B_i : i \in \mathbb{N}\} \setminus \{\mathfrak{B}_1 \cup \mathfrak{B}_2\}$ and so on. Thus continuing we have $\mathfrak{B}_1, \mathfrak{B}_2, \mathfrak{B}_3, \ldots$, each a finite or countable disjoint subcollection of $\{B_i : i \in \mathbb{N}\}$, and $\{\mathfrak{B}_1, \mathfrak{B}_2, \mathfrak{B}_3, \ldots\}$ is a disjoint collection. We claim that for some $m \leq 4^n P(n) + 1$ we have

$$E \subset \bigcup_{k=1}^{m} \bigcup_{B \in \mathfrak{B}_k} B, \text{ that is, } E \setminus \bigcup_{k=1}^{m} \bigcup_{B \in \mathfrak{B}_k} B = \emptyset.$$

Suppose $m \in \mathbb{N}$ is such that $E \setminus \bigcup_{k=1}^{m} \bigcup_{B \in \mathfrak{B}_k} B \neq \emptyset$. Then there exists $x \in E \setminus \bigcup_{k=1}^{m} \bigcup_{B \in \mathfrak{B}_k} B$. Let us show that $m \leq 4^n P(n)$. Since $E \subset \bigcup_{i \in \mathbb{N}} B_i$ as we showed in **1**, there exists $i_0 \in \mathbb{N}$ such that $x \in B_{i_0}$. Since $x \in B_{i_0}$ and $x \in E \setminus \bigcup_{k=1}^{m} \bigcup_{B \in \mathfrak{B}_k} B$, we have $B_{i_0} \notin \mathfrak{B}_k$ for $k = 1, \ldots, m$. Then for every $k = 1, \ldots, m$, B_{i_0} intersects some members of $\mathfrak{B}_k$. Let B_{k,i_k} be the first member in $\mathfrak{B}_k$ such that $B_{i_0} \cap B_{k,i_k} \neq \emptyset$. The fact that B_{i_0} was not selected to be $B_{k,1}$ implies that B_{i_0} was later than $B_{k,1}$ in the sequence $(B_i : i \in \mathbb{N})$. The fact that B_{i_0} was not selected to be $B_{k,1}, \ldots, B_{k,i_k-1}$ implies that B_{i_0} was later than $B_{k,1}, \ldots, B_{k,i_k-1}$ in the sequence $(B_i : i \in \mathbb{N})$. Thus B_{i_0} was not earlier than B_{k,i_k} in the sequence $(B_i : i \in \mathbb{N})$. Thus for the radius r_{i_0} of B_{i_0} and the radius r_{k,i_k} of B_{k,i_k}, we have $r_{k,i_k} \geq r_{i_0}$. Since $B_{i_0} \cap B_{k,i_k} \neq \emptyset$, there exists a closed ball C_k with radius $\frac{1}{2} r_{i_0}$ such that

$$C_k \subset (2 * B_{i_0}) \cap B_{k,i_k} \quad \text{for } k = 1, \ldots, m.$$

Since every $x \in \mathbb{R}^n$ is contained in at most $P(n)$ balls in the collection $\{B_i : i \in \mathbb{N}\}$, x is contained in at most $P(n)$ balls in the collection $\{B_{k,i_k} : k = 1, \ldots, m\} \subset \{B_i : i \in \mathbb{N}\}$. Then since $C_k \subset B_{k,i_k}$ for $k = 1, \ldots, m$, x is contained in at most $P(n)$ balls in the collection $\{B_{k,i_k} : k = 1, \ldots, m\}$. Now $C_k \subset \bigcup_{k=1}^{m} C_k$ and $\mathbf{1}_{C_k} \leq \mathbf{1}_{\bigcup_{k=1}^{m} C_k}$ for $k = 1, \ldots, m$. Since every $x \in \mathbb{R}^n$ is contained in at most $P(n)$ balls in the collection $\{C_k : k = 1, \ldots, m\}$, we have

$$\sum_{k=1}^{m} \mathbf{1}_{C_k} \leq P(n) \mathbf{1}_{\bigcup_{k=1}^{m} C_k}.$$

Since $C_k \subset 2 * B_{i_0}$ for $k = 1, \dots, m$, we have $\bigcup_{k=1}^m C_k \subset 2 * B_{i_0}$. Thus we have

$$\begin{aligned}(2r_{i_0})^n \mu_L^n(B(0,1)) &= \mu_L^n(2 * B_{i_0}) \geq \mu_L^n\Big(\bigcup_{k=1}^m C_k\Big) = \int_{\mathbb{R}^n} \mathbf{1}_{\bigcup_{k=1}^m C_k}\, d\mu_L^n \\ &\geq \frac{1}{P(n)} \int_{\mathbb{R}^n} \sum_{k=1}^m \mathbf{1}_{C_k}\, d\mu_L^n \geq \frac{1}{P(n)} \sum_{k=1}^m \mu_L^n(C_k) \\ &= \frac{m}{P(n)} \Big(\frac{r_{i_0}}{2}\Big)^n \mu_L^n(B(0,1)).\end{aligned}$$

Thus we have

$$m \leq 4^n P(n).$$

We have just shown that if $E \setminus \bigcup_{k=1}^m \bigcup_{B \in \mathfrak{B}_k} B \neq \emptyset$ then $m \leq 4^n P(n)$. This implies that for $m = 4^n P(n) + 1$ we have $E \setminus \bigcup_{k=1}^m \bigcup_{B \in \mathfrak{B}_k} B = \emptyset$. Let $Q(n) = 4^n P(n) + 1$. Then we have $E \setminus \bigcup_{k=1}^{Q(n)} \bigcup_{B \in \mathfrak{B}_k} B = \emptyset$. ■

[V] Vitali Covering Theorem for Radon Outer Measure on $\mathbb{R}^n$

Applying the Besicovitch Covering Theorem, we obtain a Vitali covering theorem for a Radon outer measure on $\mathbb{R}^n$. As in the Besicovitch Covering Theorem, this will be done for a Vitali covering of a bounded set in $\mathbb{R}^n$ first. Then we extend this to an arbitrary set in $\mathbb{R}^n$.

Proposition 6.25. *A Radon outer measure μ on $\mathbb{R}^n$ is a σ-finite outer measure.*

Proof. Let μ be a Radon outer measure on $\mathbb{R}^n$. Consider the measure space $(\mathbb{R}^n, \mathfrak{M}(\mu), \mu)$. Since μ is a Radon outer measure on $\mathbb{R}^n$ we have $\mathfrak{B}_{\mathbb{R}^n} \subset \mathfrak{M}(\mu)$ and for every compact set K in $\mathbb{R}^n$ we have $\mu(K) < \infty$. Note that $K \in \mathfrak{B}_{\mathbb{R}^n}$. Now a set in $\mathbb{R}^n$ is a compact set if and only if it is a bounded closed set. For every $k \in \mathbb{N}$ the set $\overline{B}(0,k) = \{x \in \mathbb{R}^n : |x| \leq k\}$ is a bounded closed set in $\mathbb{R}^n$ and is hence a compact set and then $\mu(\overline{B}(0,k)) < \infty$. Now $\mathbb{R}^n = \bigcup_{k \in \mathbb{N}} \overline{B}(0,k)$. Thus $(\mathbb{R}^n, \mathfrak{M}(\mu), \mu)$ is a σ-finite measure space and μ is a σ-finite outer measure. ■

Proposition 6.26. *Let μ be a Radon outer measure on $\mathbb{R}^n$.*
(a) *For $r \in [0, \infty)$, consider the $n-1$-dimensional spherical hypersurface defined by $S_r := \{x \in \mathbb{R}^n : |x| = r\}$ in $\mathbb{R}^n$. Then $\mu(S_r) > 0$ for only countably many values of $r \in [0, \infty)$.*
(b) *Let us write $\mathbb{R}^n = \mathbb{R}_1 \times \cdots \times \mathbb{R}_n$ and $x = (\xi_1, \dots, \xi_n) \in \mathbb{R}^n$. For each $i = 1, \dots, n$ and $a \in \mathbb{R}_i$, consider the $n-1$-dimensional parallel hyperplane in $\mathbb{R}^n$ defined by setting $H_{i,a} := \{(\xi_1, \dots, \xi_n) \in \mathbb{R}^n : \xi_i = a\}$. Then $\mu(H_{i,a}) > 0$ for only countably many values of $a \in \mathbb{R}_i$.*

Proof. The measure space $(\mathbb{R}^n, \mathfrak{M}(\mu), \mu)$ is a σ-finite measure space by Proposition 6.25.

1. For $r \in [0, \infty)$, we have $S_r \in \mathfrak{B}_{\mathbb{R}^n} \subset \mathfrak{M}(\mu)$. Now $\{S_r : r \in [0, \infty)\}$ is a disjoint collection in $\mathfrak{B}_{\mathbb{R}^n} \subset \mathfrak{M}(\mu)$ and $\bigcup_{r \in [0,\infty)} S_r = \mathbb{R}^n$. Then by Theorem 1.28, $\mu(S_r) > 0$ for only countably many values of $r \in [0, \infty)$.

2. For $a \in \mathbb{R}_i$, we have $H_{i,a} \in \mathfrak{B}_{\mathbb{R}^n} \subset \mathfrak{M}(\mu)$. Now $\{H_{i,a} : a \in \mathbb{R}_i\}$ is a disjoint collection in $\mathfrak{B}_{\mathbb{R}^n} \subset \mathfrak{M}(\mu)$ and $\bigcup_{a \in \mathbb{R}_i} H_{i,a} = \mathbb{R}^n$. Then by Theorem 1.28, $\mu(H_{i,a}) > 0$ for only countably many values of $a \in \mathbb{R}_i$.

Lemma 6.27. *Let A, B, and C be three arbitrary subsets of a non-empty set X. Then we have*

$$A \setminus C \subset (A \setminus B) \cup (B \setminus C).$$

Proof. We are to show that

$$A \cap A^c \subset (A \cap B^c) \cup (B \cap C^c).$$

Now for every point $x \in X$ we have either $x \in A$ or $x \in A^c$. Similarly we have either $x \in B$ or $x \in B^c$ and we have either $x \in C$ or $x \in C^c$. Thus X is the union of the following $2^3 = 8$ disjoint subsets of X:

$$\begin{aligned}
E_1 &= A \cap B \cap C \\
E_2 &= A \cap B \cap C^c \\
E_3 &= A \cap B^c \cap C \\
E_4 &= A \cap B^c \cap C^c \\
E_5 &= A^c \cap B \cap C \\
E_6 &= A^c \cap B \cap C^c \\
E_7 &= A^c \cap B^c \cap C \\
E_8 &= A^c \cap B^c \cap C^c.
\end{aligned}$$

In terms of these 8 sets, we have

$$\begin{aligned}
A \cap C^c &= E_2 \cup E_4 \\
A \cap B^c &= E_3 \cup E_4 \\
B \cap C^c &= E_2 \cup E_5.
\end{aligned}$$

Then we have

$$A \cap C^c = E_2 \cup E_4 \subset E_3 \cup E_4 \cup E_2 \cup E_5 = (A \cap B^c) \cup (B \cap C^c).$$

This completes the proof. ∎

Theorem 6.28. (Vitali Covering Theorem for Radon Outer Measure on $\mathbb{R}^n$) *Let μ be a Radon outer measure on $\mathbb{R}^n$. Let $A \subset \mathbb{R}^n$ and let $\mathcal{B}$ be a collection of closed balls in $\mathbb{R}^n$ satisfying the following conditions:*

$1°$ *for every $x \in A$ there exists a closed ball $\overline{B}(x, r)$ in $\mathcal{B}$.*

2° *for every* $x \in A$*, we have* $\inf\{r > 0 : \overline{B}(x, r) \in \mathcal{B}\} = 0$.
Then there exists a countable disjoint collection $\{B_i : i \in \mathbb{N}\} \subset \mathcal{B}$ *such that*

$$\mu\left(A \setminus \bigcup_{i \in \mathbb{N}} B_i\right) = 0.$$

Proof. If $\mu(A) = 0$ the theorem is trivial. Thus we assume that $\mu(A) > 0$.

1. Consider first the case that A is a bounded set. Then there exists $R > 0$ such that $A \subset B(0, R)$. By 2°, for every $x \in A$ there exists a closed ball $\overline{B}(x, r(x))$ in the collection $\mathcal{B}$ such that $\overline{B}(x, r(x)) \subset B(0, R)$ and $r(x) \leq 1$. Consider the collection $\mathcal{B}_0 = \{\overline{B}(x, r(x)) : x \in A\} \subset \mathcal{B}$. We have $\sup\{r(x) : x \in A\} \leq 1$. Thus our collection $\mathcal{B}_0$ satisfies the conditions on the collection $\mathcal{B}$ in Theorem 6.24.

Since μ is a Radon outer measure on $\mathbb{R}^n$, we have

$$\mu(A) = \inf\{\mu(V) : A \subset V, V \text{ open}\}.$$

Thus there exists an open set V in $\mathbb{R}^n$ such that $A \subset V$ and

$$\mu(V) = \left\{1 + \frac{1}{4Q(n)}\right\} \mu(A)$$

where $Q(n)$ is the constant for $\mathbb{R}^n$ in Theorem 6.24. Let $U = V \cap B(0, R)$. Then U is an open set in $\mathbb{R}^n$, $A \subset U, U \subset V$, and

$$\mu(U) = \left\{1 + \frac{1}{4Q(n)}\right\} \mu(A). \tag{1}$$

Now $\overline{U} \subset \overline{B}(0, R)$, a compact set in $\mathbb{R}^n$ and $\mu(\overline{B}(0, R)) < \infty$ since μ is a Radon outer measure on $\mathbb{R}^n$. Thus we have $\mu(U) \leq \mu(\overline{U}) < \infty$.

By Theorem 6.24 there exist $\mathcal{B}_1, \ldots, \mathcal{B}_{Q(n)}$, each a finite or countable disjoint subcollection of $\mathcal{B}_0$ such that

$$A \subset \bigcup_{i=1}^{Q(n)} \bigcup_{B \in \mathcal{B}_i} B \subset B(0, R).$$

Then $\mu(A) \leq \sum_{i=1}^{Q(n)} \mu\left(\bigcup_{B \in \mathcal{B}_i} B\right)$. Thus there exists $i_0 \in \{1, \ldots, Q(n)\}$ such that

$$\mu(A) \leq Q(n)\mu\left(\bigcup_{B \in \mathcal{B}_{i_0}} B\right). \tag{2}$$

Now $\mathcal{B}_{i_0}$ is a finite or countable disjoint collection. Let us write $\mathcal{B}_{i_0} = \{B'_j : j \in \mathbb{N}\}$. Then

$$\sum_{j \in \mathbb{N}} \mu(B'_j) = \mu\left(\bigcup_{j \in \mathbb{N}} B'_j\right) = \mu\left(\bigcup_{B \in \mathcal{B}_{i_0}} B\right) \leq \mu(B(0, R)) < \infty.$$

Thus there exists $k_1 \in \mathbb{N}$ such that

$$\mu\left(\bigcup_{j=1}^{k_1} B'_j\right) = \sum_{j=1}^{k_1} \mu(B'_j) \geq \frac{1}{2}\mu\left(\bigcup_{j\in\mathbb{N}} B'_j\right) = \frac{1}{2}\mu\left(\bigcup_{B\in\mathfrak{B}_{i_0}} B\right).$$

Then by (2) we have

$$\mu\left(\bigcup_{j=1}^{k_1} B'_j\right) \geq \frac{\mu(A)}{2Q(n)}. \tag{3}$$

Let

$$A_1 = A \setminus \bigcup_{j=1}^{k_1} B'_j. \tag{4}$$

Then

$$\begin{aligned}\mu(A_1) &\leq \mu\left(U \setminus \bigcup_{j=1}^{k_1} B'_j\right) = \mu(U) - \mu\left(\bigcup_{j=1}^{k_1} B'_j\right)\\ &\leq \left\{1 + \frac{1}{4Q(n)}\right\}\mu(A) - \frac{1}{2Q(n)}\mu(A)\\ &= \left\{1 - \frac{1}{4Q(n)}\right\}\mu(A) = u\mu(A)\end{aligned}$$

where

$$u = 1 - \frac{1}{4Q(n)} \in (0,1).$$

Now A_1 is contained in the open set $\mathbb{R}^n \setminus \bigcup_{j=1}^{k_1} B'_j$ and therefore we can find an open set U_1 such that $A_1 \subset U_1 \subset \mathbb{R}^n \setminus \bigcup_{j=1}^{k_1} B'_j$ and

$$\mu(U_1) \leq \left\{1 + \frac{1}{4Q(n)}\right\}\mu(A_1).$$

As above we can find Closed balls $B'_{k_1+1}, \ldots, B'_{k_2}$ in the collection $\mathfrak{B}_0$ such that for the set

$$A_2 = A_1 \setminus \bigcup_{j=k_1+1}^{k_2} B'_j = A \setminus \bigcup_{j=1}^{k_2} B'_j$$

we have

$$\mu(A_2) \leq u\mu(A_1) \leq u^2\mu(A).$$

Clearly $\{B'_1, \ldots, B'_{k_2}\}$ is a disjoint collection. After m steps we have

$$\mu\left(A \setminus \bigcup_{j=1}^{k_m} B'_j\right) \leq u^m \mu(A).$$

Since $u \in (0, 1)$, we have

$$\mu\left(A \setminus \bigcup_{j \in \mathbb{N}} B'_j\right) = 0. \tag{5}$$

This completes the proof for the case that A is a bounded set in $\mathbb{R}^n$.

2. Let us remove the restriction that A is a bounded set. Let us write $\mathbb{R}^n = \mathbb{R}_1 \times \cdots \times \mathbb{R}_n$ and $x = (\xi_1, \ldots, \xi_n) \in \mathbb{R}^n$. For each $i \in \{1, \ldots, n\}$, let $(a_{i,j_i} : j_i \in \mathbb{Z}) \subset \mathbb{R}_i$ be such that a_{i,j_i} is strictly increasing as $j_i \uparrow$ with $\lim_{j_i \to -\infty} a_{i,j_i} = -\infty$ and $\lim_{j_i \to \infty} a_{i,j_i} = \infty$. Consider the closed boxes in $\mathbb{R}^n$ defined by

$$C_{j_1,\ldots,j_n} = [a_{1,j_1-1}, a_{1,j_1}] \times \cdots \times [a_{n,j_n-1}, a_{n,j_n}] \quad \text{for } j_1, \ldots, j_n \in \mathbb{Z} \tag{6}$$

and their interiors, that is, the open boxes defined by

$$O_{j_1,\ldots,j_n} = (a_{1,j_1-1}, a_{1,j_1}) \times \cdots \times (a_{n,j_n-1}, a_{n,j_n}) \quad \text{for } j_1, \ldots, j_n \in \mathbb{Z} \tag{7}$$

we have

$$\bigcup_{j_1 \in \mathbb{Z}} \cdots \bigcup_{j_n \in \mathbb{Z}} C_{j_1,\ldots,j_n} = \mathbb{R}^n. \tag{8}$$

For the boundaries of these boxes we have

$$\partial O_{j_1,\ldots,j_n} = \partial C_{j_1,\ldots,j_n} = C_{j_1,\ldots,j_n} \setminus O_{j_1,\ldots,j_n} \quad \text{for } j_1, \ldots, j_n \in \mathbb{Z}. \tag{9}$$

Let us show that $(a_{i,j_i} : j_i \in \mathbb{Z}) \subset \mathbb{R}_i$ for $i = 1, \ldots, n$ can be so chosen that the measures of these boundaries are all equal to 0. Now $\partial C_{j_1,\ldots,j_n}$ consists of n pairs of faces of the closed box $C_{j_1,\ldots,j_n}$. These n pairs of faces are subsets of n pairs of $n-1$-dimensional hyperplanes given by

$$\begin{aligned}
&H_{1,a_{1,j_1-1}} = \{x \in \mathbb{R}^n : \xi_1 = a_{1,j_1-1}\}, \quad H_{1,a_{1,j_1}} = \{x \in \mathbb{R}^n : \xi_1 = a_{1,j_1}\},\\
&\vdots\\
&H_{n,a_{n,j_n-1}} = \{x \in \mathbb{R}^n : \xi_n = a_{n,j_n-1}\}, \quad H_{n,a_{n,j_n}} = \{x \in \mathbb{R}^n : \xi_n = a_{n,j_n}\}
\end{aligned}$$

According to (b) of Proposition 6.26, $\mu(H_{i,a}) > 0$ for only countably many values of $a \in \mathbb{R}_i$. Thus we can select $(a_{i,j_i} : j_i \in \mathbb{Z}) \subset \mathbb{R}_i$ such that a_{i,j_i} is strictly increasing as $j_i \uparrow$ with $\lim_{j_i \to -\infty} a_{i,j_i} = -\infty$ and $\lim_{j_i \to \infty} a_{i,j_i} = \infty$ and $\mu(H_{i,a_{i,j_i}}) = 0$. With such choice the measure of each of the $2 \times n$ faces of the closed box $C_{j_1,\ldots,j_n}$ is equal to 0 and thus we have

$$\mu(\partial C_{j_1,\ldots,j_n}) = 0 \quad \text{for } j_1, \ldots, j_n \in \mathbb{Z}. \tag{10}$$

Now let

$$A_{j_1,\dots,j_n} = A \cap O_{j_1,\dots,j_n} \quad \text{for } j_1,\dots,j_n \in \mathbb{Z}. \tag{11}$$

We have

$$\begin{aligned}
A \setminus \bigcup_{j_1\in\mathbb{Z}} \cdots \bigcup_{j_n\in\mathbb{Z}} A_{j_1,\dots,j_n} &= A \cap \mathbb{R}^n \setminus \bigcup_{j_1\in\mathbb{Z}} \cdots \bigcup_{j_n\in\mathbb{Z}} A_{j_1,\dots,j_n} \\
&= \bigcup_{j_1\in\mathbb{Z}} \cdots \bigcup_{j_n\in\mathbb{Z}} (A \cap C_{j_1,\dots,j_n}) \setminus \bigcup_{j_1\in\mathbb{Z}} \cdots \bigcup_{j_n\in\mathbb{Z}} (A \cap O_{j_1,\dots,j_n}) \\
&\subset \bigcup_{j_1\in\mathbb{Z}} \cdots \bigcup_{j_n\in\mathbb{Z}} \Big\{ (A \cap C_{j_1,\dots,j_n}) \setminus (A \cap O_{j_1,\dots,j_n}) \Big\} \\
&= \bigcup_{j_1\in\mathbb{Z}} \cdots \bigcup_{j_n\in\mathbb{Z}} A \cap \{ C_{j_1,\dots,j_n} \setminus O_{j_1,\dots,j_n} \} \\
&\subset \bigcup_{j_1\in\mathbb{Z}} \cdots \bigcup_{j_n\in\mathbb{Z}} \partial C_{j_1,\dots,j_n}.
\end{aligned}$$

Then by (10), we have

$$\begin{aligned}
\mu\left(A \setminus \bigcup_{j_1\in\mathbb{Z}} \cdots \bigcup_{j_n\in\mathbb{Z}} A_{j_1,\dots,j_n} \right) &\le \mu\left(\bigcup_{j_1\in\mathbb{Z}} \cdots \bigcup_{j_n\in\mathbb{Z}} \partial C_{j_1,\dots,j_n} \right) \\
&\le \sum_{j_1\in\mathbb{Z}} \cdots \sum_{j_n\in\mathbb{Z}} \mu(\partial C_{j_1,\dots,j_n}) \\
&= 0.
\end{aligned} \tag{12}$$

Now $A_{j_1,\dots,j_n}$ is a bounded set contained in a bounded open set $O_{j_1,\dots,j_n}$. Thus by our result in **1**, there exists a countable disjoint subcollection of $\mathfrak{B}$, $\{B_{j_1,\dots,j_n,k} : k \in \mathbb{N}\}$ contained in $O_{j_1,\dots,j_n}$ such that

$$\mu\left(A_{j_1,\dots,j_n} \setminus \bigcup_{k\in\mathbb{N}} B_{j_1,\dots,j_n,k} \right) = 0. \tag{13}$$

Consider $\bigcup_{j_1\in\mathbb{Z}} \cdots \bigcup_{j_n\in\mathbb{Z}} \{B_{j_1,\dots,j_n,k} : k \in \mathbb{N}\}$, a countable disjoint subcollection of $\mathfrak{B}$. Applying Lemma 6.27, we have

$$\begin{aligned}
&A \setminus \bigcup_{j_1\in\mathbb{Z}} \cdots \bigcup_{j_n\in\mathbb{Z}} \bigcup_{k\in\mathbb{N}} B_{j_1,\dots,j_n,k} \\
&\subset \left(A \setminus \bigcup_{j_1\in\mathbb{Z}} \cdots \bigcup_{j_n\in\mathbb{Z}} A_{j_1,\dots,j_n} \right) \cup \left(\bigcup_{j_1\in\mathbb{Z}} \cdots \bigcup_{j_n\in\mathbb{Z}} A_{j_1,\dots,j_n} \setminus \bigcup_{j_1\in\mathbb{Z}} \cdots \bigcup_{j_n\in\mathbb{Z}} \bigcup_{k\in\mathbb{N}} B_{j_1,\dots,j_n,k} \right) \\
&\subset \left(A \setminus \bigcup_{j_1\in\mathbb{Z}} \cdots \bigcup_{j_n\in\mathbb{Z}} A_{j_1,\dots,j_n} \right) \cup \left(\bigcup_{j_1\in\mathbb{Z}} \cdots \bigcup_{j_n\in\mathbb{Z}} \left(A_{j_1,\dots,j_n} \setminus \bigcup_{k\in\mathbb{N}} B_{j_1,\dots,j_n,k} \right) \right).
\end{aligned}$$

Then

$$\begin{aligned}
&\mu\left(A\setminus\bigcup_{j_1\in\mathbb{Z}}\cdots\bigcup_{j_n\in\mathbb{Z}}\bigcup_{k\in\mathbb{N}}B_{j_1,\ldots,j_n,k}\right)\\
&\le\mu\left(A\setminus\bigcup_{j_1\in\mathbb{Z}}\cdots\bigcup_{j_n\in\mathbb{Z}}A_{j_1,\ldots,j_n}\right)+\sum_{j_1\in\mathbb{Z}}\cdots\sum_{j_n\in\mathbb{Z}}\mu\left(A_{j_1,\ldots,j_n}\setminus\bigcup_{k\in\mathbb{N}}B_{j_1,\ldots,j_n,k}\right)\\
&=0
\end{aligned}$$

by (12) and (13). ∎

§7 Differentiation of Measures

[I] Measure of a Ball as a Function of the Center

Consider the Lebesgue measure space $(\mathbb{R}^n, \mathfrak{M}^n_L, \mu^n_L)$ on $\mathbb{R}^n$. With $r > 0$ fixed, consider a real-valued function f on $\mathbb{R}^n$ defined by $f(x) = \mu^n_L(C_r(x))$ for $x \in \mathbb{R}^n$. By the translation invariance of the measure space $(\mathbb{R}^n, \mathfrak{M}^n_L, \mu^n_L)$ on $\mathbb{R}^n$, the value of $\mu^n_L(C_r(x))$ depends on $r > 0$ only and does not depend on $x \in \mathbb{R}^n$ and indeed $\mu^n_L(C_r(x)) = \pi^{n/2}\{\Gamma(\frac{n}{2}+1)\}^{-1} r^n$ which is independent of $x \in \mathbb{R}^n$. Thus f is a constant function on $\mathbb{R}^n$.

Now let μ be a Radon outer measure on a set metric space (X, d) and consider the measure space $(X, \mathfrak{M}(\mu), \mu)$. With $r > 0$ fixed, consider a function f on X defined by $f(x) = \mu^n_L(C_r(x))$ for $x \in X$. Let us investigate the continuity properties and $\mathfrak{B}_X$-measurability of the function f on X.

Lemma 7.1. *Let (X, d) be a metric space. Let $x \in X$ and let $(x_k : k \in \mathbb{N})$ be a sequence in X such that $\lim_{k\to\infty} x_k = x$. With $r > 0$ fixed, let $f = \mathbf{1}_{C_r(x)}$ and $f_k = \mathbf{1}_{C_r(x_k)}$ for $k \in \mathbb{N}$. Then we have*

$$\limsup_{k\to\infty} f_k \le f \quad \textit{on } X.$$

Proof. We are to show

$$(1) \qquad \Big(\limsup_{k\to\infty} f_k\Big)(y) \le f(y) \quad \text{for all } y \in X.$$

Note that since f and f_k are characteristic functions of sets in X we have $f(y), f_k(y) \in [0, 1]$ for all $y \in X$. We consider the case $y \in C_r(x)$ and the case $y \notin C_r(x)$ separately.

1. Consider the case $y \in C_r(x)$. Then $f(y) = \mathbf{1}_{C_r(x)}(y) = 1$. Since $f_k(y) \in [0, 1]$ for all $y \in X$ and all $k \in \mathbb{N}$, we have $\big(\limsup_{k\to\infty} f_k\big)(y) \le 1 = f(y)$. This proves (1) for $y \in C_r(x)$.

2. Consider the case $y \notin C_r(x)$. In this case we have $f(y) = \mathbf{1}_{C_r(x)}(y) = 0$. Let us show that if $y \notin C_r(x)$ then there exists $N \in \mathbb{N}$ such that $y \notin C_r(x_k)$ for $k \ge N$. Now $y \notin C_r(x)$ so that $d(y, x) = r + \eta$ where $\eta > 0$. Since $\lim_{k\to\infty} x_k = x$ there exists $N \in \mathbb{N}$ such that $d(x_k, x) < \frac{\eta}{2}$ for $k \ge N$. Now $d(y, x) \le d(y, x_k) + d(x_k, y)$. Thus we have

$$d(y, x_k) \ge d(y, x) - d(x_k, y) \ge r + \eta - \tfrac{\eta}{2} = r + \tfrac{\eta}{2}$$

for $k \ge N$. Then $y \notin C_r(x_k)$ for $k \ge N$ and this implies $f_k(y) = \mathbf{1}_{C_r(x_k)}(y) = 0$ for $k \ge N$. Thus we have $\sup_{k\ge N} f_k(y) = 0$. Then $\big(\limsup_{k\to\infty} f_k\big)(y) = \lim_{N\to\infty}\big\{\sup_{k\ge N} f_k(y)\big\} = 0 = f(y)$. This proves (1) for $y \notin C_r(x)$. ■

Theorem 7.2. *Let μ be a boundedly finite Borel outer measure on a metric space (X, d).*
(a) *Let $x \in X$ and let $(x_k : k \in \mathbb{N})$ be a sequence in X such that $\lim_{k\to\infty} x_k = x$. Then with*

an arbitrarily fixed $r > 0$, we have

(1) $$\limsup_{k\to\infty} \mu\big(C_r(x_k)\big) \le \mu\big(C_r(x)\big).$$

(b) *Let us define a real-valued function f on X by*

(2) $$f(x) = \mu\big(C_r(x)\big) \quad \textit{for } x \in X.$$

Then f is upper semicontinuous on X and hence a $\mathfrak{B}_X$-measurable and therefore a $\mathfrak{M}(\mu)$-measurable function on X.

Proof. 1. Let us define real-valued functions on X by setting $f = \mathbf{1}_{C_r(x)}$ and $f_k = \mathbf{1}_{C_r(x_k)}$ for $k \in \mathbb{N}$. By Lemma 7.1 we have

(3) $$\limsup_{k\to\infty} f_k \le f \quad \text{on } X.$$

Then

(4) $$\liminf_{k\to\infty}\{1 - f_k\} \ge 1 - \limsup_{k\to\infty} f_k \ge 1 - f \quad \text{on } X.$$

By (4) and Fatou's Lemma we have

(5) $$\begin{aligned}\int_{C_{2r}(x)} \{1 - f\}\, d\mu &\le \int_{C_{2r}(x)} \liminf_{k\to\infty}\{1 - f_k\}\, d\mu \\ &\le \liminf_{k\to\infty} \int_{C_{2r}(x)} \{1 - f_k\}\, d\mu.\end{aligned}$$

Since $\lim_{k\to\infty} x_k = x$, there exists $N \in \mathbb{N}$ such that $C_r(x_k) \subset C_{2r}(x)$ for $k \ge N$. For $k \ge N$, (5) reduces to

(6) $$\begin{aligned}\mu\big(C_{2r}(x)\big) - \mu\big(C_r(x)\big) &\le \liminf_{k\to\infty}\big\{\mu\big(C_{2r}(x)\big) - \mu\big(C_r(x_k)\big)\big\} \\ &= \mu\big(C_{2r}(x)\big) + \liminf_{k\to\infty}\big\{-\mu\big(C_r(x_k)\big)\big\}.\end{aligned}$$

Subtracting $\mu\big(C_{2r}(x)\big) < \infty$ from both sides of (4), we have

$$-\mu\big(C_r(x)\big) \le \liminf_{k\to\infty}\big\{-\mu\big(C_r(x_k)\big)\big\} = -\limsup_{k\to\infty} \mu\big(C_r(x_k)\big)$$

and then

$$\mu\big(C_r(x)\big) \ge \limsup_{k\to\infty} \mu\big(C_r(x_k)\big).$$

This proves (1).

2. According to Proposition 15.83, [LRA], a real-valued function φ on a metric space (X, d) is upper semicontinuous at $x \in X$ if and only if for every sequence $(x_k : k \in \mathbb{N})$ in

X such that $\lim_{k\to\infty} x_k = x$ we have $\limsup_{k\to\infty} \varphi(x_k) \le \varphi(x)$. For the function f defined by (2), we have $\limsup_{k\to\infty} f(x_k) \le f(x)$ by (1). Thus f is upper semicontinuous on X.

According to Proposition 15.82, [LRA], an upper semicontinuous function φ on a metric space (X, d) is $\mathfrak{B}_X$-measurable function on X. Thus our function f is $\mathfrak{B}_X$-measurable function on X. ■

[II] Derivative of an Outer Measure with respect to Another Outer Measure on a Metric Space

Definition 7.3. *Let μ and ν be Borel outer measures on a metric space (X, d). The upper and lower derivative of ν with respect to μ at $x \in X$ are defined by*

$$\overline{D}_\mu \nu(x) = \begin{cases} \limsup_{r\to 0} \dfrac{\nu(C_r(x))}{\mu(C_r(x))} & \text{if } \mu(C_r(x)) > 0 \text{ for all } r \in (0,\infty) \\ \infty & \text{if } \mu(C_r(x)) = 0 \text{ for some } r \in (0,\infty) \end{cases}$$

and

$$\underline{D}_\mu \nu(x) = \begin{cases} \liminf_{r\to 0} \dfrac{\nu(C_r(x))}{\mu(C_r(x))} & \text{if } \mu(C_r(x)) > 0 \text{ for all } r \in (0,\infty) \\ \infty & \text{if } \mu(C_r(x)) = 0 \text{ for some } r \in (0,\infty) \end{cases}$$

where $C_r(x) = \{y \in X : d(y,x) \le r\}$, that is, a closed ball in X with center at x and with radius r. Note that if $\mu(C_r(x)) = 0$ for some $r > 0$ then $\mu(C_{r'}(x)) = 0$ for all $r' \in (0, r)$ by the monotonicity of the outer measure μ.

Let us note that if $\operatorname{supp}\mu = X$, that is, if $\mu(V) > 0$ for every non-empty open set V in X (see Proposition 1.44), then $\mu(C_r(x)) > 0$ for every $x \in X$ and $r \in (0,\infty)$ and thus Definition 7.3 is simplified.

Definition 7.4. *Let μ, ν, $\overline{D}_\mu\nu$ and $\underline{D}_\mu\nu$ be as in Definition 7.3. If $\overline{D}_\mu\nu(x) = \underline{D}_\mu\nu(x)$, then we define*

$$D_\mu\nu(x) := \overline{D}_\mu\nu(x) = \underline{D}_\mu\nu(x)$$

and call it the derivative of ν with respect to μ at $x \in X$. We also call $D_\mu\nu(x)$ the density of ν with respect to μ at $x \in X$. We say that ν is differentiable with respect to μ at $x \in X$ only when $D_\mu\nu(x) < \infty$.

Definition 7.5. *Let μ and ν be two outer measures on a set X. We say that ν is absolutely continuous with respect to μ and write $\nu \ll \mu$ if for any $A \subset X$ such that $\mu(X) = 0$ we have $\nu(A) = 0$.*

Definition 7.6. *Let μ and ν be Borel outer measures on a topological space X. We say μ*

and ν are mutually singular and write $\mu \perp \nu$ if there exists $B \in \mathfrak{B}_X$ such that $\mu(X \setminus B) = 0$ and $\nu(B) = 0$.

[III] Derivatives of Radon Outer Measures on $\mathbb{R}^n$

Lemma 7.7. *Let μ and ν be Radon outer measures on $\mathbb{R}^n$. For $t \in (0, \infty)$ arbitrarily fixed, let*

(1) $$A' = \{x \in \mathbb{R}^n : \underline{D}_\mu \nu(x) \leq t\}.$$

(2) $$A'' = \{x \in \mathbb{R}^n : \overline{D}_\mu \nu(x) \geq t\}.$$

Let A be an arbitrary subset of $\mathbb{R}^n$, measurable or not. Then

(3) $$A \subset A' \Rightarrow \nu(A) \leq t\,\mu(A).$$

(4) $$A \subset A'' \Rightarrow \nu(A) \geq t\,\mu(A).$$

Proof. 1. Let us prove (3). Let A be an arbitrary subset of A'. Let $\varepsilon > 0$. Since μ is a Radon outer measure, there exists an open set U such that $A \subset U$ and $\mu(U) \leq \mu(A) + \varepsilon$. Let $x \in A \subset A'$. Then $\underline{D}_\mu \nu(x) \leq t$, that is,

$$\liminf_{r \to 0} \frac{\nu\big(C_r(x)\big)}{\mu\big(C_r(x)\big)} \leq t.$$

Thus there exists $r_x > 0$ such that

(5) $$\frac{\nu\big(C_r(x)\big)}{\mu\big(C_r(x)\big)} \leq t + \varepsilon \quad \text{for all } r \in (0, r_x).$$

Select r_x so small that $C_r(x) \subset U$ for all $r \in (0, r_x)$. Let $\mathcal{B} = \{C_r(x) : x \in A, r \in (0, r_x)\}$. Then $\mathcal{B}$ is a Vitali cover of A as in Definition 6.18 and $\mathcal{B}$ is contained in the open set U. Then by Theorem 6.28 (Vitali Covering Theorem for Radon Outer Measure on $\mathbb{R}^n$), there exists a countable disjoint subcollection of $\mathcal{B}$, denoted by $\{C_i : i \in \mathbb{N}\}$, such that

(6) $$\nu\left(A \setminus \bigcup_{i \in \mathbb{N}} C_i\right) = 0.$$

Note that since $C_i \in \mathcal{B}$ we have by (5)

(7) $$\nu(C_i) \leq (t + \varepsilon)\mu(C_i).$$

By Observation 1.15, (6) implies

(8) $$\begin{aligned} \nu(A) &\leq \nu\left(\bigcup_{i \in \mathbb{N}} C_i\right) = \sum_{i \in \mathbb{N}} \nu(C_i) \leq (t + \varepsilon) \sum_{i \in \mathbb{N}} \mu(C_i) \\ &= (t + \varepsilon)\mu\left(\bigcup_{i \in \mathbb{N}} C_i\right) \leq (t + \varepsilon)\mu(U) \\ &\leq (t + \varepsilon)\{\mu(A) + \varepsilon\}. \end{aligned}$$

Since this holds for every $\varepsilon > 0$, we have $\nu(A) \leq t\,\mu(A)$. This proves (3).

2. Let us prove (4). Let $A \subset A''$. Let $\varepsilon > 0$ be arbitrarily given. Since both μ and ν are Radon outer measures on $\mathbb{R}^n$, there exists an open set Uin $\mathbb{R}^n$ such that $A \subset U$ and $\mu(U) \leq \mu(A) = \varepsilon$ and $\nu(U) \leq \nu(A) = \varepsilon$ also. Let $x \in A \subset A''$. Then $\overline{D}_\mu \nu(x) \geq t$, that is,

$$\limsup_{r \to 0} \frac{\nu\big(C_r(x)\big)}{\mu\big(C_r(x)\big)} \geq t.$$

Thus there exists $r_x > 0$ such that

$$\frac{\nu\big(C_r(x)\big)}{\mu\big(C_r(x)\big)} \geq t + \varepsilon \quad \text{for all } r \in (0, r_x). \tag{9}$$

Select r_x so small that $C_r(x) \subset U$ for all $r \in (0, r_x)$. Let $\mathcal{B} = \{C_r(x) : x \in A, r \in (0, r_x)\}$. Then $\mathcal{B}$ is a Vitali cover of A as in Definition 6.18 and $\mathcal{B}$ is contained in the open set U. Then by Theorem 6.28 (Vitali Covering Theorem for Radon Outer Measure on $\mathbb{R}^n$), there exists a countable disjoint subcollection of $\mathcal{B}$, denoted by $\{C_i : i \in \mathbb{N}\}$, such that

$$\mu\left(A \setminus \bigcup_{i \in \mathbb{N}} C_i\right) = 0. \tag{10}$$

Now (10) implies according to Observation 1.15 that

$$\mu(A) \leq \mu\left(\bigcup_{i \in \mathbb{N}} C_i\right). \tag{11}$$

Note also that since $C_i \in \mathcal{B}$, we have by (9)

$$\nu(C_i) \geq (t + \varepsilon)\mu(C_i). \tag{12}$$

Thus by (12) and (11) we have

$$\begin{aligned}
\nu(A) + \varepsilon \geq \nu(U) \geq \nu\left(\bigcup_{i \in \mathbb{N}} C_i\right) &= \sum_{i \in \mathbb{N}} \nu(C_i) \\
&\geq (t + \varepsilon) \sum_{i \in \mathbb{N}} \mu(C_i) \\
&= (t + \varepsilon)\mu\left(\bigcup_{i \in \mathbb{N}} C_i\right) \\
&\geq (t + \varepsilon)\mu(A).
\end{aligned} \tag{13}$$

Then the arbitrariness of $\varepsilon > 0$ implies that $\nu(A) \geq t\mu(A)$. This proves (4). ∎

Theorem 7.8. (Derivative of Radon Outer Measure on $\mathbb{R}^n$) *Let μ and ν be two arbitrary Radon outer measures on $\mathbb{R}^n$. Then the derivative $D_\mu \nu$ exists, is $\mathfrak{B}_{\mathbb{R}^n}$-measurable and finite*

$(\mathfrak{B}_{\mathbb{R}^n}, \mu)$*-a.e. on* $\mathbb{R}^n$*. Thus* $D_\mu \nu$ *exists, is* $\mathfrak{M}(\mu)$*-measurable and finite* $(\mathfrak{M}(\mu), \mu)$*-a.e. on* $\mathbb{R}^n$*.*

Proof. 1. Let us show first that for $r \in (0, \infty)$ arbitrarily fixed, derivative $D_\mu \nu$ exists, is $\mathfrak{B}_{\mathbb{R}^n}$-measurable and finite $(\mathfrak{B}_{\mathbb{R}^n}, \mu)$-a.e. on the closed ball $C_r(0)$. Let $\mathbb{Q}_+$ be the set of all nonnegative rational numbers. For $s, t \in \mathbb{Q}_+$ such that $0 \leq s < t$, let

$$(1) \qquad A_{s,t} = \{x \in C_r(0) : \underline{D}_\mu \nu(x) \leq s < t \leq \overline{D}_\mu \nu(x)\}$$

and for $k \in \mathbb{N}$, let

$$(2) \qquad A_k = \{x \in C_r(0) : \overline{D}_\mu \nu(x) \geq k\}.$$

Note that since $C_r(0)$ is a compact set in $\mathbb{R}^n$ and since μ and ν are Radon outer measures on $\mathbb{R}^n$, we have $\mu(C_r(0)) < \infty$ and $\nu(C_r(0)) < \infty$. By Lemma 7.7, we have

$$(3) \qquad t\,\mu(A_{s,t}) \leq \nu(A_{s,t}) \leq s\,\mu(A_{s,t}) \leq s\,\mu(C_r(0)) < \infty,$$

$$(4) \qquad k\,\mu(A_k) \leq \nu(A_k) \leq \nu(C_r(0)) < \infty.$$

Since $s < t$, (3) implies that $\mu(A_{s,t}) = 0$. Now for any two nonnegative extended real numbers α and β, we have $\alpha < \beta$ if and only if there exist $s, t \in \mathbb{Q}_+$ such that $\alpha \leq s < t \leq \beta$. Thus we have

$$\begin{aligned} &\{x \in C_r(0) : \underline{D}_\mu \nu(x) < \overline{D}_\mu \nu(x)\} \\ &= \bigcup_{s \in \mathbb{Q}_+} \bigcup_{t \in \mathbb{Q}_+} \{x \in C_r(0) : \underline{D}_\mu \nu(x) \leq s < t \leq \overline{D}_\mu \nu(x)\} \\ &= \bigcup_{s \in \mathbb{Q}_+} \bigcup_{t \in \mathbb{Q}_+} A_{s,t,r}. \end{aligned}$$

Then by the countable subadditivity of the outer measure μ we have

$$\mu\{x \in C_r(0) : \underline{D}_\mu \nu(x) < \overline{D}_\mu \nu(x)\} = \mu\left(\bigcup_{s \in \mathbb{Q}_+} \bigcup_{t \in \mathbb{Q}_+} A_{s,t}\right) \leq \sum_{s \in \mathbb{Q}_+} \sum_{t \in \mathbb{Q}_+} \mu(A_{s,t}) = 0.$$

Since μ is a Borel regular outer measure on $\mathbb{R}^n$, for every $A \subset \mathbb{R}^n$ there exists $B \in \mathfrak{B}_{\mathbb{R}^n}$ such that $A \subset B$ and $\mu(A) = \mu(B)$. Thus there exists a null set N_1 in the measure space $(\mathbb{R}^n, \mathfrak{B}_{\mathbb{R}^n}, \mu)$ such that

$$(5) \qquad \underline{D}_\mu \nu(x) = \overline{D}_\mu \nu(x) \quad \text{for } x \in N_1^c \cap C_r(0).$$

Now $\{x \in C_r(0) : \overline{D}_\mu \nu(x) = \infty\} \subset A_k$ for every $k \in \mathbb{N}$. Thus by (4) we have

$$\mu\{x \in C_r(0) : \overline{D}_\mu \nu(x) = \infty\} \leq \mu(A_k) \leq \frac{1}{k} \nu(C_r(0))$$

and letting $k \to \infty$ we have

$$\mu\{x \in C_r(0) : \overline{D}_\mu \nu(x) = \infty\} = 0.$$

By the Borel regularity of μ there exists a null set N_1 in measure space $(\mathbb{R}^n, \mathfrak{B}_{\mathbb{R}^n}, \mu)$ such that

$$\overline{D}_\mu \nu(x) < \infty \quad \text{for } x \in N_2^c \cap C_r(0). \tag{6}$$

Let $N = N_1 \cup N_2$. Then N is a null set in $(\mathbb{R}^n, \mathfrak{B}_{\mathbb{R}^n}, \mu)$ and we have

$$\underline{D}_\mu \nu(x) = \overline{D}_\mu \nu(x) < \infty \quad \text{for } x \in N^c \cap C_r(0). \tag{7}$$

This shows that $D_\mu \nu(x)$ exists and is finite for $x \in N^c$. To show the $\mathfrak{B}_{\mathbb{R}^n}$-measurability of $D_\mu \nu$, let $(r_k : k \in \mathbb{N})$ be a sequence of positive numbers such that $\lim_{k \to \infty} r_k = 0$. Then we have

$$D_\mu \nu(x) = \lim_{r \to 0} \frac{\nu(C_r(x))}{\mu(C_r(x))} = \lim_{k \to \infty} \frac{\nu(C_{r_k}(x))}{\mu(C_{r_k}(x))} \quad \text{for } x \in N^c \cap C_r(0).$$

According to Theorem 7.2, $\mu(C_{r_k}(x))$ and $\nu(C_{r_k}(x))$ are $\mathfrak{B}_{\mathbb{R}^n}$-measurable functions of x. Then so is the quotient and the limit of the quotients.

2. We have $\mathbb{R}^n = \bigcup_{k \in \mathbb{N}} C_k(0)$. By our result in **1**, $D_\mu \nu$ exists, is $\mathfrak{B}_{\mathbb{R}^n}$-measurable and finite $(\mathfrak{B}_{\mathbb{R}^n}, \mu)$-a.e. on the closed ball $C_k(0)$ for every $k \in \mathbb{N}$. Then since a union of countably many null sets is again a null set, $D_\mu \nu$ exists, is $\mathfrak{B}_{\mathbb{R}^n}$-measurable and finite $(\mathfrak{B}_{\mathbb{R}^n}, \mu)$-a.e. on $\mathbb{R}^n$.

3. Since μ is a Radon outer measure on $\mathbb{R}^n$, we have $\mathfrak{B}_{\mathbb{R}^n} \subset \mathfrak{M}(\mu)$. Thus $D_\mu \nu$ exists, is $\mathfrak{M}(\mu)$-measurable and finite $(\mathfrak{M}(\mu), \mu)$-a.e. on $\mathbb{R}^n$. ∎

Proposition 7.9. Let μ and ν be two Borel-regular outer measures on a topological space X. Then for every $A \subset X$, there exists a Borel set B such that $A \subset B$, $\mu(A) = \mu(B)$, and $\nu(A) = \nu(B)$.

Proof. If μ and ν are Borel-regular outer measures on a topological space X, then for every $A \subset X$ there exists a Borel set B_1 such that $A \subset B_1$ and $\mu(A) = (B_1)$ and similarly there exists a Borel set B_2 such that $A \subset B_2$ and $\nu(A) = \nu(B_2)$. Let $B = B_1 \cap B_2$. Then B is a Borel set and $A \subset B$. Now $\mu(A) \le \mu(B) \le \mu(B_1) = \mu(A)$ by the monotonicity of the outer measure μ so that $\mu(A) = \mu(B)$ and similarly $\nu(A) \le \nu(B) \le \nu(B_2) = \nu(A)$ so that $\nu(A) = \nu(B)$. ∎

Theorem 7.10. (Integral of the Derivative of a Radon Outer Measure on $\mathbb{R}^n$) *Let μ and ν be Radon outer measures on $\mathbb{R}^n$. Then for every $E \in \mathfrak{M}(\mu)$, we have*

$$\int_E D_\mu \nu(x)\, \mu(dx) \le \nu(E).$$

(Note that E need not be in $\mathfrak{M}(\nu)$ since no integration with respect to ν is involved.)

Proof. By Theorem 7.8, $D_\mu \nu$ is a nonnegative extended real-valued $\mathfrak{B}_{\mathbb{R}^n}$-measurable function on $\mathbb{R}^n$ and hence it is both $\mathfrak{M}(\mu)$ measurable and $\mathfrak{M}(\nu)$ measurable on $\mathbb{R}^n$. Note that μ is a measure on the σ-algebra $\mathfrak{M}(\mu)$ and hence μ is a measure on $\mathfrak{B}_{\mathbb{R}^n}$ and similarly ν is a measure on the σ-algebra $\mathfrak{M}(\nu)$ and hence ν is a measure on $\mathfrak{B}_{\mathbb{R}^n}$. Now let $E \in \mathfrak{M}(\mu)$. Since E is a subset of $\mathbb{R}^n$, by Proposition 7.9 there exists $B \in \mathfrak{B}_{\mathbb{R}^n}$ such that $E \subset B$, $\mu(E) = \mu(B)$ and $\nu(E) = \nu(B)$. Since $E \subset B$ and $\mu(E) = \mu(B)$, we have

$$(1) \qquad \int_E D_\mu \nu(x)\, \mu(dx) = \int_B D_\mu \nu(x)\, \mu(dx).$$

With $t \in (1, \infty)$ arbitrarily fixed, let us decompose B into a disjoint collection in $\mathfrak{B}_{\mathbb{R}^n}$ by setting

$$\begin{aligned}
B_p &= \{x \in B : D_\mu \nu(x) \in [t^p, t^{p+1})\} \quad \text{for } p \in \mathbb{Z}, \\
B_{-\infty} &= \{x \in B : D_\mu \nu(x) = 0\}, \\
B_\infty &= \{x \in B : D_\mu \nu(x) = \infty\}.
\end{aligned}$$

Since $D_\mu \nu(x) < \infty$ for $(\mathfrak{B}_{\mathbb{R}^n}, \mu)$-a.e. $x \in \mathbb{R}^n$ by Theorem 7.8, we have $\mu(B_\infty) = 0$. Thus we have

$$\begin{aligned}
(2) \qquad \int_B D_\mu \nu \, d\mu &= \int_{B_{-\infty}} D_\mu \nu \, d\mu + \sum_{p \in \mathbb{Z}} \int_{B_p} D_\mu \nu \, d\mu + \int_{E_\infty} B_\mu \nu \, d\mu \\
&= \sum_{p \in \mathbb{Z}} \int_{B_p} D_\mu \nu \, d\mu \le \sum_{p \in \mathbb{Z}} t^{p+1} \mu(B_p) \\
&\le \sum_{p \in \mathbb{Z}} t^{p+1} \frac{1}{t^p} \nu(B_p) \quad \text{by (4) of Lemma 7.7} \\
&= t \sum_{p \in \mathbb{Z}} \nu(B_p) \\
&= t\, \nu\Big(\bigcup_{p \in \mathbb{Z}} B_p\Big) \quad \text{since } \nu \text{ is a measure on } \mathfrak{B}_{\mathbb{R}^n} \\
&\le t\, \nu(B) = t\, \nu(E).
\end{aligned}$$

By (1) and (2), we have $\int_E D_\mu \nu(x)\, \mu(dx) \le \nu(E)$. ∎

Theorem 7.11. *Let μ and ν be Radon outer measures on $\mathbb{R}^n$. Then $\nu \ll \mu$ if and only if*

$$(1) \qquad \int_E D_\mu \nu(x)\, \mu(dx) = \nu(E) \quad \textit{for every } E \in \mathfrak{M}(\mu).$$

Proof. 1. Suppose (1) holds. Let $A \subset \mathbb{R}^n$. By Proposition 7.9 there exists $B \in \mathfrak{B}_{\mathbb{R}^n}$ such that $A \subset B$, $\mu(A) = \mu(B)$ and $\nu(A) = \nu(B)$. Now $B \in \mathfrak{B}_{\mathbb{R}^n} \subset \mathfrak{M}(\mu)$ so that by (1) we have

$$(2) \qquad \int_B D_\mu \nu(x)\,\mu(dx) = \nu(B) = \nu(A).$$

Thus if $\mu(A) = 0$, then $\mu(B) = \mu(A) = 0$ and then $\int_B D_\mu \nu(x)\,\mu(dx) = 0$. Then by (2) we have $\nu(A) = 0$. This shows that $\nu \ll \mu$.

2. Conversely suppose $\nu \ll \mu$. Let us show that (1) holds. By Theorem 7.8, for every $E \in \mathfrak{M}(\mu)$ we have

$$(3) \qquad \int_E D_\mu \nu(x)\,\mu(dx) \le \nu(E).$$

Thus it remains to show that $\nu \ll \mu$ implies

$$(4) \qquad \nu(E) \le \int_E D_\mu \nu(x)\,\mu(dx).$$

According to Theorem 7.8, for any two Radon outer measures μ and ν on $\mathbb{R}^n$, we have
1° $D_\mu \nu$ exists, is $\mathfrak{M}(\mu)$-measurable, and $D_\mu \nu < \infty$, $(\mathfrak{M}(\mu), \mu)$-a.e. on $\mathbb{R}^n$,
2° $D_\nu \mu$ exists, is $\mathfrak{M}(\nu)$-measurable, and $D_\nu \mu < \infty$, $(\mathfrak{M}(\nu), \nu)$-a.e. on $\mathbb{R}^n$.
Note also that from Definition 7.3 and Definition 7.4,

$$D_\nu \mu(x) = \frac{1}{D_\mu \nu(x)} \quad \text{whenever both exist.}$$

Let $E \in \mathfrak{M}(\mu)$. With $t \in (1, \infty)$ arbitrarily fixed, let us decompose E into a disjoint collection in $\mathfrak{M}(\mu)$ by setting

$$\begin{aligned} E_p &= \{x \in E : D_\mu \nu(x) \in [t^p, t^{p+1})\} \quad \text{for } p \in \mathbb{Z}, \\ E_{-\infty} &= \{x \in E : D_\mu \nu(x) = 0\}, \\ E_\infty &= \{x \in E : D_\mu \nu(x) = \infty\}. \end{aligned}$$

Since $D_\mu \nu < \infty$, $(\mathfrak{M}(\mu), \mu)$-a.e. on $\mathbb{R}^n$ by 1°, we have $\mu(E_\infty) = 0$. Then $\nu \ll \mu$ implies

$$(5) \qquad \nu(E_\infty) = 0.$$

On $E_{-\infty}$ we have $D_\mu \nu = 0$ and hence

$$D_\nu \mu(x) = \frac{1}{D_\mu \nu(x)} = 0 \quad \text{for } x \in E_{-\infty}.$$

But $D_\nu \mu < \infty$, $(\mathfrak{M}(\nu), \nu)$-a.e. on $\mathbb{R}^n$ by 2°. Thus we have

$$\text{(6)} \qquad \nu(E_{-\infty}) = 0.$$

Now we have

$$\begin{aligned}
\nu(E) &= \nu(E_{-\infty}) + \sum_{p \in \mathbb{Z}} \nu(E_p) + \nu(E_\infty) \\
&= \sum_{p \in \mathbb{Z}} \nu(E_p) \quad \text{by (5) and (6)} \\
&\leq \sum_{p \in \mathbb{Z}} t^{p+1} \mu(E_p) \quad \text{by (3) of Lemma 7.7} \\
&= t \sum_{p \in \mathbb{Z}} t^p \mu(E_p) \leq t \sum_{p \in \mathbb{Z}} \int_{E_p} D_\mu \nu \, d\mu \quad \text{by the definition of } E_p \\
&= t \int_E D_\mu \nu \, d\mu \quad \text{since } D_\mu \nu = 0 \text{ on } E_{-\infty} \text{ and } \mu(E_\infty) = 0.
\end{aligned}$$

Thus letting $t \downarrow 1$, we have $\nu(E) \leq \int_E D_\mu \nu \, d\mu$. This proves (4) and completes the proof. ∎

According to Theorem 7.8, for any two Radon outer measures μ and ν on $\mathbb{R}^n$ the derivative $D_\mu \nu$ exists, is $\mathfrak{M}(\mu)$-measurable and finite $(\mathfrak{M}(\mu), \mu)$-a.e. on $\mathbb{R}^n$. We show next that $\nu \ll \mu$ if and only if $D_\mu \nu$ is finite $(\mathfrak{M}(\nu), \nu)$-a.e. on $\mathbb{R}^n$.

Theorem 7.12. *Let μ and ν be Radon outer measures on $\mathbb{R}^n$. Then $\nu \ll \mu$ if and only if*

$$\text{(1)} \qquad D_\mu \nu(x) < \infty \quad \textit{for } (\mathfrak{M}(\nu), \nu)\textit{-a.e. } x \in \mathbb{R}^n.$$

Proof. By Theorem 7.8, we have

$$\text{(2)} \qquad D_\mu \nu(x) < \infty \quad \text{for } (\mathfrak{M}(\mu), \mu)\text{-a.e. } x \in \mathbb{R}^n.$$

1. Suppose $\nu \ll \mu$. Now according to (2) we have

$$\text{(3)} \qquad \mu\{x \in \mathbb{R}^n : D_\mu \nu(x) = \infty\} = 0.$$

Since $D_\mu \nu$ is $\mathfrak{B}_{\mathbb{R}^n}$-measurable according to Theorem 7.8, we have

$$\{x \in \mathbb{R}^n : D_\mu \nu(x) = \infty\} \in \mathfrak{B}_{\mathbb{R}^n} \subset \mathfrak{M}(\nu).$$

Since $\nu \ll \mu$, (3) implies

$$\nu\{x \in \mathbb{R}^n : D_\mu \nu(x) = \infty\} = 0.$$

Thus we have $D_\mu\nu(x) < \infty$ for $(\mathfrak{B}_{\mathbb{R}^n}, \nu)$-a.e. $x \in \mathbb{R}^n$ and certainly for $(\mathfrak{M}(\nu), \nu)$-a.e. $x \in \mathbb{R}^n$ since $\mathfrak{B}_{\mathbb{R}^n} \subset \mathfrak{M}(\nu)$.

2. Conversely suppose (1) holds. To show that $\nu \ll \mu$ we show that if $A \subset \mathbb{R}^n$ is such that $\mu(A) = 0$ then $\nu(A) = 0$. Now by Proposition 7.9, there exists $B \in \mathfrak{B}_{\mathbb{R}^n}$ such that $A \subset B$, $\mu(A) = \mu(B)$ and $\nu(A) = \nu(B)$. To show $\nu(A) = 0$ we show $\nu(B) = 0$. Now

$$\begin{aligned} B &= \{x \in B : D_\mu\nu(x) < \infty\} \cup \{x \in B : D_\mu\nu(x) = \infty\} \\ &= \bigcup_{k \in \mathbb{N}} \{x \in B : D_\mu\nu(x) \leq k\} \cup \{x \in B : D_\mu\nu(x) = \infty\}. \end{aligned}$$

Since $D_\mu\nu$ is $\mathfrak{B}_{\mathbb{R}^n}$-measurable according to Theorem 7.8, these sets are all $\mathfrak{B}_{\mathbb{R}^n}$-measurable. By our assumption of (1), we have $\nu\{x \in B : D_\mu\nu(x) = \infty\} = 0$. Thus we have

$$\begin{aligned} \nu(B) &= \nu\left(\bigcup_{k \in \mathbb{N}} \{x \in B : D_\mu\nu(x)\right) \leq \sum_{k \in \mathbb{N}} \nu\{x \in B : D_\mu\nu(x) \leq k\} \\ &= \sum_{k \in \mathbb{N}} \nu\{x \in B : \underline{D}_\mu\nu(x) \leq k\} \leq \sum_{k \in \mathbb{N}} k\,\mu\{x \in B : \underline{D}_\mu\nu(x) \leq k\} \\ &= 0 \end{aligned}$$

where the second inequality is by (3) of Lemma 7.7 and the last equality is by the fact that we have $\mu\{x \in B : \underline{D}_\mu\nu(x) \leq k\} \leq \mu(B) = \mu(A) = 0$. ∎

Theorem 7.13. (Lebesgue Decomposition Theorem for Radon Outer Measures on $\mathbb{R}^n$) *Let μ and ν be Radon outer measures on $\mathbb{R}^n$.*

(a) *ν can be decomposed as*

$$\nu = \nu_{ac} + \nu_s,$$

where ν_{ac} and ν_s are Radon outer measures on $\mathbb{R}^n$ such that

$$\nu_{ac} \ll \mu \quad \text{and} \quad \nu_s \perp \mu.$$

(b) *Furthermore we have*

$$D_\mu\nu = D_\mu\nu_{ac} \quad \text{and} \quad D_\mu\nu_s = 0, (\mathfrak{B}_{\mathbb{R}^n}, \mu)\text{-a.e. on } \mathbb{R}^n$$

and consequently, for every $E \in \mathfrak{B}_{\mathbb{R}^n}$ we have

$$\nu(E) = \int_E D_\mu\nu\, d\mu + \nu_s(E).$$

Proof. 1. Let μ and ν be Radon outer measures on $\mathbb{R}^n$. We assume first that $\mu(\mathbb{R}^n) < \infty$ and $\nu(\mathbb{R}^n) < \infty$. Let

(1) $$\mathfrak{E} = \{E \in \mathfrak{B}_{\mathbb{R}^n} : \mu(E^c) = 0\} \subset \mathfrak{B}_{\mathbb{R}^n}.$$

Let us show that there exists $B \in \mathfrak{E}$ such that $\nu(B)$ is the minimum of $\nu(E)$ for all $E \in \mathfrak{E}$, that is, there exists $B \in \mathfrak{E}$ such that

(2) $$\nu(B) = \inf_{E \in \mathfrak{E}} \nu(E) \in [0, \infty).$$

Now for every $k \in \mathbb{N}$ there exists $B_k \in \mathfrak{E}$ such that $\nu(B_k) \leq \inf_{E \in \mathfrak{E}} \nu(E) + \frac{1}{k}$. Then define $B := \bigcap_{k \in \mathbb{N}} B_k$. Since $B_k \in \mathfrak{E} \subset \mathfrak{B}_{\mathbb{R}^n}$, we have $B \in \mathfrak{B}_{\mathbb{R}^n}$. Also $B_k \in \mathfrak{E}$ implies $\mu(B_k^c) = 0$ and then $\mu(B^c) = \mu\big(\big(\bigcap_{k \in \mathbb{N}} B_k\big)^c\big) = \mu\big(\bigcup_{k \in \mathbb{N}} B_k^c\big) \leq \sum_{k \in \mathbb{N}} \mu(B_k^c) = 0$. Thus $B \in \mathfrak{E}$. Now since $B \subset B_k$ for every $k \in \mathbb{N}$, we have $\nu(B) \leq \nu(B_k) \leq \inf_{E \in \mathfrak{E}} \nu(E) + \frac{1}{k}$ for every $k \in \mathbb{N}$. This implies that $\nu(B) \leq \inf_{E \in \mathfrak{E}} \nu(E)$. On the other hand, since $B \in \mathfrak{E}$ we have $\nu(B) \geq \inf_{E \in \mathfrak{E}} \nu(E)$. Therefore we have $\nu(B) = \inf_{E \in \mathfrak{E}} \nu(E)$. This proves (2).

We show next that the set $B \in \mathfrak{E}$ characterized by (2) has the following property:

(3) $$A \subset \mathbb{R}^n, A \subset B, \mu(A) = 0 \Rightarrow \nu(A) = 0.$$

Now since μ and ν are Radon outer measures on $\mathbb{R}^n$, μ and ν are Borel regular outer measures on $\mathbb{R}^n$ by Observation 3.43. Then by Proposition 7.9, there exists $E \in \mathfrak{B}_{\mathbb{R}^n}$ such that $A \subset E$, $\mu(A) = \mu(E)$ and $\nu(A) = \nu(E)$. Consider the set $B \setminus E$. Let us show that $B \setminus E \in \mathfrak{E}$. Now since $B, E \in \mathfrak{B}_{\mathbb{R}^n}$, we have $B \setminus E \in \mathfrak{B}_{\mathbb{R}^n}$. We have also

$$\mu((B \setminus E)^c) = \mu((B \cap E^c)^c) = \mu(B^c \cup E) \leq \mu(B^c) + \mu(E) = 0,$$

where the last equality is from the fact that $\mu(B^c) = 0$ implied by the fact that $B \in \mathfrak{E}$ and by the fact that $\mu(E) = \mu(A) = 0$. Now that $B \setminus E \in \mathfrak{B}_{\mathbb{R}^n}$ and $\mu((B \setminus E)^c) = 0$, we have

$$B \setminus E \in \mathfrak{E}.$$

Now $A \subset E$ and $A \subset B$ so that $A \subset E \cap B$. Then $\nu(E) = \nu(A) \leq \nu(E \cap B)$. On the other hand $\nu(E \cap B) \leq \nu(E)$. Thus $\nu(E) = \nu(E \cap B)$. Then since $B \setminus E = B \setminus (E \cap B)$, we have

$$\nu(B \setminus E) = \nu(B) - \nu(E \cap B) = \nu(B) - \nu(E) = \nu(B) - \nu(A).$$

If $\nu(A) > 0$, then $\nu(B \setminus E) < \nu(B)$. Since $B, B \setminus E \in \mathfrak{E}$, this contradicts (2). Thus $\nu(A) > 0$ leads to a contradiction. Therefore we have $\nu(A) = 0$. This proves (3).

1.1. With the set B defined above, let us define two outer measures on $\mathbb{R}^n$ by letting

(4) $$\nu_{ac} := \nu|_B \quad \text{and} \quad \nu_s := \nu|_{B^c}.$$

Since ν is a Radon outer measure on $\mathbb{R}^n$, ν is a Borel regular outer measure on $\mathbb{R}^n$ by Observation 3.43. Since $B, B^c \in \mathfrak{B}_{\mathbb{R}^n} \subset \mathfrak{M}(\nu)$ and $\nu(B) < \infty$ and $\nu(B^c) < \infty$, the two outer measures on $\mathbb{R}^n$ defined by (4) are Radon outer measures on $\mathbb{R}^n$ by Corollary 3.40. Clearly we have $\nu = \nu_{ac} + \nu_s$.

1.2. Let us show that $\nu_{ac} \ll \mu$, that is, for any $A \subset \mathbb{R}^n$, if $\mu(A) = 0$ then $\nu_{ac}(A) = 0$. Let $A \subset \mathbb{R}^n$ and suppose $\mu(A) = 0$. We have $\nu_{ac}(A) = \nu|_B(A) = \nu(A \cap B)$. Now

$A \cap B \subset A$ and $\mu(A \cap B) \le \mu(A) = 0$. Then since $A \cap B \subset B$ and $\mu(A \cap B) = 0$, we have $\nu(A \cap B) = 0$ by (3). Thus we have $\nu_{ac}(A) = 0$.

1.3. Let us show that $\nu_s \perp \mu$. Now for our $B \in \mathfrak{B}_{\mathbb{R}^n}$, we have

$$\nu_s(B) = \nu|_{B^c}(B) = \nu(B \cap B^c) = \nu(\emptyset) = 0.$$

On the other hand since $B \in \mathfrak{E}$ we have $\mu(B^c) = 0$. This shows that $\nu_s \perp \mu$.

1.4. Let us prove (b). For $\alpha > 0$, let

$$C_\alpha = \{x \in B : D_\mu \nu_s(x) \ge \alpha\}.$$

Since $C \subset B$ we have $\nu_s(C_\alpha) = \nu|_{B^c}(C_\alpha) = \nu(C_\alpha \cap B^c) = \nu(\emptyset) = 0$. By (4) of Lemma 7.7, we have $\alpha\, \mu(C_\alpha) \le \nu_s(C_\alpha) = 0$ and hence $\mu(C_\alpha) = 0$ for every $\alpha > 0$. Thus we have

$$D_\mu \nu_s = 0,\ (\mathfrak{B}_{\mathbb{R}^n}, \mu)\text{-a.e. on } \mathbb{R}^n. \tag{5}$$

Since $\nu = \nu_{ac} + \nu_s$ we have $D_\mu \nu = D_\mu \nu_{ac} + D_\mu \nu_s$. Then by (5) we have

$$D_\mu \nu = D_\mu \nu_{ac},\ (\mathfrak{B}_{\mathbb{R}^n}, \mu)\text{-a.e. on } \mathbb{R}^n. \tag{6}$$

Then for every $E \in \mathfrak{B}_{\mathbb{R}^n}$, we have

$$\nu(E) = \nu_{ac}(E) + \nu_s(E) = \int_E D_\mu \nu \, d\mu + \nu_s(E). \tag{7}$$

This completes the proof of the theorem for the case that $\mu(\mathbb{R}^n) < \infty$ and $\nu(\mathbb{R}^n) < \infty$.

2. Let us remove the assumption that $\mu(\mathbb{R}^n) < \infty$ and $\nu(\mathbb{R}^n) < \infty$. For $j \in \mathbb{N}$, let $K_j = \{x \in \mathbb{R}^n : d(x, 0) \le j\}$. Then K_j is a compact set in $\mathbb{R}^n$. Since μ and ν are Radon outer measures on $\mathbb{R}^n$ we have $\mu(K_j) < \infty$ and $\nu(K_j) < \infty$. Now $(K_j : j \in \mathbb{N})$ is an increasing sequence in $\mathfrak{B}_{\mathbb{R}^n}$ and $\mathbb{R}^n = \bigcup_{j \in \mathbb{N}} K_j$. Let $R_1 = K_1$ and $R_j = K_j \setminus K_{j-1}$ for $j \ge 2$. Then $(R_j : j \in \mathbb{N})$ is a disjoint sequence in $\mathfrak{B}_{\mathbb{R}^n}$ with $\bigcup_{j \in \mathbb{N}} R_j = \mathbb{R}^n$ and moreover $\mu(R_j) \le \mu(K_j) < \infty$ and $\nu(R_j) \le \nu(K_j) < \infty$ for $j \in \mathbb{N}$. Let

$$\mu_j = \mu|_{R_j} \quad \text{and} \quad \nu_j = \nu|_{R_j}. \tag{8}$$

By Corollary 3.40, μ_j and ν_j are Radon outer measures on $\mathbb{R}^n$ and moreover $\mu_j(\mathbb{R}^n) < \infty$ and $\nu_j(\mathbb{R}^n) < \infty$. We have also

$$\mu = \sum_{j \in \mathbb{N}} \mu_j \quad \text{and} \quad \nu = \sum_{j \in \mathbb{N}} \nu_j. \tag{9}$$

For each $j \in \mathbb{N}$, since μ_j and ν_j are finite Radon outer measures on $\mathbb{R}^n$, by our results in **1**, ν_j can be decomposed as

$$\nu_j = \nu_{j,ac} + \nu_{j,s},$$

where $\nu_{j,ac}$ and $\nu_{j,s}$ are Radon outer measures on $\mathbb{R}^n$ such that

$$\nu_{j,ac} \ll \mu_j \quad \text{and} \quad \nu_{j,s} \perp \mu_j.$$

Let

$$\nu_{ac} = \sum_{j\in\mathbb{N}} \nu_{j,ac} \quad\text{and}\quad \nu_s = \sum_{j\in\mathbb{N}} \nu_{j,s}.$$

We have

$$\nu = \sum_{j\in\mathbb{N}} \nu_j = \sum_{j\in\mathbb{N}} \{\nu_{j,ac} + \nu_{j,s}\} = \sum_{j\in\mathbb{N}} \nu_{j,ac} + \sum_{j\in\mathbb{N}} \nu_{j,s} = \nu_{ac} + \nu_s.$$

To show that $\nu_{ac} \ll \mu$, suppose $A \subset \mathbb{R}^n$ and $\mu(A) = 0$. Since $\mu(A) = \sum_{j\in\mathbb{N}} \mu_j(A)$, we have $\mu_j(A) = 0$ for every $j \in \mathbb{N}$. Since $\nu_{j,ac} \ll \mu_j$, $\mu_j(A) = 0$ implies $\nu_{j,ac}(A) = 0$. Then $\nu_{ac}(A) = \sum_{j\in\mathbb{N}} \nu_{j,ac}(A) = 0$. This shows that $\nu_{ac} \ll \mu$.

Let us show that $\nu_s \perp \mu$. Since $\nu_{j,s} \perp \mu_j$, R_j can be given as the union of two disjoint members of $\mathfrak{B}_{\mathbb{R}^n}$, B_j and C_j, such that $\mu_j(C_j) = 0$ and $\nu_{j,s}(B_j) = 0$. Let $B = \bigcup_{j\in\mathbb{N}} B_j \in \mathfrak{B}_{\mathbb{R}^n}$ and $C = \bigcup_{j\in\mathbb{N}} C_j \in \mathfrak{B}_{\mathbb{R}^n}$. Since $\{R_j : j \in \mathbb{N}\}$ is a disjoint collection, B and C are disjoint. Then we have $\mu(C) = \sum_{j\in\mathbb{N}} \mu_j(C_j) = 0$ and $\nu_s(B) = \sum_{j\in\mathbb{N}} \nu_{j,s}(B_j) = 0$. This shows that $\nu_s \perp \mu$. ∎

[IV] Derivatives of Borel Regular Outer Measures on a Metric Space

Commentary. Here we consider the class of boundedly finite Borel regular outer measures μ on a metric space (X, d) satisfying the following conditions:
1° μ satisfies the Vitali condition.
2° $\mu(A) = \inf\{\mu(V) : A \subset V, V \text{ is open}\}$ for every $A \subset X$.
This class includes Radon outer measures on $\mathbb{R}^n$. This can be shown as follows. To start with, a Radon outer measure μ on $\mathbb{R}^n$ is a Borel regular outer measure on $\mathbb{R}^n$ by Observation 3.43. Condition 2° is a defining condition for a Radon outer measure by Definition 3.37. If $E \subset \mathbb{R}^n$ is a bounded set then its closure $\overline{E}$ is a bounded closed set in $\mathbb{R}^n$ so that $\mu(\overline{E}) < \infty$ by Definition 3.37. Thus $\mu(E) \le \mu(\overline{E}) < \infty$. This shows that μ is boundedly finite. Finally by Theorem 6.28 (Vitali Covering Theorem for Radon Outer Measures on $\mathbb{R}^n$), μ satisfies condition 1°.
Therefore our results in [III] for Radon outer measures on $\mathbb{R}^n$ are all particular cases of what we prove in [IV].

Lemma 7.14. *Let μ and ν be Borel outer measures on a metric space (X, d) satisfying the following conditions:*
1° *μ satisfies the Vitali condition.*
2° *$\mu(A) = \inf\{\mu(V) : A \subset V, V \text{ is open}\}$ for every $A \subset X$.*
3° *$\nu(A) = \inf\{\nu(V) : A \subset V, V \text{ is open}\}$ for every $A \subset X$.*
For $t \in (0, \infty)$ arbitrarily fixed, let

(1) $$A' = \{x \in X : \underline{D}_\mu \nu(x) \le t\}.$$

(2) $$A'' = \{x \in X : \overline{D}_\mu \nu(x) \ge t\}.$$

Let A be an arbitrary subset of X, measurable or not. Then

$$A \subset A' \Rightarrow \nu(A) \leq t\,\mu(A). \tag{3}$$

$$A \subset A'' \Rightarrow \nu(A) \geq t\,\mu(A). \tag{4}$$

Proof. 1. Let us prove (3). Let A be an arbitrary subset of A'. Let $\varepsilon > 0$. Since μ satisfies condition 2°, there exists an open set U such that $A \subset U$ and $\mu(U) \leq \mu(A) + \varepsilon$. Let $x \in A \subset A'$. Then $\underline{D}_\mu \nu(x) \leq t$, that is,

$$\liminf_{r \to 0} \frac{\nu(C_r(x))}{\mu(C_r(x))} \leq t.$$

Thus there exists $r_x > 0$ such that

$$\frac{\nu(C_r(x))}{\mu(C_r(x))} \leq t + \varepsilon \quad \text{for all } r \in (0, r_x). \tag{5}$$

Select r_x so small that $C_r(x) \subset U$ for all $r \in (0, r_x)$. Let $\mathcal{B} = \{C_r(x) : x \in A, r \in (0, r_x)\}$. Then for every $a \in A$ we have $\inf\{r > 0 : C(a,r) \in \mathcal{B}\} = 0$. Then since μ satisfies condition 1°, there exists a countable disjoint subcollection of $\mathcal{B}$, denoted by $\{C_i : i \in \mathbb{N}\}$, such that

$$\nu\left(A \setminus \bigcup_{i \in \mathbb{N}} C_i\right) = 0. \tag{6}$$

Note that since $C_i \in \mathcal{B}$ we have by (5)

$$\nu(C_i) \leq (t + \varepsilon)\mu(C_i). \tag{7}$$

By Observation 1.15, (6) implies

$$\begin{aligned} \nu(A) \leq \nu\left(\bigcup_{i \in \mathbb{N}} C_i\right) &= \sum_{i \in \mathbb{N}} \nu(C_i) \leq (t + \varepsilon) \sum_{i \in \mathbb{N}} \mu(C_i) \\ &= (t + \varepsilon)\mu\left(\bigcup_{i \in \mathbb{N}} C_i\right) \leq (t + \varepsilon)\mu(U) \\ &\leq (t + \varepsilon)\{\mu(A) + \varepsilon\}. \end{aligned} \tag{8}$$

Since this holds for every $\varepsilon > 0$, we have $\nu(A) \leq t\,\mu(A)$. This proves (3).

2. Let us prove (4). Let $A \subset A''$. Let $\varepsilon > 0$ be arbitrarily given. By 2° and 3°, there exists an open set Uin X such that $A \subset U$ and $\mu(U) \leq \mu(A) = \varepsilon$ and $\nu(U) \leq \nu(A) = \varepsilon$ also. Let $x \in A \subset A''$. Then $\overline{D}_\mu \nu(x) \geq t$, that is,

$$\limsup_{r \to 0} \frac{\nu(C_r(x))}{\mu(C_r(x))} \geq t.$$

Thus there exists $r_x > 0$ such that

$$\frac{\nu(C_r(x))}{\mu(C_r(x))} \geq t + \varepsilon \quad \text{for all } r \in (0, r_x). \tag{9}$$

Select r_x so small that $C_r(x) \subset U$ for all $r \in (0, r_x)$. Let $\mathcal{B} = \{C_r(x) : x \in A, r \in (0, r_x)\}$. Then for every $a \in A$ we have $\inf\{r > 0 : C(a, r) \in \mathcal{B}\} = 0$. Then since μ satisfies condition $1°$, there exists a countable disjoint subcollection of $\mathcal{B}$, denoted by $\{C_i : i \in \mathbb{N}\}$, such that

$$\mu\left(A \setminus \bigcup_{i \in \mathbb{N}} C_i\right) = 0. \tag{10}$$

Now (10) implies according to Observation 1.15 that

$$\mu(A) \leq \mu\left(\bigcup_{i \in \mathbb{N}} C_i\right). \tag{11}$$

Note also that since $C_i \in \mathcal{B}$, we have by (9)

$$\nu(C_i) \geq (t + \varepsilon)\mu(C_i). \tag{12}$$

Thus by (12) and (11) we have

$$\begin{aligned} \nu(A) + \varepsilon \geq \nu(U) \geq \nu\left(\bigcup_{i \in \mathbb{N}} C_i\right) &= \sum_{i \in \mathbb{N}} \nu(C_i) \\ &\geq (t + \varepsilon) \sum_{i \in \mathbb{N}} \mu(C_i) \\ &= (t + \varepsilon)\mu\left(\bigcup_{i \in \mathbb{N}} C_i\right) \\ &\geq (t + \varepsilon)\mu(A). \end{aligned} \tag{13}$$

Then the arbitrariness of $\varepsilon > 0$ implies that $\nu(A) \geq t\mu(A)$. This proves (4). ∎

Theorem 7.15. (Derivative of Borel Regular Outer Measures on a Metric Space) *Let μ and ν be two boundedly finite Borel regular outer measures on a metric space (X, d) satisfying the following conditions:*

$1°$ *μ satisfies the Vitali condition.*

$2°$ *$\mu(A) = \inf\{\mu(V) : A \subset V, V \text{ is open}\}$ for every $A \subset X$.*

$3°$ *$\nu(A) = \inf\{\nu(V) : A \subset V, V \text{ is open}\}$ for every $A \subset X$.*

Then the derivative $D_\mu \nu$ exists, is $\mathfrak{B}_X$-measurable and finite $(\mathfrak{B}_X, \mu)$-a.e. on X. Thus $D_\mu \nu$ exists, is $\mathfrak{M}(\mu)$-measurable and finite $(\mathfrak{M}(\mu), \mu)$-a.e. on X.

Proof. 1. Let us show first that for $r \in (0, \infty)$ arbitrarily fixed, derivative $D_\mu \nu$ exists, is $\mathfrak{B}_X$-measurable and finite $(\mathfrak{B}_X, \mu)$-a.e. on the closed ball $C_r(x_0)$ where $x_0 \in X$ is arbitrarily chosen. Let $\mathbb{Q}_+$ be the set of all nonnegative rational numbers. For $s, t \in \mathbb{Q}_+$ such that $0 \le s < t$, let

(1) $$A_{s,t} = \{x \in C_r(x_0) : \underline{D}_\mu \nu(x) \le s < t \le \overline{D}_\mu \nu(x)\}$$

and for $k \in \mathbb{N}$, let

(2) $$A_k = \{x \in C_r(x_0) : \overline{D}_\mu \nu(x) \ge k\}.$$

Since μ and ν are both boundedly finite Borel regular outer measures on X, we have $\mu(C_r(x_0)) < \infty$ and $\nu(C_r(x_0)) < \infty$. By Lemma 7.14, we have

(3) $$t\,\mu(A_{s,t}) \le \nu(A_{s,t}) \le s\,\mu(A_{s,t}) \le s\,\mu(C_r(x_0)) < \infty,$$

(4) $$k\,\mu(A_k) \le \nu(A_k) \le \nu(C_r(x_0)) < \infty.$$

Since $s < t$, (3) implies that $\mu(A_{s,t}) = 0$. Now for any two nonnegative extended real numbers α and β, we have $\alpha < \beta$ if and only if there exist $s, t \in \mathbb{Q}_+$ such that $\alpha \le s < t \le \beta$. Thus we have

$$\begin{aligned}
&\{x \in C_r(x_0) : \underline{D}_\mu \nu(x) < \overline{D}_\mu \nu(x)\} \\
&= \bigcup_{s \in \mathbb{Q}_+} \bigcup_{t \in \mathbb{Q}_+} \{x \in C_r(x_0) : \underline{D}_\mu \nu(x) \le s < t \le \overline{D}_\mu \nu(x)\} \\
&= \bigcup_{s \in \mathbb{Q}_+} \bigcup_{t \in \mathbb{Q}_+} A_{s,t,r}.
\end{aligned}$$

Then by the countable subadditivity of the outer measure μ we have

$$\mu\{x \in C_r(x_0) : \underline{D}_\mu \nu(x) < \overline{D}_\mu \nu(x)\} = \mu\left(\bigcup_{s \in \mathbb{Q}_+} \bigcup_{t \in \mathbb{Q}_+} A_{s,t}\right) \le \sum_{s \in \mathbb{Q}_+} \sum_{t \in \mathbb{Q}_+} \mu(A_{s,t}) = 0.$$

Since μ is a Borel regular outer measure on X, for every $A \subset X$ there exists $B \in \mathfrak{B}_X$ such that $A \subset B$ and $\mu(A) = \mu(B)$. Thus there exists a null set N_1 in the measure space $(X, \mathfrak{B}_X, \mu)$ such that

(5) $$\underline{D}_\mu \nu(x) = \overline{D}_\mu \nu(x) \quad \text{for } x \in N_1^c \cap C_r(x_0).$$

Now $\{x \in C_r(x_0) : \overline{D}_\mu \nu(x) = \infty\} \subset A_k$ for every $k \in \mathbb{N}$. Thus by (4) we have

$$\mu\{x \in C_r(x_0) : \overline{D}_\mu \nu(x) = \infty\} \le \mu(A_k) \le \frac{1}{k}\nu(C_r(x_0))$$

and letting $k \to \infty$ we have

$$\mu\{x \in C_r(x_0) : \overline{D}_\mu \nu(x) = \infty\} = 0.$$

By the Borel regularity of μ there exists a null set N_1 in measure space $(X, \mathfrak{B}_X, \mu)$ such that

$$\overline{D}_\mu \nu(x) < \infty \quad \text{for } x \in N_2^c \cap C_r(x_0). \tag{6}$$

Let $N = N_1 \cup N_2$. Then N is a null set in $(X, \mathfrak{B}_X, \mu)$ and we have

$$\underline{D}_\mu \nu(x) = \overline{D}_\mu \nu(x) < \infty \quad \text{for } x \in N^c \cap C_r(x_0). \tag{7}$$

This shows that $D_\mu \nu(x)$ exists and is finite for $x \in N^c$. To show the $\mathfrak{B}_X$-measurability of $D_\mu \nu$, let $(r_k : k \in \mathbb{N})$ be a sequence of positive numbers such that $\lim_{k\to\infty} r_k = 0$. Then we have

$$D_\mu \nu(x) = \lim_{r\to 0} \frac{\nu(C_r(x))}{\mu(C_r(x))} = \lim_{k\to\infty} \frac{\nu(C_{r_k}(x))}{\mu(C_{r_k}(x))} \quad \text{for } x \in N^c \cap C_r(x_0).$$

According to Proposition 7.2, $\mu(C_{r_k}(x))$ and $\mu(C_{r_k}(x))$ are $\mathfrak{B}_X$-measurable functions of x. Then so is the quotient and the limit of the quotients.

2. We have $X = \bigcup_{k\in\mathbb{N}} C_k(0)$. By our result in **1**, $D_\mu \nu$ exists, is $\mathfrak{B}_X$-measurable and finite $(\mathfrak{B}_X, \mu)$-a.e. on the closed ball $C_k(0)$ for every $k \in \mathbb{N}$. Then since a union of countably many null sets is again a null set, $D_\mu \nu$ exists, is $\mathfrak{B}_X$-measurable and finite $(\mathfrak{B}_X, \mu)$-a.e. on X.

3. Since μ is a Borel regular outer measure on X, we have $\mathfrak{B}_X \subset \mathfrak{M}(\mu)$. Thus $D_\mu \nu$ exists, is $\mathfrak{M}(\mu)$-measurable and finite $(\mathfrak{M}(\mu), \mu)$-a.e. on X. ∎

Theorem 7.16. (Integral of the Derivative of a Borel Regular Outer Measure on a Metric Space) *Let μ and ν be two boundedly finite Borel regular outer measures on a metric space (X, d) satisfying the following conditions:*
1° *μ satisfies the Vitali condition.*
2° *$\mu(A) = \inf\{\mu(V) : A \subset V, V$ is open$\}$ for every $A \subset X$.*
3° *$\nu(A) = \inf\{\nu(V) : A \subset V, V$ is open$\}$ for every $A \subset X$.*
Then for every $E \in \mathfrak{M}(\mu)$, we have

$$\int_E D_\mu \nu(x)\, \mu(dx) \le \nu(E).$$

(Note that E need not be in $\mathfrak{M}(\nu)$ since no integration with respect to ν is involved.)

Proof. By Theorem 7.15, $D_\mu \nu$ is a nonnegative extended real-valued $\mathfrak{B}_X$-measurable function on X and hence it is both $\mathfrak{M}(\mu)$ measurable and $\mathfrak{M}(\nu)$ measurable on X. Note that μ is a measure on the σ-algebra $\mathfrak{M}(\mu)$ and hence μ is a measure on $\mathfrak{B}_X$ and similarly ν is a measure on the σ-algebra $\mathfrak{M}(\nu)$ and hence ν is a measure on $\mathfrak{B}_X$. Now let $E \in \mathfrak{M}(\mu)$.

Since E is a subset of X, by Proposition 7.9 there exists $B \in \mathfrak{B}_X$ such that $E \subset B$, $\mu(E) = \mu(B)$ and $\nu(E) = \nu(B)$. Since $E \subset B$ and $\mu(E) = \mu(B)$, we have

$$\text{(1)} \qquad \int_E D_\mu \nu(x)\, \mu(dx) = \int_B D_\mu \nu(x)\, \mu(dx).$$

With $t \in (1, \infty)$ arbitrarily fixed, let us decompose B into a disjoint collection in $\mathfrak{B}_X$ by setting

$$B_p = \{x \in B : D_\mu \nu(x) \in [t^p, t^{p+1})\} \quad \text{for } p \in \mathbb{Z},$$
$$B_{-\infty} = \{x \in B : D_\mu \nu(x) = 0\},$$
$$B_\infty = \{x \in B : D_\mu \nu(x) = \infty\}.$$

Since $D_\mu \nu(x) < \infty$ for $(\mathfrak{B}_X, \mu)$-a.e. $x \in X$ by Theorem 7.15, we have $\mu(B_\infty) = 0$. Thus we have

$$\begin{aligned}
\text{(2)} \qquad \int_B D_\mu \nu\, d\mu &= \int_{B_{-\infty}} D_\mu \nu\, d\mu + \sum_{p \in \mathbb{Z}} \int_{B_p} D_\mu \nu\, d\mu + \int_{E_\infty} B_\mu \nu\, d\mu \\
&= \sum_{p \in \mathbb{Z}} \int_{B_p} D_\mu \nu\, d\mu \le \sum_{p \in \mathbb{Z}} t^{p+1} \mu(B_p) \\
&\le \sum_{p \in \mathbb{Z}} t^{p+1} \frac{1}{t^p} \nu(B_p) \quad \text{by (4) of Lemma 7.14} \\
&= t \sum_{p \in \mathbb{Z}} \nu(B_p) \\
&= t\, \nu\Big(\bigcup_{p \in \mathbb{Z}} B_p\Big) \quad \text{since } \nu \text{ is a measure on } \mathfrak{B}_X \\
&\le t\, \nu(B) = t\, \nu(E).
\end{aligned}$$

By (1) and (2), we have $\int_E D_\mu \nu(x)\, \mu(dx) \le \nu(E)$. ∎

Theorem 7.17. *Let μ and ν be two boundedly finite Borel regular outer measures on a metric space (X, d) satisfying the following conditions:*
1° μ satisfies the Vitali condition.
2° $\mu(A) = \inf\{\mu(V) : A \subset V, V \text{ is open}\}$ for every $A \subset X$.
3° $\nu(A) = \inf\{\nu(V) : A \subset V, V \text{ is open}\}$ for every $A \subset X$.
Then $\nu \ll \mu$ if and only if

$$\text{(1)} \qquad \int_E D_\mu \nu(x)\, \mu(dx) = \nu(E) \quad \text{for every } E \in \mathfrak{M}(\mu).$$

Proof. 1. Suppose (1) holds. Let $A \subset X$. By Proposition 7.9 there exists $B \in \mathfrak{B}_X$ such that $A \subset B$, $\mu(A) = \mu(B)$ and $\nu(A) = \nu(B)$. Now $B \in \mathfrak{B}_X \subset \mathfrak{M}(\mu)$ so that by (1) we have

$$\int_B D_\mu\nu(x)\,\mu(dx) = \nu(B) = \nu(A). \tag{2}$$

Thus if $\mu(A) = 0$, then $\mu(B) = \mu(A) = 0$ and then $\int_B D_\mu\nu(x)\,\mu(dx) = 0$. Then by (2) we have $\nu(A) = 0$. This shows that $\nu \ll \mu$.

2. Conversely suppose $\nu \ll \mu$. Let us show that (1) holds. By Theorem 7.8, for every $E \in \mathfrak{M}(\mu)$ we have

$$\int_E D_\mu\nu(x)\,\mu(dx) \le \nu(E). \tag{3}$$

Thus it remains to show that $\nu \ll \mu$ implies

$$\nu(E) \le \int_E D_\mu\nu(x)\,\mu(dx). \tag{4}$$

According to Theorem 7.15, for any two boundedly finite Borel regular outer measures on X satisfying conditions 1°, 2° and 3°, we have
(a) $D_\mu\nu$ exists, is $\mathfrak{M}(\mu)$-measurable, and $D_\mu\nu < \infty$, $(\mathfrak{M}(\mu), \mu)$-a.e. on X,
(b) $D_\nu\mu$ exists, is $\mathfrak{M}(\nu)$-measurable, and $D_\nu\mu < \infty$, $(\mathfrak{M}(\nu), \nu)$-a.e. on X.
Note also that from Definition 7.3 and Definition 7.4,

$$D_\nu\mu(x) = \frac{1}{D_\mu\nu(x)} \quad \text{whenever both exist.}$$

Let $E \in \mathfrak{M}(\mu)$. With $t \in (1, \infty)$ arbitrarily fixed, let us decompose E into a disjoint collection in $\mathfrak{M}(\mu)$ by setting

$$\begin{aligned} E_p &= \{x \in E : D_\mu\nu(x) \in [t^p, t^{p+1})\} \quad \text{for } p \in \mathbb{Z}, \\ E_{-\infty} &= \{x \in E : D_\mu\nu(x) = 0\}, \\ E_\infty &= \{x \in E : D_\mu\nu(x) = \infty\}. \end{aligned}$$

Since $D_\mu\nu < \infty$, $(\mathfrak{M}(\mu), \mu)$-a.e. on X by (a), we have $\mu(E_\infty) = 0$. Then $\nu \ll \mu$ implies

$$\nu(E_\infty) = 0. \tag{5}$$

On $E_{-\infty}$ we have $D_\mu\nu = 0$ and hence

$$D_\nu\mu(x) = \frac{1}{D_\mu\nu(x)} = 0 \quad \text{for } x \in E_{-\infty}.$$

But $D_\nu\mu < \infty$, $(\mathfrak{M}(\nu), \nu)$-a.e. on X by (b). Thus we have

(6) $$\nu(E_{-\infty}) = 0.$$

Now we have

$$\begin{aligned}
\nu(E) &= \nu(E_{-\infty}) + \sum_{p\in\mathbb{Z}} \nu(E_p) + \nu(E_\infty) \\
&= \sum_{p\in\mathbb{Z}} \nu(E_p) \quad \text{by (5) and (6)} \\
&\le \sum_{p\in\mathbb{Z}} t^{p+1}\mu(E_p) \quad \text{by (3) of Lemma 7.14} \\
&= t\sum_{p\in\mathbb{Z}} t^p\mu(E_p) \le t\sum_{p\in\mathbb{Z}} \int_{E_p} D_\mu\nu\, d\mu \quad \text{by the definition of } E_p \\
&= t\int_E D_\mu\nu\, d\mu \quad \text{since } D_\mu\nu = 0 \text{ on } E_{-\infty} \text{ and } \mu(E_\infty) = 0.
\end{aligned}$$

Thus letting $t \downarrow 1$, we have $\nu(E) \le \int_E D_\mu\nu\, d\mu$. This proves (4) and completes the proof. ∎

According to Theorem 7.15, for any boundedly finite Borel regular outer measures μ and ν on a metric space (X, d) satisfying conditions 1°, 2° and 3°, the derivative $D_\mu\nu$ exists, is $\mathfrak{M}(\mu)$-measurable and finite $(\mathfrak{M}(\mu), \mu)$-a.e. on X. We show next that $\nu \ll \mu$ if and only if $D_\mu\nu$ is finite $(\mathfrak{M}(\nu), \nu)$-a.e. on X.

Theorem 7.18. *Let μ and ν be two boundedly finite Borel regular outer measures on a metric space (X, d) satisfying the following conditions:*
1° *μ satisfies the Vitali condition.*
2° $\mu(A) = \inf\{\mu(V) : A \subset V, V$ *is open*$\}$ *for every* $A \subset X$.
3° $\nu(A) = \inf\{\nu(V) : A \subset V, V$ *is open*$\}$ *for every* $A \subset X$.
Then $\nu \ll \mu$ if and only if

(1) $$D_\mu\nu(x) < \infty \quad \textit{for } (\mathfrak{M}(\nu), \nu)\textit{-a.e. } x \in X.$$

Proof. By Theorem 7.15, we have

(2) $$D_\mu\nu(x) < \infty \quad \text{for } (\mathfrak{M}(\mu), \mu)\text{-a.e. } x \in X.$$

1. Suppose $\nu \ll \mu$. Now according to (2) we have

(3) $$\mu\{x \in X : D_\mu\nu(x) = \infty\} = 0.$$

Since $D_\mu \nu$ is $\mathfrak{B}_X$-measurable according to Theorem 7.15, we have

$$\{x \in X : D_\mu \nu(x) = \infty\} \in \mathfrak{B}_X \subset \mathfrak{M}(\nu).$$

Since $\nu \ll \mu$, (3) implies

$$\nu\{x \in X : D_\mu \nu(x) = \infty\} = 0.$$

Thus we have $D_\mu \nu(x) < \infty$ for $(\mathfrak{B}_X, \nu)$-a.e. $x \in X$ and certainly for $(\mathfrak{M}(\nu), \nu)$-a.e. $x \in X$ since $\mathfrak{B}_X \subset \mathfrak{M}(\nu)$.

2. Conversely suppose (1) holds. To show that $\nu \ll \mu$ we show that if $A \subset X$ is such that $\mu(A) = 0$ then $\nu(A) = 0$. Now by Proposition 7.9, there exists $B \in \mathfrak{B}_X$ such that $A \subset B$, $\mu(A) = \mu(B)$ and $\nu(A) = \nu(B)$. To show $\nu(A) = 0$ we show $\nu(B) = 0$. Now

$$\begin{aligned} B &= \{x \in B : D_\mu \nu(x) < \infty\} \cup \{x \in B : D_\mu \nu(x) = \infty\} \\ &= \bigcup_{k \in \mathbb{N}} \{x \in B : D_\mu \nu(x) \leq k\} \cup \{x \in B : D_\mu \nu(x) = \infty\}. \end{aligned}$$

Since $D_\mu \nu$ is $\mathfrak{B}_X$-measurable according to Theorem 7.15, these sets are all $\mathfrak{B}_X$-measurable. By our assumption of (1), we have $\nu\{x \in B : D_\mu \nu(x) = \infty\} = 0$. Thus we have

$$\begin{aligned} \nu(B) &= \nu\left(\bigcup_{k \in \mathbb{N}} \{x \in B : D_\mu \nu(x)\right) \leq \sum_{k \in \mathbb{N}} \nu\{x \in B : D_\mu \nu(x) \leq k\} \\ &= \sum_{k \in \mathbb{N}} \nu\{x \in B : \underline{D}_\mu \nu(x) \leq k\} \leq \sum_{k \in \mathbb{N}} k\, \mu\{x \in B : \underline{D}_\mu \nu(x) \leq k\} \\ &= 0 \end{aligned}$$

where the second inequality is by (3) of Lemma 7.14 and the last equality is by the fact that we have $\mu\{x \in B : \underline{D}_\mu \nu(x) \leq k\} \leq \mu(B) = \mu(A) = 0$. ∎

Theorem 7.19. (Lebesgue Decomposition Theorem for Borel Regular Outer Measures on a Metric Space) *Let μ and ν be two boundedly finite Borel regular outer measures on a metric space (X, d) satisfying the following conditions:*

1° *μ satisfies the Vitali condition.*
2° *$\mu(A) = \inf\{\mu(V) : A \subset V, V$ is open$\}$ for every $A \subset X$.*
3° *$\nu(A) = \inf\{\nu(V) : A \subset V, V$ is open$\}$ for every $A \subset X$.*

(a) *ν can be decomposed as*

$$\nu = \nu_{ac} + \nu_s,$$

where ν_{ac} and ν_s are Borel regular outer measures on X such that

$$\nu_{ac} \ll \mu \quad \text{and} \quad \nu_s \perp \mu.$$

(b) *Furthermore we have*

$$D_\mu \nu = D_\mu \nu_{ac} \quad \text{and} \quad D_\mu \nu_s = 0, \mu \text{ a.e.}$$

and consequently, for every $E \in \mathfrak{B}_X$ we have

$$\nu(E) = \int_E D_\mu \nu \, d\mu + \nu_s(E).$$

Proof. 1. Let μ and ν be Radon outer measures on X. We assume first that $\mu(X) < \infty$ and $\nu(X) < \infty$. Let

$$\mathfrak{E} = \{E \in \mathfrak{B}_X : \mu(E^c) = 0\} \subset \mathfrak{B}_X. \tag{1}$$

Let us show that there exists $B \in \mathfrak{E}$ such that $\nu(B)$ is the minimum of $\nu(E)$ for all $E \in \mathfrak{E}$, that is, there exists $B \in \mathfrak{E}$ such that

$$\nu(B) = \inf_{E \in \mathfrak{E}} \nu(E) \in [0, \infty). \tag{2}$$

Now for every $k \in \mathbb{N}$ there exists $B_k \in \mathfrak{E}$ such that $\nu(B_k) \leq \inf_{E \in \mathfrak{E}} \nu(E) + \frac{1}{k}$. Then define $B := \bigcap_{k \in \mathbb{N}} B_k$. Since $B_k \in \mathfrak{E} \subset \mathfrak{B}_X$, we have $B \in \mathfrak{B}_X$. Also $B_k \in \mathfrak{E}$ implies $\mu(B_k^c) = 0$ and then $\mu(B^c) = \mu\big(\big(\bigcap_{k \in \mathbb{N}} B_k\big)^c\big) = \mu\big(\bigcup_{k \in \mathbb{N}} B_k^c\big) \leq \sum_{k \in \mathbb{N}} \mu(B_k^c) = 0$. Thus $B \in \mathfrak{E}$. Now since $B \subset B_k$ for every $k \in \mathbb{N}$, we have $\nu(B) \leq \nu(B_k) \leq \inf_{E \in \mathfrak{E}} \nu(E) + \frac{1}{k}$ for every $k \in \mathbb{N}$. This implies that $\nu(B) \leq \inf_{E \in \mathfrak{E}} \nu(E)$. On the other hand, since $B \in \mathfrak{E}$ we have $\nu(B) \geq \inf_{E \in \mathfrak{E}} \nu(E)$. Therefore we have $\nu(B) = \inf_{E \in \mathfrak{E}} \nu(E)$. This proves (2).

We show next that the set $B \in \mathfrak{E}$ characterized by (2) has the following property:

$$A \subset X, A \subset B, \mu(A) = 0 \Rightarrow \nu(A) = 0. \tag{3}$$

Now since μ and ν are Borel regular outer measures on X, by Proposition 7.9 there exists $E \in \mathfrak{B}_X$ such that $A \subset E$, $\mu(A) = \mu(E)$ and $\nu(A) = \nu(E)$. Consider the set $B \setminus E$. Let us show that $B \setminus E \in \mathfrak{E}$. Now since $B, E \in \mathfrak{B}_X$, we have $B \setminus E \in \mathfrak{B}_X$. We have also

$$\mu\big((B \setminus E)^c\big) = \mu\big((B \cap E^c)^c\big) = \mu(B^c \cup E) \leq \mu(B^c) + \mu(E) = 0,$$

where the last equality is from the fact that $\mu(B^c) = 0$ implied by the fact that $B \in \mathfrak{E}$ and by the fact that $\mu(E) = \mu(A) = 0$. Now that $B \setminus E \in \mathfrak{B}_X$ and $\mu\big((B \setminus E)^c\big) = 0$, we have

$$B \setminus E \in \mathfrak{E}.$$

Now $A \subset E$ and $A \subset B$ so that $A \subset E \cap B$. Then $\nu(E) = \nu(A) \leq \nu(E \cap B)$. On the other hand $\nu(E \cap B) \leq \nu(E)$. Thus $\nu(E) = \nu(E \cap B)$. Then since $B \setminus E = B \setminus (E \cap B)$, we have

$$\nu(B \setminus E) = \nu(B) - \nu(E \cap B) = \nu(B) - \nu(E) = \nu(B) - \nu(A).$$

If $\nu(A) > 0$, then $\nu(B \setminus E) < \nu(B)$. Since $B, B \setminus E \in \mathfrak{E}$, this contradicts (2). Thus $\nu(A) > 0$ leads to a contradiction. Therefore we have $\nu(A) = 0$. This proves (3).

1.1. With the set B defined above, let us define two outer measures on X by letting

$$\nu_{ac} := \nu|_B \quad \text{and} \quad \nu_s := \nu|_{B^c}. \tag{4}$$

Since ν is a Borel regular outer measure on X and $B, B^c \in \mathfrak{B}_X$, ν_{ac} and ν_s are Borel regular outer measures on X by Theorem 3.24. Clearly we have $\nu = \nu_{ac} + \nu_s$.

1.2. Let us show that $\nu_{ac} \ll \mu$, that is, for any $A \subset X$, if $\mu(A) = 0$ then $\nu_{ac}(A) = 0$. Let $A \subset X$ and suppose $\mu(A) = 0$. We have $\nu_{ac}(A) = \nu|_B(A) = \nu(A \cap B)$. Now $A \cap B \subset A$ and $\mu(A \cap B) \leq \mu(A) = 0$. Then since $A \cap B \subset B$ and $\mu(A \cap B) = 0$, we have $\nu(A \cap B) = 0$ by (3). Thus we have $\nu_{ac}(A) = 0$.

1.3. Let us show that $\nu_s \perp \mu$. Now for our $B \in \mathfrak{B}_X$, we have

$$\nu_s(B) = \nu|_{B^c}(B) = \nu(B \cap B^c) = \nu(\emptyset) = 0.$$

On the other hand since $B \in \mathfrak{E}$ we have $\mu(B^c) = 0$. This shows that $\nu_s \perp \mu$.

1.4. Let us prove (b). For $\alpha > 0$, let

$$C_\alpha = \{x \in B : D_\mu \nu_s(x) \geq \alpha\}.$$

Since $C \subset B$ we have $\nu_s(C_\alpha) = \nu|_{B^c}(C_\alpha) = \nu(C_\alpha \cap B^c) = \nu(\emptyset) = 0$. By (4) of Lemma 7.14, we have $\alpha\, \mu(C_\alpha) \leq \nu_s(C_\alpha) = 0$ and hence $\mu(C_\alpha) = 0$ for every $\alpha > 0$. Thus we have

$$D_\mu \nu_s = 0, (\mathfrak{B}_X, \mu)\text{-a.e. on} X. \tag{5}$$

Since $\nu = \nu_{ac} + \nu_s$ we have $D_\mu \nu = D_\mu \nu_{ac} + D_\mu \nu_s$. Then by (5) we have

$$D_\mu \nu = D_\mu \nu_{ac}, (\mathfrak{B}_X, \mu)\text{-a.e. on} X. \tag{6}$$

Then for every $E \in \mathfrak{B}_X$, we have

$$\nu(E) = \nu_{ac}(E) + \nu_s(E) = \int_E D_\mu \nu \, d\mu + \nu_s(E). \tag{7}$$

This completes the proof of the theorem under the condition that $\mu(X) < \infty$ and $\nu(X) < \infty$.

2. Let us remove the assumption that $\mu(X) < \infty$ and $\nu(X) < \infty$. Let $x_0 \in X$ be arbitrarily chosen. For $j \in \mathbb{N}$, let $K_j = \{x \in X : d(x, x_0) \leq j\}$. Since μ and ν are boundedly finite Borel regular outer measures on X, we have $\mu(K_j) < \infty$ and $\nu(K_j) < \infty$. Now $(K_j : j \in \mathbb{N})$ is an increasing sequence in $\mathfrak{B}_X$ and $X = \bigcup_{j \in \mathbb{N}} K_j$. Let $R_1 = K_1$ and $R_j = K_j \setminus K_{j-1}$ for $j \geq 2$. Then $(R_j : j \in \mathbb{N})$ is a disjoint sequence in $\mathfrak{B}_X$ with $\bigcup_{j \in \mathbb{N}} R_j = X$ and moreover $\mu(R_j) \leq \mu(K_j) < \infty$ and $\nu(R_j) \leq \nu(K_j) < \infty$ for $j \in \mathbb{N}$. Let

$$\mu_j = \mu|_{R_j} \quad \text{and} \quad \nu_j = \nu|_{R_j}. \tag{8}$$

By Theorem 3.24, μ_j and ν_j are Borel regular outer measures on X and moreover we have $\mu_j(X) < \infty$ and $\nu_j(X) < \infty$. We have also

$$\mu = \sum_{j \in \mathbb{N}} \mu_j \quad \text{and} \quad \nu = \sum_{j \in \mathbb{N}} \nu_j. \tag{9}$$

For each $j \in \mathbb{N}$, since μ_j and ν_j are finite Borel regular outer measures on X, by our results in **1**, ν_j can be decomposed as

$$\nu_j = \nu_{j,ac} + \nu_{j,s},$$

where $\nu_{j,ac}$ and $\nu_{j,s}$ are Borel regular outer measures on X such that

$$\nu_{j,ac} \ll \mu_j \quad \text{and} \quad \nu_{j,s} \perp \mu_j.$$

Let

$$\nu_{ac} = \sum_{j \in \mathbb{N}} \nu_{j,ac} \quad \text{and} \quad \nu_s = \sum_{j \in \mathbb{N}} \nu_{j,s}.$$

We have

$$\nu = \sum_{j \in \mathbb{N}} \nu_j = \sum_{j \in \mathbb{N}} \{\nu_{j,ac} + \nu_{j,s}\} = \sum_{j \in \mathbb{N}} \nu_{j,ac} + \sum_{j \in \mathbb{N}} \nu_{j,s} = \nu_{ac} + \nu_s.$$

To show that $\nu_{ac} \ll \mu$, suppose $A \subset X$ and $\mu(A) = 0$. Since $\mu(A) = \sum_{j \in \mathbb{N}} \mu_j(A)$, we have $\mu_j(A) = 0$ for every $j \in \mathbb{N}$. Since $\nu_{j,ac} \ll \mu_j$, $\mu_j(A) = 0$ implies $\nu_{j,ac}(A) = 0$. Then $\nu_{ac}(A) = \sum_{j \in \mathbb{N}} \nu_{j,ac}(A) = 0$. This shows that $\nu_{ac} \ll \mu$.

Let us show that $\nu_s \perp \mu$. Since $\nu_{j,s} \perp \mu_j$, R_j can be given as the union of two disjoint members of $\mathfrak{B}_X$, B_j and C_j, such that $\mu_j(C_j) = 0$ and $\nu_{j,s}(B_j) = 0$. Let $B = \bigcup_{j \in \mathbb{N}} B_j \in \mathfrak{B}_X$ and $C = \bigcup_{j \in \mathbb{N}} C_j \in \mathfrak{B}_X$. Since $\{R_j : j \in \mathbb{N}\}$ is a disjoint collection, B and C are disjoint. Then we have $\mu(C) = \sum_{j \in \mathbb{N}} \mu_j(C_j) = 0$ and $\nu_s(B) = \sum_{j \in \mathbb{N}} \nu_{j,s}(B_j) = 0$. This shows that $\nu_s \perp \mu$. ∎

§8 Averaging Operators and Differentiation of Integrals

[I] The Averaging Operator

Proposition 8.1. *Let (X, d) be a metric space and let μ be a boundedly finite Borel outer measure on X. Consider the measure space $(X, \mathfrak{M}(\mu), \mu)$. Let $f \in \mathcal{L}^1_{loc}(X, \mathfrak{M}(\mu), \mu)$. With $r \in (0, \infty)$ fixed, let us define a function φ on X by setting. Let ν be a set function on $\mathfrak{M}(\mu)$ defined by setting*

$$\varphi(x) = \int_{C_r(x)} f \, d\mu \quad \text{for } x \in X.$$

Then φ is a $\mathfrak{B}_X$-measurable and hence a $\mathfrak{M}(\mu)$-measurable function on X.

Proof. 1. Consider first the case that $f \in \mathcal{L}^1_{loc}(X, \mathfrak{M}(\mu), \mu)$ and $f \geq 0$ on X. Let us define a set function ν on $\mathfrak{M}(\mu)$ by setting

$$\nu(E) = \int_E f \, d\mu \quad \text{for } E \in \mathfrak{M}(\mu).$$

Then ν is a measure on the σ-algebra $\mathfrak{M}(\mu)$. Since μ is a Borel outer measure we have $\mathfrak{B}_X \subset \mathfrak{M}(\mu)$. Thus ν is a Borel measure. Let $E \in \mathfrak{M}(\mu)$ be a bounded set in X. Since $f \in \mathcal{L}^1_{loc}(X, \mathfrak{M}(\mu), \mu)$, we have $\nu(E) = \int_E f \, d\mu < \infty$. Thus ν is a boundedly finite Borel measure. Then by (b) of Theorem 7.2, $\nu(C_r(x))$ is an upper semicontinuous function on X and hence a $\mathfrak{B}_X$-measurable function on X. But $\nu(C_r(x)) = \int_{C_r(x)} f \, d\mu = \varphi(x)$ for $x \in X$. Thus $\varphi(x)$ is a $\mathfrak{B}_X$-measurable function of $x \in X$.

2. For $f \in\in \mathcal{L}^1_{loc}(X, \mathfrak{M}(\mu), \mu)$, let us write $f = f^+ - f^-$. Then we have

$$\varphi(x) = \int_{C_r(x)} f \, d\mu = \int_{C_r(x)} f^+ \, d\mu - \int_{C_r(x)} f^- \, d\mu.$$

We showed in **1** that $\int_{C_r(x)} f^+ \, d\mu$ and $\int_{C_r(x)} f^- \, d\mu$ are $\mathfrak{B}_X$-measurable functions of $x \in X$. Then so is their difference $\varphi(x)$. ■

The next theorem is identical with Proposition 8.1 in its content. However it is proved without Theorem 7.2 and in fact it contains Theorem 7.2 as a particular case.

Theorem 8.2. *Let (X, d) be a metric space and let μ be a boundedly finite Borel outer measure on X. Consider the measure space $(X, \mathfrak{M}(\mu), \mu)$. Let $f \in \mathcal{L}^1_{loc}(X, \mathfrak{M}(\mu), \mu)$. With $r > 0$ fixed, define a function φ on X by setting*

(1)
$$\varphi(x) = \int_{C_r(x)} f \, d\mu \quad \text{for } x \in X.$$

Then φ is a $\mathfrak{B}_X$-measurable and hence a $\mathfrak{M}(\mu)$-measurable function on X. In particular with $f = 1$ on X, $\mu(C_r(x))$ is a $\mathfrak{B}_X$-measurable function of $x \in X$.

Proof. 1. Let $f \in \mathcal{L}^1_{loc}(X, \mathfrak{M}(\mu), \mu)$ and assume further that $f \geq 0$ on X. We show that in this case φ is an upper semicontinuous function on X and hence φ is a $\mathfrak{B}_X$-measurable function on X.

According to Proposition 15.83, [LRA], a real-valued function φ on a metric space (X, d) is upper semicontinuous at $x \in X$ if and only if for every sequence $(x_k : k \in \mathbb{N})$ in X such that $\lim_{k\to\infty} x_k = x$ we have $\limsup_{k\to\infty} \varphi(x_k) \leq \varphi(x)$. Consider the function φ on X defined by (1). Let $x \in X$ and let $(x_k : k \in \mathbb{N})$ be a sequence in X such that $\lim_{k\to\infty} x_k = x$. Let D be a closed ball containing x with sufficiently large radius so that $C_r(x) \subset D$ and $C_r(x_k) \subset D$ for all $k \in \mathbb{N}$. Since $f \in \mathcal{L}^1_{loc}(X, \mathfrak{M}(\mu), \mu)$, f is μ-integrable on D. Then

$$(2) \qquad \begin{aligned} \limsup_{k\to\infty} \varphi(x_k) &= \limsup_{k\to\infty} \int_{C_r(x_k)} f \, d\mu \\ &= \limsup_{k\to\infty} \int_D \mathbf{1}_{C_r(x_k)} f \, d\mu \\ &\leq \int_D \limsup_{k\to\infty} \mathbf{1}_{C_r(x_k)} f \, d\mu \end{aligned}$$

where the inequality is by Generalized Fatou's Lemma (Theorem 9.19, [LRA]) which is applicable since $\mathbf{1}_{C_r(x_k)} f \leq \mathbf{1}_D f$ which is μ-integrable on D. Now according to Lemma 7.1, we have $\limsup_{k\to\infty} \mathbf{1}_{C_r(x_k)} \leq \mathbf{1}_{C_r(x)}$ on X. Then since $f \geq 0$ on X we have

$$(3) \qquad \limsup_{k\to\infty} \mathbf{1}_{C_r(x_k)} f \leq \mathbf{1}_{C_r(x)} f \quad \text{on } X.$$

Substituting (3) in (2), we have

$$\limsup_{k\to\infty} \varphi(x_k) \leq \int_D \mathbf{1}_{C_r(x)} f \, d\mu = \int_{C_r(x)} f \, d\mu = \varphi(x).$$

This proves the upper semicontinuity of φ at $x \in X$ by Proposition 15.83, [LRA]. The upper semicontinuity of φ on X implies that φ is $\mathfrak{B}_X$-measurable on X by Proposition 15.82, [LRA].

2. For $f \in \mathcal{L}^1_{loc}(X, \mathfrak{M}(\mu), \mu)$, let us write $f = f^+ - f^-$. By our result above, the functions $\varphi^+(x) = \int_{C_r(x)} f^+ \, d\mu$ and $\varphi^-(x) = \int_{C_r(x)} f^- \, d\mu$ for $x \in X$ are $\mathfrak{B}_X$-measurable. Then $\varphi(x) = \varphi^+(x) - \varphi^-(x)$ for $x \in X$ is $\mathfrak{B}_X$-measurable. ∎

Theorem 8.3. *Let (X, d) be a metric space and let μ be a boundedly finite Borel outer measure on X satisfying the condition that $\mu(O) > 0$ for every non-empty open set O in X. Consider the measure space $(X, \mathfrak{M}(\mu), \mu)$. Let $f \in \mathcal{L}^1_{loc}(X, \mathfrak{M}(\mu), \mu)$. With $r > 0$ fixed, define a function $A_r f$ on X by setting*

$$(1) \qquad A_r f(x) = \frac{1}{\mu(C_r(x))} \int_{C_r(x)} f \, d\mu \quad \textit{for } x \in X.$$

Then $A_r f$ is a $\mathfrak{B}_X$-measurable and hence a $\mathfrak{M}(\mu)$-measurable function on X.

Proof. According to Theorem 8.2, the function $\mu(C_r(x))$ for $x \in X$ is a $\mathfrak{B}_X$-measurable function on X and the function $\varphi(x) = \int_{C_r(x)} f\,d\mu$ for $x \in X$ is a $\mathfrak{B}_X$-measurable function on X. Thus the quotient $A_r f(x)$ for $x \in X$ is a $\mathfrak{B}_X$-measurable function on X. ∎

[II] Continuity of the Averaging Operator

The Lebesgue outer measure $(\mu_L^n)^*$ on $\mathbb{R}^n$ has the following properties:
1° $(\mu_L^n)^*(O) > 0$ for every non-empty open set O in $\mathbb{R}^n$.
2° $(\mu_L^n)^*(C_r(x))$ is a continuous function of $x \in \mathbb{R}^n$ for each fixed $r \in (0, \infty)$. Actually by the translation invariance of $(\mu_L^n)^*$, $(\mu_L^n)^*(C_r(x))$ is constant on $\mathbb{R}^n$.
3° $(\mu_L^n)^*(C_r(x)) = \Gamma\left(\frac{n}{2}+1\right)^{-1}\pi^{\frac{n}{2}}r_n$ so that $(\mu_L^n)^*(C_r(x))$ is a continuous function of $r \in (0, \infty)$ for each fixed $x \in \mathbb{R}^n$.

We consider a boundedly finite Borel outer measure μ on a metric space (X, d) satisfying the following conditions:
1° $\mu(O) > 0$ for every non-empty open set O in X.
2° $\mu(C_r(x))$ is a continuous function of $x \in X$ for each fixed $r \in (0, \infty)$.
3° $\mu(C_r(x))$ is a continuous function of $r \in (0, \infty)$ for each fixed $x \in X$.

Notations. Let (X, d) be a metric space. For $x \in X$ and $r \in (0, \infty)$, we write

$$S(x, r) = \{y \in X : d(x, y) = r\}.$$

Observation 8.4. Let (X, d) be a metric space.
(a) For $x_0 \in X$ and $r_0 \in (0, \infty)$ arbitrarily fixed, we have

$$\lim_{x \to x_0} \mathbf{1}_{C(x,r_0)}(y) = \mathbf{1}_{C(x_0,r_0)}(y) \quad \text{for } y \in S(x_0, r_0)^c.$$

(b) For $r_0 \in (0, \infty)$ and $x_0 \in X$ arbitrarily fixed, we have

$$\lim_{r \to r_0} \mathbf{1}_{C(x_0,r)}(y) = \mathbf{1}_{C(x_0,r_0)}(y) \quad \text{for } y \in S(x_0, r_0)^c.$$

Proof. 1. Let us prove (a). Note that $S(x_0, r_0)^c = B(x_0, r_0) \cup C(x_0, r_0)^c$.

1.1. Let $y \in B(x_0, r_0)$ be arbitrarily chosen. Let $\delta = r_0 - d(y, x_0)$. Then for any $x \in B(x_0, \delta)$ we have

$$d(y, x) \le d(y, x_0) + d(x_0, x) < d(y, x_0) + \delta \le r_0$$

so that $y \in B(x, r_0) \subset C(x, r_0)$. Thus $\mathbf{1}_{C(x,r_0)}(y) = 1$ for every $x \in B(x_0, \delta)$. Then

$$\lim_{x \to x_0} \mathbf{1}_{C(x,r_0)}(y) = 1 = \mathbf{1}_{B(x_0,r_0)}(y) = \mathbf{1}_{C(x_0,y_0)}(y).$$

1.2. Let $y \in C(x_0, r_0)^c$ be arbitrarily chosen. Let $\delta = d(y, x_0) - r_0 \in (0, \infty)$. Then for any $x \in B(x_0, \delta)$ we have

$$d(y, x) \geq d(y, x_0) - d(x_0, x) > r_0 + \delta - \delta = r_0$$

so that $y \notin C(x, r_0)$. Thus $\mathbf{1}_{C(x,r_0)}(y) = 0$ for every $x \in B(x_0, \delta)$. Then we have

$$\lim_{x \to x_0} \mathbf{1}_{C(x,r_0)}(y) = 0 = \mathbf{1}_{C(x_0,r_0)}(y).$$

2. Let us prove (b). Recall that $S(x_0, r_0)^c = B(x_0, r_0) \cup C(x_0, r_0)^c$.

2.1. Let $y \in B(x_0, r_0)$ be arbitrarily chosen. If $y = x_0$, then $y \in B(x_0, r)$ for every $r \in (0, \infty)$ so that $\mathbf{1}_{C(x_0,r)}(y) = 1$ for every $r \in (0, \infty)$ and thus

$$\lim_{r \to r_0} \mathbf{1}_{C(x_0,r)}(y) = 1 = \mathbf{1}_{C(x_0,r_0)}(y).$$

If $y \in B(x_0, r_0)$ but $y \neq x_0$, let $\delta = r_0 - d(y, x_0) \in (0, r_0)$. Then $y \in B(x_0, r) \subset C(c_0, r)$ for every $r \in (r_0 - \delta, \infty)$ so that

$$\lim_{r \to r_0} \mathbf{1}_{C(x_0,r)}(y) = 1 = \mathbf{1}_{C(x_0,r_0)}(y).$$

2.2. Let $y \in C(x_0, r_0)^c$ be arbitrarily chosen. Let $\delta = d(y, x_0) - r_0 \in (0, \infty)$ so that $y \notin C(x_0, r)$ for any $r \in (0, r_0 + \delta)$. Thus we have

$$\lim_{r \to r_0} \mathbf{1}_{C(x_0,r)}(y) = 0 = \mathbf{1}_{C(x_0,r_0)}(y). \blacksquare$$

Lemma 8.5. *Let μ be a boundedly finite Borel outer measure on a metric space (X, d) satisfying the condition that $\mu(C_r(x))$ is a continuous function of $r \in (0, \infty)$ for each fixed $x \in X$. Then $\mu(S_r(x)) = 0$ for every $x \in X$ and $r \in (0, \infty)$.*

Proof. Observe that

$$\begin{aligned}
S_r(x) &= C_r(x) \setminus B_r(x), \\
B_r(x) &= \bigcup_{k \in \mathbb{N}} C_{r-\frac{1}{k}}(x) = \lim_{k \to \infty} C_{r-\frac{1}{k}}(x), \\
S_r(x) &= C_r(x) \setminus \lim_{k \to \infty} C_{r-\frac{1}{k}}(x), \\
\mu(S_r(x)) &= \mu\Big(C_r(x) \setminus \lim_{k \to \infty} C_{r-\frac{1}{k}}(x)\Big).
\end{aligned}$$

Now we have

$$\mu\Big(\lim_{k \to \infty} C_{r-\frac{1}{k}}(x)\Big) = \lim_{k \to \infty} \mu(C_{r-\frac{1}{k}}(x)) \leq \mu(C_r(x)) < \infty.$$

Thus we have

$$= \mu\big(C_r(x)\big) - \mu\Big(\lim_{k\to\infty} C_{r-\frac{1}{k}}(x)\Big)$$
$$= \mu\big(C_r(x)\big) - \lim_{k\to\infty} \mu\big(C_{r-\frac{1}{k}}(x)\big).$$

By the continuity of $\mu\big(C_r(x)\big)$ as a function of $r \in (0, \infty)$, we have

$$\lim_{k\to\infty} \mu\big(C_{r-\frac{1}{k}}(x)\big) = \mu\big(C_r(x)\big).$$

Then we have $\mu\big(S_r(x)\big) = 0$. ■

Lemma 8.6. *Let μ be a boundedly finite Borel outer measure on a metric space (X, d) satisfying the condition that $\mu\big(C_r(x)\big)$ is a continuous function of $r \in (0, \infty)$ for each fixed $x \in X$.*
(a) *For fixed $x_0 \in X$ and $r_0 \in (0, \infty)$, there exists a null set N in $(X, \mathfrak{B}_X, \mu)$ such that*

$$\lim_{x\to x_0} \mathbf{1}_{C(x,r_0)}(y) = \mathbf{1}_{C(x_0,r_0)}(y) \quad \textit{for } y \in N^c.$$

(b) *For fixed $r_0 \in (0, \infty)$ and $x_0 \in X$, there exists a null set N in $(X, \mathfrak{B}_X, \mu)$ such that*

$$\lim_{r\to r_0} \mathbf{1}_{C(x_0,r)}(y) = \mathbf{1}_{C(x_0,r_0)}(y) \quad \textit{for } y \in N^c.$$

Proof. According to Observation 8.4, we have

$$\lim_{x\to x_0} \mathbf{1}_{C(x,r_0)}(y) = \mathbf{1}_{C(x_0,r_0)}(y) \quad \text{for } y \in S(x_0, r_0)^c,$$

and

$$\lim_{r\to r_0} \mathbf{1}_{C(x_0,r)}(y) = \mathbf{1}_{C(x_0,r_0)}(y) \quad \text{for } y \in S(x_0, r_0)^c.$$

According to Lemma 8.5, $S(x_0, r_0)$ is a null set in $(X, \mathfrak{B}_X, \mu)$. ■

Theorem 8.7. *Let μ be a boundedly finite Borel outer measure on a metric space (X, d) satisfying the following condition:*
1° *$\mu(O) > 0$ for every non-empty open set O in X.*
Consider the measure space $(X, \mathfrak{M}(\mu), \mu)$ and let $f \in \mathcal{L}^1_{loc}(X, \mathfrak{M}(\mu), \mu)$. Let us define a function on $(0, \infty) \times X$ by setting

$$A_r f(x) = \frac{1}{\mu\big(C_r(x)\big)} \int_{C_r(x)} f \, d\mu \quad \textit{for } (r, x) \in (0, \infty) \times X.$$

(a) *Suppose μ satisfies the condition:*
2° *$\mu\big(C_r(x)\big)$ is a continuous function of $x \in X$ for each fixed $r \in (0, \infty)$.*

Then $A_r f(x)$ is a continuous function of $x \in X$ for each fixed $r \in (0, \infty)$.
(b) *Suppose μ satisfies the condition:*
3° $\mu(C_r(x))$ *is a continuous function of $r \in (0, \infty)$ for each fixed $x \in X$.*
Then $A_r f(x)$ is a continuous function of $r \in (0, \infty)$ for each fixed $x \in X$.

Proof. 1. To prove (a), let us assume the conditions 1° and 2°. Let $r \in (0, \infty)$ be fixed. To show that $A_r f(x)$ is a continuous function of $x \in X$ we show that for every $x_0 \in X$, we have

$$\lim_{x \to x_0} A_r f(x) = A_r f(x_0). \tag{1}$$

Since $A_r f(x) = \frac{1}{\mu(C_r(x))} \int_{C_r(x)} f \, d\mu$, it suffices to show

$$\lim_{x \to x_0} \frac{1}{\mu(C_r(x))} = \frac{1}{\mu(C_r(x_0))} \tag{2}$$

and

$$\lim_{x \to x_0} \int_{C_r(x)} f \, d\mu = \int_{C_r(x_0)} f \, d\mu. \tag{3}$$

Now (2) is immediate from condition 2°. Thus it remains to show (3). Let $a \in X$ be arbitrarily chosen and fixed. Let $\delta > 0$ and $R > d(x_0, a) + \delta + r$. Then for every $x \in C_\delta(x_0)$, we have $C_r(x) \subset C_R(a)$ so that

$$\int_{C_r(x)} f \, d\mu = \int_{C_R(a)} \mathbf{1}_{C_r(x)} f \, d\mu.$$

Then we have

$$\begin{aligned} \lim_{x \to x_0} \int_{C_r(x)} f \, d\mu &= \lim_{x \to x_0} \int_{C_R(a)} \mathbf{1}_{C_r(x)} f \, d\mu \\ &= \int_{C_R(a)} \lim_{x \to x_0} \mathbf{1}_{C_r(x)} f \, d\mu \\ &= \int_{C_r(x_0)} f \, d\mu, \end{aligned}$$

where the second equality is by the Dominated Convergence Theorem and the last equality is by (a) of Lemma 8.6. This proves (3).

2. To prove (b), let us assume the conditions 1° and 3°. Let $x \in X$ be fixed. To show that $A_r f(x)$ is a continuous function of $r \in (0, \infty)$ we show that for every $r_0 \in (0, \infty)$, we have

$$\lim_{r \to r_0} A_r f(x) = A_{r_0} f(x). \tag{4}$$

Since $A_r f(x) = \frac{1}{\mu(C_r(x))} \int_{C_r(x)} f \, d\mu$, it suffices to show

(5) $$\lim_{r \to r_0} \frac{1}{\mu(C_r(x))} = \frac{1}{\mu(C_{r_0}(x))}$$

and

(6) $$\lim_{r \to r_0} \int_{C_r(x)} f \, d\mu = \int_{C_{r_0}(x)} f \, d\mu.$$

Here (5) is immediate from condition 3° and it remains to show (6). Let $a \in X$ be arbitrarily chosen and fixed. Let $\delta \in (0, r_0)$ and $R > r_0 + \delta + d(x, a)$. Then for every $r \in (r_0 - \delta, r_0 + \delta)$, we have $C_r(x) \subset C_R(a)$ so that

$$\int_{C_r(x)} f \, d\mu = \int_{C_R(a)} \mathbf{1}_{C_r(x)} f \, d\mu.$$

Now $|\mathbf{1}_{C_r(x)} f| \leq |\mathbf{1}_{C_R(a)} f|$, which is μ-integrable on $C_R(a)$. By (b) of Observation 8.4, we have also $\lim_{r \to r_0} \mathbf{1}_{C_r(x)} = \mathbf{1}_{C_{r_0}(x)}$ for $(\mathfrak{B}_X, \mu)$-a.e. $x \in X$. Thus by the Dominated Convergence Theorem, we have

$$\begin{aligned} \lim_{r \to r_0} \int_{C_r(x)} f \, d\mu &= \lim_{r \to r_0} \int_{C_R(a)} \mathbf{1}_{C_r(x)} f \, d\mu \\ &= \int_{C_R(a)} \lim_{r \to r_0} \mathbf{1}_{C_r(x)} f \, d\mu \\ &= \int_{C_R(a)} \mathbf{1}_{C_{r_0}(x)} f \, d\mu \\ &= \int_{C_{r_0}(x)} f \, d\mu. \end{aligned}$$

This proves (6). ∎

[III] Measurability of the Limit Inferior and Limit Superior of the Averaging Operators

Observation 8.8. Let $(X, \mathfrak{A}, \mu)$ be a measure space and let $(f_n : n \in \mathbb{N})$ be a sequence of extended real-valued $\mathfrak{A}$-measurable functions on a set $D \in \mathfrak{A}$. Then $\inf_{n \in \mathbb{N}} f_n$, $\sup_{n \in \mathbb{N}} f_n$, $\liminf_{n \to \infty} f_n$, $\limsup_{n \to \infty} f_n$ are all $\mathfrak{A}$-measurable functions on D and if $\lim_{n \to \infty} f_n$ exists, that is, if $\liminf_{n \to \infty} f_n = \limsup_{n \to \infty} f_n$ on D, then $\lim_{n \to \infty} f_n$ is a $\mathfrak{A}$-measurable function on D.

The situation is different when instead of a sequence $(f_n : n \in \mathbb{N})$ we have a collection of $\mathfrak{A}$-measurable functions indexed by real numbers, say $\{f_r : r \in I\}$ where I is an open interval in $\mathbb{R}$.

1. Let $\{\xi : r \in I\} \subset \overline{\mathbb{R}}$ and let $r_0 \in \overline{I}$. We say that $\lim_{r \to r_0} \xi_r = \xi \in \mathbb{R}$ if for every $\varepsilon > 0$ there exists $\delta > 0$ such that $|\xi_r - \xi| < \varepsilon$ for all $r \in (r_0 - \delta, r_0) \cup (r_0, r_0 + \delta)$. We say that $\lim_{r \to r_0} \xi_r = \infty$ if for every $M > 0$ there exists $\delta > 0$ such that $\xi_r > M$ for all $r \in (r_0 - \delta, r_0) \cup (r_0, r_0 + \delta)$. Similarly we say that $\lim_{r \to r_0} \xi_r = -\infty$ if for every $M > 0$ there exists $\delta > 0$ such that $\xi_r < -M$ for all $r \in (r_0 - \delta, r_0) \cup (r_0, r_0 + \delta)$.

2. It follows that $\lim_{r \to r_0} \xi_r = \xi \in \overline{\mathbb{R}}$ if and only if for every sequence $(r_n : n \in \mathbb{N})$ in $I \setminus \{r_0\}$ such that $\lim_{n \to \infty} r_n = r_0$ we have $\lim_{n \to \infty} \xi_{r_n} = \xi$.

3. Thus if $\{f_r : r \in I\}$ is a collection of extended real-valued $\mathfrak{A}$-measurable functions on $D \in \mathfrak{A}$ and $\lim_{r \to r_0} f_r(x) = f(x)$ for $x \in D$, then for every sequence $(r_n : n \in \mathbb{N})$ in $I \setminus \{r_0\}$ such that $\lim_{n \to \infty} r_n = r_0$ we have $\lim_{n \to \infty} f_{r_n}(x) = f(x)$ for $x \in D$. Thus $\lim_{r \to r_0} f_r$ is the limit of a sequence $(f_{r_n} : n \in \mathbb{N})$ of $\mathfrak{A}$-measurable functions on D and hence $\lim_{r \to r_0} f_r$ is $\mathfrak{A}$-measurable on D.

4. However $\inf_{r \in I} f_r$ and $\sup_{r \in I} f_r$, being the supremum and the infimum of an uncountable collection of $\mathfrak{A}$-measurable functions on D, need not be $\mathfrak{A}$-measurable on D. Similarly $\liminf_{r \to r_0} f_r = \lim_{\delta \to 0} \inf_{(r_0 - \delta, r_0) \cup (r_0, r_0 + \delta)} f_r$ and $\limsup_{r \to r_0} f_r = \lim_{\delta \to 0} \sup_{(r_0 - \delta, r_0) \cup (r_0, r_0 + \delta)} f_r$ need not be $\mathfrak{A}$-measurable on D.

Proposition 8.9. *Let $(X, \mathfrak{A}, \mu)$ be a measure space and let I be an open interval in $\mathbb{R}$ and $r_0 \in \overline{I}$. Let $\{f_r : r \in I\}$ be a collection of extended real-valued $\mathfrak{A}$-measurable functions on a set $D \in \mathfrak{A}$. Assume that $f_r(x)$ is a continuous function of $r \in I$ for each $x \in D$. Then $\inf_{r \in I} f_r$, $\sup_{r \in I} f_r$, $\liminf_{r \to r_0} f_r$, and $\limsup_{r \to r_0} f_r$ are $\mathfrak{A}$-measurable functions on D.*

Proof. Let $\{r_n : n \in \mathbb{N}\}$ be a countable dense subset of I. For instance $\{r_n : n \in \mathbb{N}\}$ may be the set of all rational numbers contained in I. Since $f_r(x)$ is a continuous function of $r \in I$ for each $x \in D$, we have

$$\inf_{r \in I} f_r(x) = \inf_{n \in \mathbb{N}} f_{r_n}(x).$$

Since $(f_{r_n} : n \in \mathbb{N})$ is a sequence of $\mathfrak{A}$-measurable functions on D, $\inf_{n \in \mathbb{N}} f_{r_n}$ is a $\mathfrak{A}$-measurable function on D. Thus $\inf_{r \in I} f_r$ is a $\mathfrak{A}$-measurable function on D. Similarly $\sup_{r \in I} f_r$ is a $\mathfrak{A}$-measurable function on D.

Now we have

$$\liminf_{r \to r_0} f_r = \lim_{\delta \to 0} \inf_{(r_0 - \delta, r_0) \cup (r_0, r_0 + \delta)} f_r.$$

Let $\{r_n : n \in \mathbb{N}\}$ be a countable dense subset of $(r_0 - \delta, r_0) \cup (r_0, r_0 + \delta)$. Since $f_r(x)$ is a continuous function of $r \in I$ for each $x \in D$, we have

$$\inf_{(r_0 - \delta, r_0) \cup (r_0, r_0 + \delta)} f_r(x) = \inf_{n \in \mathbb{N}} f_{r_n}(x).$$

Since $(f_{r_n} : n \in \mathbb{N})$ is a sequence of $\mathfrak{A}$-measurable functions on D, $\inf_{n\in\mathbb{N}} f_{r_n}$ is a $\mathfrak{A}$-measurable function on D. Thus $\inf_{(r_0-\delta,r_0)\cup(r_0,r_0+\delta)} f_r$ is a $\mathfrak{A}$-measurable function on D. Then by **3** of Observation 8.8, $\liminf_{r\to r_0} f_r$ is a $\mathfrak{A}$-measurable function on D. Similarly $\limsup_{r\to r_0} f_r$ is a $\mathfrak{A}$-measurable function on D. ∎

Theorem 8.10. *Let μ be a boundedly finite Borel outer measure on a metric space (X, d) satisfying the following conditions:*
1° $\mu(O) > 0$ for every non-empty open set O in X.
2° $\mu(C_r(x))$ is a continuous function of $r \in (0, \infty)$ for each fixed $x \in X$.
Let $f \in \mathcal{L}^1_{loc}(X, \mathfrak{M}(\mu), \mu)$ and consider

$$A_r f(x) = \frac{1}{\mu(C_r(x))} \int_{C_r(x)} f \, d\mu \quad \text{for } (r, x) \in (0, \infty) \times X.$$

(a) *For every open interval $I \subset (0, \infty)$, $\inf_{r\in I} A_r f$ and $\sup_{r\in I} A_r f$ are $\mathfrak{B}_X$-measurable functions on X.*
(b) *For every open interval $I \subset (0, \infty)$ and $r_0 \in \overline{I}$, $\liminf_{r\to r_0} A_r f$ and $\limsup_{r\to r_0} A_r f$ are $\mathfrak{B}_X$-measurable functions on X.*

Proof. By Theorem 8.3, $A_r f(x)$ is a $\mathfrak{B}_X$-measurable functions of $x \in X$ for each fixed $r \in (0, \infty)$. By (b) of Theorem 8.7, $A_r f(x)$ is a continuous function of $r \in (0, \infty)$ for each fixed $x \in X$. Then by Proposition 8.9, $\inf_{r\in I} A_r f$, $\sup_{r\in I} A_r f$, $\liminf_{r\to r_0} A_r f$ and $\limsup_{r\to r_0} A_r f$ are $\mathfrak{B}_X$-measurable functions on X. ∎

Theorem 8.11. *Let (X, d) be a separable metric space and a boundedly compact metric space. Let μ be a Radon outer measure on X. Consider the measure space $(X, \mathfrak{M}(\mu), \mu)$ and let $f \in \mathcal{L}^1_{loc}(X, \mathfrak{B}_X, \mu)$ and $f \geq 0$ on X. Define a set function ν on $\mathfrak{M}(\mu)$ by setting*

$$\nu(E) = \int_E f \, d\mu \quad \text{for } E \in \mathfrak{M}(\mu) \tag{1}$$

and then define a set function $\widetilde{\nu}$ on $\mathfrak{P}(X)$ by setting

$$\widetilde{\nu}(A) = \inf \{\nu(E) : A \subset E \in \mathfrak{M}(\mu)\} \quad \text{for } A \in \mathfrak{P}(X). \tag{2}$$

Then $\widetilde{\nu}$ is a Radon outer measure on X.

Proof. 1. The set function ν on $\mathfrak{M}(\mu)$ defined by (1) is a measure on the σ-algebra $\mathfrak{M}(\mu)$. Thus we have a measure space $(X, \mathfrak{M}(\mu), \nu)$. By Theorem 1.38, the set function $\widetilde{\nu}$ defined by (2) has the following properties:
(a) $\widetilde{\nu}$ is an outer measure on X.

(b) $\widetilde{\nu} = \nu$ on $\mathfrak{M}(\mu)$.
(c) $\mathfrak{B}_X \subset \mathfrak{M}(\mu) \subset \mathfrak{M}(\widetilde{\nu})$.
(d) $\widetilde{\nu}$ is a regular outer measure on X.
Let us show that $\widetilde{\nu}$ is a Radon outer measure on X. Since X is a separable metric space and a boundedly compact metric space, to show that $\widetilde{\nu}$ is a Radon outer measure it suffices to show that $\widetilde{\nu}$ is a boundedly finite Borel regular outer measure according to Theorem 3.39.

2. Let us show that $\widetilde{\nu}$ is boundedly finite, that is, for every bounded set $A \in \mathfrak{P}(X)$ we have $\widetilde{\nu}(A) < \infty$. Now let $A \in \mathfrak{P}(X)$ be a bounded set in X. Then $A \subset B_r(x)$ for some $x \in X$ and $r > 0$. Since $B_r(x) \in \mathfrak{B}_X \subset \mathfrak{M}(\mu)$, we have $\nu\big(B_r(x)\big) = \int_{B_r(x)} f\,d\mu$ by (1). Since $B_r(x)$ is a bounded set in X and since $f \in \mathcal{L}^1_{loc}(X, \mathfrak{B}_X, \mu)$, we have $\int_{B_r(x)} f\,d\mu < \infty$. Thus $\nu\big(B_r(x)\big) < \infty$. Then we have $\widetilde{\nu}(A) < \infty$ by (2). This shows that $\widetilde{\nu}$ is boundedly finite.

3. By (a) and (c), $\widetilde{\nu}$ is a Borel outer measure on X. Let us show that $\widetilde{\nu}$ is Borel regular, that is, for any $A \in \mathfrak{P}(X)$ there exists $B \in \mathfrak{B}_X$ such that $A \subset B$ and $\widetilde{\nu}(A) = \widetilde{\nu}(B)$. By (2)

$$\begin{aligned}\widetilde{\nu}(A) &= \inf\{\nu(E) : A \subset E \in \mathfrak{M}(\mu)\}\\ &\le \inf\{\nu(B) : A \subset B \in \mathfrak{B}_X\},\end{aligned}$$

since $\mathfrak{B}_X \subset \mathfrak{M}(\mu)$. Since $\widetilde{\nu}$ is a regular outer measure on X by (d), for every $B \in \mathfrak{B}_X \subset \mathfrak{P}(X)$ there exists $E \in \mathfrak{M}(\mu)$ such that $B \subset E$ and $\widetilde{\nu}(B) = \widetilde{\nu}(E)$. Since $\mathfrak{B}_X \subset \mathfrak{M}(\mu)$, we have $\widetilde{\nu}(B) = \nu(B)$ by (b) and similarly $\widetilde{\nu}(E) = \nu(E)$. Thus for every $B \in \mathfrak{B}_X$ there exists $E \in \mathfrak{M}(\mu)$ such that $B \subset E$ and $\nu(B) = \nu(E)$. Therefore we have

$$(3) \qquad \widetilde{\nu}(A) = \inf\{\nu(B) : A \subset B \in \mathfrak{B}_X\} \quad \text{for } A \in \mathfrak{P}(X).$$

Let $A \subset X$. Then by (3), for every $k \in \mathbb{N}$ there exists $B_k \in \mathfrak{B}_X$ such that $A \subset B_k$ and $\nu(B_k) \le \widetilde{\nu}(A) + \frac{1}{k}$. Let $B = \bigcap_{k\in\mathbb{N}} B_k$. Then $B \in \mathfrak{B}_X$ and $A \subset B$. By (3) we have $\widetilde{\nu}(A) \le \nu(B) \le \nu(B_k) \le \widetilde{\nu}(A) + \frac{1}{k}$ for every $k \in \mathbb{N}$. Thus $\widetilde{\nu}(A) = \nu(B)$. By (b), $\widetilde{\nu}(B) = \nu(B)$. Then we have $B \in \mathfrak{B}_X$ such that $A \subset B$ and $\widetilde{\nu}(A) = \widetilde{\nu}(B)$. ∎

[IV] Differentiation of Integrals with Respect to Radon Measure on $\mathbb{R}^n$

Theorem 8.12. (Lebesgue Differentiation Theorem for Integrals with Respect to Radon Measure on $\mathbb{R}^n$) *Let μ be a Radon outer measure on $\mathbb{R}^n$ such that $\mu(O) > 0$ for every non-empty open set O in $\mathbb{R}^n$. Consider the measure space $(\mathbb{R}^n, \mathfrak{M}(\mu), \mu)$. Let $f \in \mathcal{L}^1_{loc}\big(\mathbb{R}^n, \mathfrak{M}(\mu), \mu\big)$. Then we have*

$$(1) \qquad \lim_{r\to 0} \frac{1}{\mu\big(C_r(x)\big)} \int_{C_r(x)} f\,d\mu = f(x) \quad \textit{for } \big(\mathfrak{M}(\mu), \mu\big)\textit{-a.e. } x \in \mathbb{R}^n.$$

Proof. 1. Let $f \in \mathcal{L}^1_{loc}(\mathbb{R}^n, \mathfrak{M}(\mu), \mu)$ and $f \geq 0$ on $\mathbb{R}^n$. According to Theorem 8.11, if we define a set function ν on $\mathfrak{M}(\mu)$ by setting

$$\nu(E) = \int_E f(x)\,\mu(dx) \quad \text{for } E \in \mathfrak{M}(\mu), \tag{2}$$

and then define a set function $\widetilde{\nu}$ on $\mathfrak{P}(\mathbb{R}^n)$ by setting

$$\widetilde{\nu}(A) = \inf\{\nu(E) : A \subset E \in \mathfrak{M}(\mu)\} \quad \text{for } A \in \mathfrak{P}(\mathbb{R}^n), \tag{3}$$

then $\widetilde{\nu}$ is a Radon outer measure on $\mathbb{R}^n$.

Let us show that $\widetilde{\nu} \ll \mu$. Thus let $A \in \mathfrak{P}(\mathbb{R}^n)$ and $\mu(A) = 0$. We are to show that $\widetilde{\nu}(A) = 0$. Now since μ is a Radon outer measure on $\mathbb{R}^n$, μ is a Borel regular outer measure on $\mathbb{R}^n$ by Observation 3.43. Then for our $A \in \mathfrak{P}(\mathbb{R}^n)$ there exists $E \in \mathfrak{B}_{\mathbb{R}^n}$ such that $A \subset E$ and $\mu(A) = \mu(E)$. Since $E \in \mathfrak{B}_{\mathbb{R}^n} \subset \mathfrak{M}(\mu)$, (2) and $\mu(E) = \mu(A) = 0$ imply that $\nu(E) = \int_E f(x)\,\mu(dx) = 0$. Then (3) implies that $\widetilde{\nu}(A) = 0$. This proves $\widetilde{\nu} \ll \mu$.

Since μ and $\widetilde{\nu}$ are Radon outer measures on $\mathbb{R}^n$, by Theorem 7.8 the derivative $D_\mu\widetilde{\nu}$ exists, is $\mathfrak{M}(\mu)$-measurable and finite $(\mathfrak{M}(\mu), \mu)$-a.e. on $\mathbb{R}^n$. Now $\widetilde{\nu} \ll \mu$ implies according to Theorem 7.11 that for every $E \in \mathfrak{M}(\mu)$ we have

$$\int_E D_\mu\widetilde{\nu}(x)\,\mu(dx) = \widetilde{\nu}(E).$$

On the other hand for every $E \in \mathfrak{M}(\mu)$, (3) and (2) imply

$$\widetilde{\nu}(E) = \nu(E) = \int_E f(x)\,\mu(dx).$$

Thus we have

$$\int_E D_\mu\widetilde{\nu}(x)\,\mu(dx) = \int_E f(x)\,\mu(dx) \quad \text{for every } E \in \mathfrak{M}(\mu). \tag{4}$$

Since f is a $\mathfrak{M}(\mu)$-measurable function on $\mathbb{R}^n$, the fact that the equality (4) holds for every $E \in \mathfrak{M}(\mu)$ implies that

$$D_\mu\widetilde{\nu}(x) = f(x) \quad \text{for } (\mathfrak{M}(\mu), \mu)\text{-a.e. } x \in \mathbb{R}^n. \tag{5}$$

Now by Definition 7.3 and Definition 7.4, whenever $D_\mu\widetilde{\nu}(x)$ exists it is given by

$$D_\mu\widetilde{\nu}(x) = \lim_{r\to 0} \frac{\widetilde{\nu}(C_r(x))}{\mu(C_r(x))} = \lim_{r\to 0} \frac{1}{\mu(C_r(x))} \int_{C_r(x)} f\,d\mu. \tag{6}$$

By (5) and (6), we have (1).

2. When $f \in \mathcal{L}^1_{loc}(\mathbb{R}^n, \mathfrak{M}(\mu), \mu)$ we decompose f as $f = f^+ - f^-$ and apply our result in **1** to f^+ and f^-. ∎

Definition 8.13. *Let μ be a Radon outer measure on $\mathbb{R}^n$ such that $\mu(O) > 0$ for every non-empty open set O in $\mathbb{R}^n$. Consider the measure space $(\mathbb{R}^n, \mathfrak{M}(\mu), \mu)$. Let $f \in \mathcal{L}^1_{loc}(\mathbb{R}^n, \mathfrak{M}(\mu), \mu)$.*
(a) *A point $x \in \mathbb{R}^n$ is called a Lebesgue point of f with respect to μ if we have*

$$\lim_{r \to 0} \frac{1}{\mu(C_r(x))} \int_{C_r(x)} |f - f(x)| \, d\mu = 0.$$

(b) *We call the set of all Lebesgue points of f with respect to μ the Lebesgue set of f with respect to μ and write $\Lambda_\mu(f)$ for it.*

Observation 8.14. Let μ be a Radon outer measure on $\mathbb{R}^n$ such that $\mu(O) > 0$ for every non-empty open set O in $\mathbb{R}^n$. Let $f \in \mathcal{L}^1_{loc}(\mathbb{R}^n, \mathfrak{M}(\mu), \mu)$. Then for every $x \in \Lambda_\mu(f)$, we have

$$\lim_{r \to 0} \frac{1}{\mu(C_r(x))} \int_{C_r(x)} f \, d\mu = f(x). \tag{1}$$

Proof. For any $x \in \mathbb{R}^n$, we have

$$\begin{aligned}
&\left| \frac{1}{\mu(C_r(x))} \int_{C_r(x)} f \, d\mu - f(x) \right| \\
&= \left| \frac{1}{\mu(C_r(x))} \int_{C_r(x)} f(y) \, \mu(dy) - \frac{1}{\mu(C_r(x))} \int_{C_r(x)} f(x) \, \mu(dy) \right| \\
&\le \frac{1}{\mu(C_r(x))} \int_{C_r(x)} |f(y) - f(x)| \, \mu(dy).
\end{aligned}$$

If x is a Lebesgue point of f with respect to μ, then

$$\begin{aligned}
&\lim_{r \to 0} \left| \frac{1}{\mu(C_r(x))} \int_{C_r(x)} f \, d\mu - f(x) \right| \\
&\le \lim_{r \to 0} \frac{1}{\mu(C_r(x))} \int_{C_r(x)} |f(y) - f(x)| \, \mu(dy) = 0.
\end{aligned}$$

This implies that

$$\lim_{r \to 0} \left\{ \frac{1}{\mu(C_r(x))} \int_{C_r(x)} f \, d\mu - f(x) \right\} = 0.$$

This implies (1). ∎

Theorem 8.15. *Let μ be a Radon outer measure on $\mathbb{R}^n$ such that $\mu(O) > 0$ for every non-empty open set O in $\mathbb{R}^n$. Consider the measure space $(\mathbb{R}^n, \mathfrak{M}(\mu), \mu)$. Let $f \in \mathcal{L}^1_{loc}(\mathbb{R}^n, \mathfrak{M}(\mu), \mu)$. Then the Lebesgue set of f with respect to μ, $\Lambda_\mu(f) \in \mathfrak{M}(\mu)$ and moreover we have $\mu(\Lambda_\mu(f)^c) = 0$.*

Proof. Theorem 8.17 below shows among other things that there exists a null set N in $(\mathbb{R}^n, \mathfrak{M}(\mu), \mu)$ such that every $x \in N^c$ is a Lebesgue point of f with respect to μ. Thus we have $N^c \subset \Lambda_\mu(f)$ and hence $\Lambda_\mu(f)^c \subset N$. Since $(\mathbb{R}^n, \mathfrak{M}(\mu), \mu)$ is a complete measure space, an arbitrary subset of a null set in $(\mathbb{R}^n, \mathfrak{M}(\mu), \mu)$ is a member of $\mathfrak{M}(\mu)$. Thus we have $\Lambda_\mu(f)^c \in \mathfrak{M}(\mu)$. This implies that $\Lambda_\mu(f) \in \mathfrak{M}(\mu)$. We have also $\mu(\Lambda_\mu(f)^c) \le \mu(N) = 0$. ∎

Definition 8.16. *Let (X, d) be a metric space and let $(X, \mathfrak{A}, \mu)$ be a measure space. We write $\mathcal{L}^1_{loc}(\mathbb{R}^n, \mathfrak{M}(\mu), \mu)$ for the collection of all extended real-valued $\mathfrak{A}$-measurable functions f that are locally μ-integrable, that is, $\int_E |f|\, d\mu < \infty$ for every bounded set $E \in \mathfrak{A}$. For $p \in (0, \infty)$, we write $\mathcal{L}^p_{loc}(\mathbb{R}^n, \mathfrak{M}(\mu), \mu)$ for the collection of all extended real-valued $\mathfrak{A}$-measurable functions f that are locally pth order μ-integrable, that is, $\int_E |f|^p\, d\mu < \infty$ for every bounded set $E \in \mathfrak{A}$. Thus for an extended real-valued $\mathfrak{A}$-measurable function f on X, we have*

$$f \in \mathcal{L}^p_{loc}(\mathbb{R}^n, \mathfrak{M}(\mu), \mu) \Leftrightarrow |f|^p \in \mathcal{L}^1_{loc}(\mathbb{R}^n, \mathfrak{M}(\mu), \mu).$$

Theorem 8.17. *Let μ be a Radon outer measure on $\mathbb{R}^n$ such that $\mu(O) > 0$ for every non-empty open set O in $\mathbb{R}^n$. Consider the measure space $(\mathbb{R}^n, \mathfrak{M}(\mu), \mu)$. Let $p \in (0, \infty)$ and let $f \in \mathcal{L}^p_{loc}(\mathbb{R}^n, \mathfrak{M}(\mu), \mu)$. Then there exists a null set N in $(\mathbb{R}^n, \mathfrak{M}(\mu), \mu)$ such that*

$$\lim_{r \to 0} \frac{1}{\mu(C_r(x))} \int_{C_r(x)} |f - f(x)|^p\, d\mu = 0 \quad \text{for } x \in N^c.$$

Proof. 1. Let us show first that

$$(1) \qquad f \in \mathcal{L}^p_{loc}(\mathbb{R}^n, \mathfrak{M}(\mu), \mu), a \in \mathbb{R} \Rightarrow f - a \in \mathcal{L}^p_{loc}(\mathbb{R}^n, \mathfrak{M}(\mu), \mu).$$

Let $f \in \mathcal{L}^p_{loc}(\mathbb{R}^n, \mathfrak{M}(\mu), \mu)$ and $a \in \mathbb{R}$. Since $f - a \in \mathcal{L}^p_{loc}(\mathbb{R}^n, \mathfrak{M}(\mu), \mu)$ if and only if $|f - a|^p \in \mathcal{L}^1_{loc}(\mathbb{R}^n, \mathfrak{M}(\mu), \mu)$, it suffices to show that $|f - a|^p \in \mathcal{L}^1_{loc}(\mathbb{R}^n, \mathfrak{M}(\mu), \mu)$. Now for $p \in (0, 1)$ and $\alpha, \beta \in \overline{\mathbb{R}}$, whenever $\alpha + \beta$ is defined we have

$$|\alpha + \beta|^p \le 2^p\{|\alpha|^p + |\beta|^p\}.$$

(See Lemma 16.7, [LRA].) Thus for every $x \in \mathbb{R}^n$, we have

$$(2) \qquad |f(x) - a|^p \le 2^p\{|f(x)|^p + |a|^p\}.$$

Since $f \in \mathcal{L}^p_{loc}(\mathbb{R}^n, \mathfrak{M}(\mu), \mu)$, we have $|f|^p \in \mathcal{L}^1_{loc}(\mathbb{R}^n, \mathfrak{M}(\mu), \mu)$. Let us show that $|a|^p \in \mathcal{L}^1_{loc}(\mathbb{R}^n, \mathfrak{M}(\mu), \mu)$. Let $E \in \mathfrak{M}(\mu)$ be a bounded set. Then $\overline{E}$ is a bounded closed set and hence a compact set in $\mathbb{R}^n$. Then since μ is a Radon outer measure on $\mathbb{R}^n$ we have $\mu(\overline{E}) < \infty$. Thus we have

$$\int_E |a|^p\, d\mu \le \int_{\overline{E}} |a|^p\, d\mu = |a|^p \mu(\overline{E}) < \infty.$$

This shows that $|a|^p \in \mathcal{L}^1_{loc}(\mathbb{R}^n, \mathfrak{M}(\mu), \mu)$. The fact that $|f|^p \in \mathcal{L}^1_{loc}(\mathbb{R}^n, \mathfrak{M}(\mu), \mu)$ and $|a|^p \in \mathcal{L}^1_{loc}(\mathbb{R}^n, \mathfrak{M}(\mu), \mu)$ implies that $2^p\{|f|^p + |a|^p\} \in \mathcal{L}^1_{loc}(\mathbb{R}^n, \mathfrak{M}(\mu), \mu)$. Then by (2) we have $|f - a|^p \in \mathcal{L}^1_{loc}(\mathbb{R}^n, \mathfrak{M}(\mu), \mu)$. This completes the proof of (1).

2. Let us assume that our $f \in \mathcal{L}^p_{loc}(\mathbb{R}^n, \mathfrak{M}(\mu), \mu)$ is real-valued. Let $\{a_i : i \in \mathbb{N}\}$ be a countable dense subset of $\mathbb{R}$. By (1) we have $f - a_i \in \mathcal{L}^p_{loc}(\mathbb{R}^n, \mathfrak{M}(\mu), \mu)$ for every $i \in \mathbb{N}$. Thus $|f - a_i|^p \in \mathcal{L}^1_{loc}(\mathbb{R}^n, \mathfrak{M}(\mu), \mu)$ for every $i \in \mathbb{N}$. Then by Theorem 8.12, there exists a null set N_i in $(\mathbb{R}^n, \mathfrak{M}(\mu), \mu)$ such that

$$\lim_{r \to 0} \frac{1}{\mu(C_r(x))} \int_{C_r(x)} |f - a_i|^p \, d\mu = |f(x) - a_i|^p \quad \text{for } x \in N_i^c.$$

Let $N = \bigcup_{i \in \mathbb{N}} N_i$, a null set in $(\mathbb{R}^n, \mathfrak{M}(\mu), \mu)$. Then we have

$$(3) \qquad \lim_{r \to 0} \frac{1}{\mu(C_r(x))} \int_{C_r(x)} |f - a_i|^p \, d\mu = |f(x) - a_i|^p \quad \text{for } i \in \mathbb{N} \text{ and } x \in N^c.$$

Let $x \in N^c$ be fixed. Since $f(x) \in \mathbb{R}$ and since $\{a_i : i \in \mathbb{N}\}$ is a dense subset of $\mathbb{R}$ for an arbitrary $\varepsilon > 0$ there exists $i \in \mathbb{N}$ such that

$$(4) \qquad |f(x) - a_i|^p < \frac{\varepsilon}{2^p}.$$

Now for any $y \in \mathbb{R}^n$ we have

$$|f(y) - f(x)|^p \leq 2^p\{|f(y) - a_i|^p + |a_i - f(x)|^p\}.$$

This implies

$$\frac{1}{\mu(C_r(x))} \int_{C_r(x)} |f - f(x)|^p \, d\mu$$
$$\leq 2^p \left\{ \frac{1}{\mu(C_r(x))} \int_{C_r(x)} |f - a_1|^p \, d\mu + |a_i - f(x)|^p \right\}.$$

Then

$$\limsup_{r \to 0} \frac{1}{\mu(C_r(x))} \int_{C_r(x)} |f - f(x)|^p \, d\mu$$
$$\leq 2^p \left\{ \limsup_{r \to 0} \frac{1}{\mu(C_r(x))} \int_{C_r(x)} |f - a_1|^p \, d\mu + |a_i - f(x)|^p \right\}$$
$$= 2^p\{|f(x) - a_i|^p + |a_i - f(x)|^p\}$$
$$< 2\varepsilon$$

where the equality is by (3) and the last inequality is by (4). Since this holds for every $\varepsilon > 0$, we have

$$\limsup_{r \to 0} \frac{1}{\mu(C_r(x))} \int_{C_r(x)} |f - f(x)|^p \, d\mu = 0.$$

This implies

$$\liminf_{r\to 0} \frac{1}{\mu(C_r(x))}\int_{C_r(x)} |f - f(x)|^p \, d\mu = 0,$$

and hence we have

$$\lim_{r\to 0} \frac{1}{\mu(C_r(x))}\int_{C_r(x)} |f - f(x)|^p \, d\mu = 0 \quad \text{for } x \in N^c. \tag{5}$$

3. Let us remove the assumption that our $f \in \mathcal{L}^p_{loc}(\mathbb{R}^n, \mathfrak{M}(\mu), \mu)$ is real-valued. Now $f \in \mathcal{L}^p_{loc}(\mathbb{R}^n, \mathfrak{M}(\mu), \mu)$ implies that f is finite a.e. on $\mathbb{R}^n$, that is, there exists a null set N_0 in $(\mathbb{R}^n, \mathfrak{M}(\mu), \mu)$ such that $f(x) \in \mathbb{R}$ for $x \in N_0^c$. Let us define a function g on $\mathbb{R}^n$ by setting

$$g(x) = \begin{cases} f(x) & \text{for } x \in N_0^c, \\ 0 & \text{for } x \in N_0. \end{cases} \tag{6}$$

Then g is real-valued on $\mathbb{R}^n$, $g(x) = f(x)$ for $(\mathfrak{M}(\mu), \mu)$-a.e. $x \in \mathbb{R}^n$ and thus we have $g \in \mathcal{L}^p_{loc}(\mathbb{R}^n, \mathfrak{M}(\mu), \mu)$. Thus by our result above, that is, (5), there exists a null set N_1 in $(\mathbb{R}^n, \mathfrak{M}(\mu), \mu)$ such that

$$\lim_{r\to 0} \frac{1}{\mu(C_r(x))}\int_{C_r(x)} |g - g(x)|^p \, d\mu = 0 \quad \text{for } x \in N_1^c. \tag{7}$$

Let $N = N_0 \cup N_1$, a null set in $(\mathbb{R}^n, \mathfrak{M}(\mu), \mu)$. If $x \in N^c$ then $x \in N_0^c$ and thus $g(x) = f(x)$ by (6). If $x \in N^c$ then $x \in N_1^c$. Thus by (7) we have

$$\lim_{r\to 0} \frac{1}{\mu(C_r(x))}\int_{C_r(x)} |g - f(x)|^p \, d\mu = 0 \quad \text{for } x \in N^c. \tag{8}$$

Since $f = g$, $(\mathfrak{M}(\mu), \mu)$-a.e. on $\mathbb{R}^n$, we have

$$\frac{1}{\mu(C_r(x))}\int_{C_r(x)} |f - f(x)|^p \, d\mu = \frac{1}{\mu(C_r(x))}\int_{C_r(x)} |g - f(x)|^p \, d\mu. \tag{9}$$

Now (9) and (8) imply

$$\lim_{r\to 0} \frac{1}{\mu(C_r(x))}\int_{C_r(x)} |f - f(x)|^p \, d\mu = 0 \quad \text{for } x \in N^c.$$

This completes the proof. ∎

If the Radon outer measure μ on $\mathbb{R}^n$ in Theorem 8.17 is in fact the Lebesgue outer measure on $\mathbb{R}^n$ then we have the following stronger result. This is a consequence of the translation invariance of the Lebesgue measure on $\mathbb{R}^n$.

Corollary 8.18. *Consider the n-dimensional Lebesgue measure space $(\mathbb{R}^n, \mathfrak{M}_L^n, \mu_L^n)$. Let $f \in \mathcal{L}_{loc}^p(\mathbb{R}^n, \mathfrak{M}_L^n, \mu_L^n)$ where $p \in (0, \infty)$. For $x \in \mathbb{R}^n$, let $D_r(x)$ be a closed ball with diameter $r > 0$ and containing X (but not necessarily centered at x). Then there exists a null set N in $(\mathbb{R}^n, \mathfrak{M}_L^n, \mu_L^n)$ such that*

$$\lim_{r\to 0} \frac{1}{\mu(D_r(x))} \int_{D_r(x)} |f - f(x)|^p \, d\mu_L^n = 0 \quad \textit{for } x \in N^c.$$

Proof. The Lebesgue outer measure on $\mathbb{R}^n$ is a Radon outer measure on $\mathbb{R}^n$. Thus by Theorem 8.17 there exists a null set N in $(\mathbb{R}^n, \mathfrak{M}_L^n, \mu_L^n)$ such that

$$(1) \qquad \lim_{r\to 0} \frac{1}{\mu(C_r(x))} \int_{C_r(x)} |f - f(x)|^p \, d\mu_L^n = 0 \quad \text{for } x \in N^c.$$

Consider $D_r(x)$. Let y be the center of $D_r(x)$. Then $D_r(x) = C_{\frac{r}{2}}(y)$ and moreover

$$\mu_L^n(D_r(x)) = \mu_L^n(C_{\frac{r}{2}}(y)) = \mu_L^n(C_{\frac{r}{2}}(x)) = \left(\frac{1}{2}\right)^n \mu_L^n(C_r(x))$$

where the second equality is by the translation invariance of μ_L^n. Note that $D_r(x) \subset C_r(x)$. Now let $x \in N^c$. Then we have

$$\begin{aligned}
&\lim_{r\to 0} \frac{1}{\mu(D_r(x))} \int_{D_r(x)} |f - f(x)|^p \, d\mu_L^n \\
&\leq \lim_{r\to 0} \frac{1}{\left(\frac{1}{2}\right)^n \mu(C_r(x))} \int_{C_r(x)} |f - f(x)|^p \, d\mu_L^n \\
&= 2^n \lim_{r\to 0} \frac{1}{\mu(C_r(x))} \int_{C_r(x)} |f - f(x)|^p \, d\mu_L^n \\
&= 0
\end{aligned}$$

by (1). This completes the proof. ∎

Definition 8.19. *Let μ be a Radon outer measure on $\mathbb{R}^n$ such that $\mu(O) > 0$ for every non-empty open set O in $\mathbb{R}^n$. Let $f \in \mathcal{L}_{loc}^1(\mathbb{R}^n, \mathfrak{M}(\mu), \mu)$. Let us define a function f^* on $\mathbb{R}^n$ by setting*

$$f^*(x) = \begin{cases} \displaystyle\lim_{r\to 0} \frac{1}{\mu(C_r(x))} \int_{C_r(x)} f \, d\mu & \textit{if it exists,} \\ 0 & \textit{otherwise.} \end{cases}$$

We call the function f^* the precise representation of f with respect to μ.

Proposition 8.20. *Let μ be a Radon outer measure on $\mathbb{R}^n$ such that $\mu(O) > 0$ for every non-empty open set O in $\mathbb{R}^n$. Let $f \in \mathcal{L}_{loc}^1(\mathbb{R}^n, \mathfrak{M}(\mu), \mu)$ and let f^* be the precise representation*

of f with respect to μ.
(a) *There exists a null set N in $(\mathbb{R}^n, \mathfrak{M}(\mu), \mu)$ such that $f^*(x) = f(x)$ for $x \in N^c$.*
(b) *f^* is $\mathfrak{M}(\mu)$-measurable on $\mathbb{R}^n$ and $f^* \in \mathcal{L}^1_{loc}(\mathbb{R}^n, \mathfrak{M}(\mu), \mu)$.*
(c) *If $f, g \in \mathcal{L}^1_{loc}(\mathbb{R}^n, \mathfrak{M}(\mu), \mu)$ and if there exists a null set N in $(\mathbb{R}^n, \mathfrak{M}(\mu), \mu)$ such that $f(x) = g(x)$ for $x \in N^c$, then $f^*(x) = g^*(x)$ for every $x \in \mathbb{R}^n$.*
(d) *$f^*(x) = (f^*)^*(x)$ for every $x \in \mathbb{R}^n$.*

Proof. 1. To prove (a), note that by Theorem 8.12 there exists a null set N in $(\mathbb{R}^n, \mathfrak{M}(\mu), \mu)$ such that

$$\lim_{r \to 0} \frac{1}{\mu(C_r(x))} \int_{C_r(x)} f \, d\mu = f(x) \quad \text{for } x \in N^c.$$

Thus by Definition 8.19 we have $f^*(x) = f(x)$ for $x \in N^c$.

2. Let us prove (b). By (a) there exists a null set N in $(\mathbb{R}^n, \mathfrak{M}(\mu), \mu)$ such that $f^*(x) = f(x)$ for $x \in N^c$. Since f is $\mathfrak{M}(\mu)$-measurable on $\mathbb{R}^n$ and $(\mathbb{R}^n, \mathfrak{M}(\mu), \mu)$ is a complete measure space, this implies that f^* is $\mathfrak{M}(\mu)$-measurable on $\mathbb{R}^n$. Then the fact that $f \in \mathcal{L}^1_{loc}(\mathbb{R}^n, \mathfrak{M}(\mu), \mu)$ and $f^*(x) = f(x)$ for $x \in N^c$ implies that we have $f^* \in \mathcal{L}^1_{loc}(\mathbb{R}^n, \mathfrak{M}(\mu), \mu)$.

3. To prove (c), suppose $f, g \in \mathcal{L}^1_{loc}(\mathbb{R}^n, \mathfrak{M}(\mu), \mu)$ and there exists a null set N in $(\mathbb{R}^n, \mathfrak{M}(\mu), \mu)$ such that $f(x) = g(x)$ for $x \in N^c$. This implies that for every $x \in \mathbb{R}^n$ we have

$$(1) \qquad A_r f(x) := \frac{1}{\mu(C_r(x))} \int_{C_r(x)} f \, d\mu = \frac{1}{\mu(C_r(x))} \int_{C_r(x)} g \, d\mu := A_r g(x).$$

Then for any $x \in \mathbb{R}^n$, if $\lim_{r \to 0} A_r f(x)$ does not exist then $\lim_{r \to 0} A_r g(x)$ does not exist so that $f^*(x) = 0 = g^*(x)$, and if $\lim_{r \to 0} A_r f(x)$ exists then so does $\lim_{r \to 0} A_r g(x)$ and the limits are equal so that $f^*(x) = g^*(x)$. Thus for every $x \in \mathbb{R}^n$ we have $f^*(x) = g^*(x)$.

4. If $f \in \mathcal{L}^1_{loc}(\mathbb{R}^n, \mathfrak{M}(\mu), \mu)$ then $f^* \in \mathcal{L}^1_{loc}(\mathbb{R}^n, \mathfrak{M}(\mu), \mu)$ by (b) and then $(f^*)^*$ is defined. According to (a) there exists a null set N in $(\mathbb{R}^n, \mathfrak{M}(\mu), \mu)$ such that $f(x) = f^*(x)$ for $x \in N^c$. Then by (c) we have $f^*(x) = (f^*)^*(x)$ for every $x \in \mathbb{R}^n$. ∎

§9 Lebesgue Differentiation Theorem for Integrals

[I] Differentiation of Integrals with Respect to Radon Measures on $\mathbb{R}^n$ Revisited

Definition 9.1. *Let μ be a Radon outer measure on $\mathbb{R}^n$ such that for every non-empty open set $O \subset \mathbb{R}^n$ we have $\mu(O) > 0$. Consider the measure space $(\mathbb{R}^n, \mathfrak{M}(\mu), \mu)$. For $g \in \mathcal{L}^1_{loc}(\mathbb{R}^n, \mathfrak{M}(\mu), \mu)$ such that $g \geq 0$ on $\mathbb{R}^n$, let us define*

$$\overline{g}(x) = \limsup_{r \to 0} \frac{1}{\mu\big(C_r(x)\big)} \int_{C_r(x)} g(y)\, \mu(dy) \quad \text{for } x \in \mathbb{R}^n.$$

Remark 9.2. Note that the condition that $\mu(O) > 0$ for every non-empty open set O in $\mathbb{R}^n$ is equivalent to the condition that $\mu\big(B_r(x)\big) > 0$ for every $x \in \mathbb{R}^n$ and $r \in (0, \infty)$. With this condition, $\frac{1}{\mu(C_r(x))}$ is always defined. Since $\mu\big(C_r(x)\big)$ is a $\mathfrak{B}_{\mathbb{R}^n}$-measurable function of $x \in \mathbb{R}^n$ by Theorem 7.2, so is $\frac{1}{\mu(C_r(x))}$. By Theorem 8.2, $\int_{C_r(x)} g(y)\, \mu(dy)$ is a $\mathfrak{B}_{\mathbb{R}^n}$-measurable function of $x \in \mathbb{R}^n$. Thus $\frac{1}{\mu(C_r(x))} \int_{C_r(x)} g(y)\, \mu(dy)$ is a $\mathfrak{B}_{\mathbb{R}^n}$-measurable function of $x \in \mathbb{R}^n$ for each fixed $r \in (0, \infty)$. However $\overline{g}(x)$ being $\limsup_{r \to 0}$, does not limit superior of a sequence, need not be $\mathfrak{B}_{\mathbb{R}^n}$-measurable. The $\mathfrak{B}_{\mathbb{R}^n}$-measurability of $\frac{1}{\mu(C_r(x))} \int_{C_r(x)} g(y)\, \mu(dy)$ implies its $\mathfrak{M}(\mu)$-measurability. By the same reason as above, $\overline{g}(x)$ need not be $\mathfrak{M}(\mu)$-measurable. In what follows we do not consider integration of $\overline{g}(x)$ and thus the measurability of $\overline{g}(x)$ is not needed. We show next that there exists a null set N in the measure space $\big(\mathbb{R}^n, \mathfrak{B}_{\mathbb{R}^n}, \mu\big)$ such that $\overline{g}(x) \leq g(x)$ for $x \in N^c$.

Observation 9.3. Let $(X, \mathfrak{A}, \mu)$ be a measure space. Let $A, B \in \mathfrak{A}$. If $\mu(A \setminus B) = 0$, then $\mu(A) \leq \mu(B)$.

Proof. Let us write $A = (A \setminus B) \cup (A \cap B)$. Then $\mu(A) = \mu(A \setminus B) + \mu(A \cap B)$. If $\mu(A \setminus B) = 0$ then $\mu(A) = \mu(A \cap B) \leq \mu(B)$. ∎

Observation 9.4. Let $(X, \mathfrak{A}, \mu)$ be a measure space. Let $A \in \mathfrak{A}$ with $\mu(A) < \infty$. Let $(A_n : n \in \mathbb{N})$ be a decreasing sequence in $\mathfrak{A}$ such that $A_n \supset A$ for every $n \in \mathbb{N}$ and $\mu(A) = \lim_{n \to \infty} \mu(A_n)$. ($A = \lim_{n \to \infty} A_n$ is not assumed.) Let f be an extended real-valued $\mathfrak{A}$-measurable and μ-integrable function on A_1. Then we have

$$\lim_{n \to \infty} \int_{A_n} f\, d\mu = \int_A f\, d\mu.$$

Proof. Since f is μ-integrable on A_1, for every $\varepsilon > 0$ there exists $\delta > 0$ such that

$$(1) \qquad \int_E |f|\, d\mu < \varepsilon \quad \text{for every } E \subset A_1, E \in \mathfrak{A} \text{ with } \mu(E) < \delta.$$

Now since $\mu(A_n \setminus A) = \mu(A_n) - \mu(A)$ and $\lim_{n\to\infty} \mu(A_n \setminus A) = \lim_{n\to\infty} \mu(A_n) - \mu(A) = 0$, there exists $N \in \mathbb{N}$ such that $\mu(A_n \setminus A) < \delta$ for $n \geq N$. Then for $n \geq N$, we have by (1)

$$\left| \int_{A_n} f\,d\mu - \int_A f\,d\mu \right| = \left| \int_{A_n \setminus A} f\,d\mu \right| \leq \int_{A_n \setminus A} |f|\,d\mu < \varepsilon.$$

This shows that $\lim_{n\to\infty} \left| \int_{A_n} f\,d\mu - \int_A f\,d\mu \right| = 0$ and hence $\lim_{n\to\infty} \int_{A_n} f\,d\mu = \int_A f\,d\mu$. ■

Proposition 9.5. *Let μ be a Radon outer measure on $\mathbb{R}^n$ such that for every non-empty open set $O \subset \mathbb{R}^n$ we have $\mu(O) > 0$. Consider the measure space $\left(\mathbb{R}^n, \mathfrak{M}(\mu), \mu\right)$. Let $g \in \mathcal{L}^1_{loc}\left(\mathbb{R}^n, \mathfrak{M}(\mu), \mu\right)$ be such that $g \geq 0$ on $\mathbb{R}^n$. Then there exists a null set N in the measure space $\left(\mathbb{R}^n, \mathfrak{B}_{\mathbb{R}^n}, \mu\right)$ such that*

$$\overline{g}(x) \leq g(x) \quad \text{for } x \in N^c. \tag{1}$$

(Note that since μ is a Borel outer measure on $\mathbb{R}^n$ we have $\mathfrak{B}_{\mathbb{R}^n} \subset \mathfrak{M}(\mu)$.)

Proof. 1. Let $g \in \mathcal{L}^1_{loc}\left(\mathbb{R}^n, \mathfrak{M}(\mu), \mu\right)$ and $g \geq 0$ on $\mathbb{R}^n$. Consider the particular case that g satisfies the additional condition that g is constant outside of a bounded closed set in $\mathbb{R}^n$, that is, there exists a bounded closed set $K \subset \mathbb{R}^n$ such that

$$g(x) = c \quad \text{for } x \in K^c \tag{2}$$

where $c \geq 0$. Let

$$A = \{x \in \mathbb{R}^n : \overline{g}(x) > g(x)\}. \tag{3}$$

If we show that outer measure $\mu(A) = 0$, then since μ is a Borel regular outer measure there exists a set $N \in \mathfrak{B}_{\mathbb{R}^n}$ such that $A \subset N$ and $\mu(A) = \mu(N)$ so that $\mu(N) = 0$ and $\overline{g}(x) \leq g(x)$ for $x \in N^c$ and (1) is proved. It remains to show that $\mu(A) = 0$.

Let us show first that A is a bounded set and indeed $A \subset K$. Now since K^c is an open set, for every $x \in K^c$ we have $C_r(x) \subset K^c$ for all sufficiently small $r > 0$. Then $g = c$ on $C_r(x)$ for all sufficiently small $r > 0$ and this implies

$$\overline{g}(x) = \limsup_{r\to 0} \frac{1}{\mu\left(C_r(x)\right)} \int_{C_r(x)} g(y)\,\mu(dy) = c = g(x).$$

Thus $\overline{g} = g$ on K^c and hence $A \subset \left(K^c\right)^c = K$.

Let $\mathbb{Q}_+$ be the set of all nonnegative rational numbers. For $q \in \mathbb{Q}_+$ let

$$A_q = \{x \in \mathbb{R}^n : \overline{g}(x) > q > g(x)\} \subset A. \tag{4}$$

Then $A = \bigcup_{q\in\mathbb{Q}_+} A_q$. By the countable subadditivity of the outer measure μ, to show $\mu(A) = 0$ it suffices to show $\mu(A_q) = 0$ for every $q \in \mathbb{Q}_+$. Now since $A_q \subset A \subset K$ and

K is a bounded set, there exists a bounded open set O such that $A_q \subset O$. Let us construct a Vitali cover of the set A_q which is contained in the bounded open set O. Now if $x \in A_q$ then $\overline{g}(x) > q$, that is,

$$\limsup_{r \to 0} \frac{1}{\mu(C_r(x))} \int_{C_r(x)} g(y)\, \mu(dy) > q.$$

Then we can select a sequence of positive numbers $\big(r_j(x) : j \in \mathbb{N}\big)$ such that $r_j(x) \downarrow 0$ and

$$\int_{C_{r_j(x)}(x)} g(y)\, \mu(dy) > q\, \mu\big(C_{r_j(x)}(x)\big). \tag{5}$$

We choose $r_j(x)$ so small that $C_{r_j(x)}(x) \subset O$. Let $\mathcal{B} = \big\{C_{r_j(x)}(x) : j \in \mathbb{N}, x \in A_q\big\}$. Then $\mathcal{B}$ is a Vitali cover of A_q as defined in Definition 6.18 and $\mathcal{B}$ is contained in the bounded open set O. By Theorem 6.28 (Vitali Covering Theorem for Radon Outer Measures on $\mathbb{R}^n$) there exists a countable disjoint subcollection of $\mathcal{B}$ which we denote by $\big\{C(x_i, r_i) : i \in \mathbb{N}\big\}$ such that

$$\mu\left(A_q \setminus \bigcup_{i \in \mathbb{N}} C(x_i, r_i)\right) = 0. \tag{6}$$

By Observation 9.3, (6) implies

$$\mu(A_q) \le \mu\left(\bigcup_{i \in \mathbb{N}} C(x_i, r_i)\right). \tag{7}$$

Since the disjoint collection $\big\{C(x_i, r_i) : i \in \mathbb{N}\big\}$ is contained in the bounded open set O, we have

$$\begin{aligned} \int_O g(y)\, \mu(dy) &\ge \int_{\bigcup_{i \in \mathbb{N}} C(x_i, r_i)} g(y)\, \mu(dy) = \sum_{i \in \mathbb{N}} \int_{C(x_i, r_i)} g(y)\, \mu(dy) \\ &> q \sum_{i \in \mathbb{N}} \mu\big(C(x_i, r_i)\big) = q\, \mu\left(\bigcup_{i \in \mathbb{N}} C(x_i, r_i)\right) \\ &\ge q\, \mu(A_q), \end{aligned} \tag{8}$$

where the second inequality is by (5) and the last inequality is by (7). Since μ is a Radon outer measure, we have $\mu(A_q) = \inf\{\mu(V) : A_q \subset V, V \text{ is open}\}$. Since A_q is contained in a bounded open set O, there exists a decreasing sequence of bounded open sets $(O_k : k \in \mathbb{N})$ such that $O_k \supset A_q$ and $\lim_{k \to \infty} \mu(O_k) = \mu(A_q)$. According to Observation 9.4, this implies

$$\lim_{k \to \infty} \int_{O_k} g(y)\, \mu(dy) = \int_{A_q} g(y)\, \mu(dy). \tag{9}$$

Since (8) holds for an arbitrary bounded open set O containing A_q, we have

$$\int_{O_k} g(y)\,\mu(dy) \geq q\,\mu(A_q). \tag{10}$$

By (9) and (10), we have

$$\int_{A_q} g(y)\,\mu(dy) \geq q\,\mu(A_q). \tag{11}$$

Now since $g < q$ on A_q, if $\mu(A_q) > 0$ then $\int_{A_q} g(y)\,\mu(dy) < q\,\mu(A_q)$ contradicting (11). Thus we must have $\mu(A_q) = 0$. Since this holds for every $q \in \mathbb{Q}_+$ and since $A = \bigcup_{q\in\mathbb{Q}_+} A_q$, we have $\mu(A) = 0$. This prove (1) for g satisfying the additional condition that g is constant outside of a bounded closed set.

2. Now consider the general case that $g \in \mathcal{L}^1_{loc}(\mathbb{R}^n, \mathfrak{M}^n_L, \mu)$ and $g \geq 0$ on $\mathbb{R}^n$. Let $K_\ell := C(0,\ell)$ for $\ell \in \mathbb{N}$. Then $(K_\ell : \ell \in \mathbb{N})$ is an increasing sequence of bounded closed sets in $\mathbb{R}^n$ with $\bigcup_{\ell\in\mathbb{N}} K_\ell = \mathbb{R}^n$. Let us define a sequence of functions $(g_\ell : \ell \in \mathbb{N})$ by setting

$$g_\ell = \begin{cases} g & \text{on } K_\ell, \\ 0 & \text{on } K_\ell^c. \end{cases}$$

Then $g_\ell \in \mathcal{L}^1_{loc}(\mathbb{R}^n, \mathfrak{M}^n_L, \mu)$, $g_\ell \geq 0$ on $\mathbb{R}^n$ and g_ℓ is constant outside of a bounded closed set K_ℓ. Thus by our result in **1**, there exists a null set N_ℓ in $(\mathbb{R}^n, \mathfrak{B}_{\mathbb{R}^n}, \mu)$ such that

$$\overline{g}_\ell \leq g_\ell \quad \text{on } \mathbb{R}^n \setminus N_\ell. \tag{12}$$

Now $g_\ell = g$ on K_ℓ. For $x \in K_{\ell-1}$ and $r \in (0,1)$ we have $C_r(x) \subset K_\ell$. This fact and the fact that $g_\ell = g$ on K_ℓ imply that $\overline{g}_\ell(x) = \overline{g}(x)$ for $x \in K_{\ell-1}$. Then from (12) we have

$$\overline{g} \leq g \quad \text{on } K_{\ell-1} \setminus N_\ell. \tag{13}$$

Let $N = \bigcup_{\ell\in\mathbb{N}} N_\ell$, a null set in $(\mathbb{R}^n, \mathfrak{B}_{\mathbb{R}^n}, \mu)$. Then since $\bigcup_{\ell\in\mathbb{N}} K_{\ell-1} = \mathbb{R}^n$, (13) implies

$$\overline{g} \leq g \quad \text{on } \mathbb{R}^n \setminus N. \tag{14}$$

This completes the proof. ∎

Theorem 9.6. (Lebesgue Differentiation Theorem for Integrals with Respect to a Radon Outer Measure on $\mathbb{R}^n$) *Let μ be a Radon outer measure on $\mathbb{R}^n$ such that for every non-empty open set $O \subset \mathbb{R}^n$ we have $\mu(O) > 0$. Consider the measure space $(\mathbb{R}^n, \mathfrak{M}(\mu), \mu)$. Then for every $f \in \mathcal{L}^1_{loc}(\mathbb{R}^n, \mathfrak{M}(\mu), \mu)$ there exists a null set N in the measure space $(\mathbb{R}^n, \mathfrak{B}_{\mathbb{R}^n}, \mu)$ such that*

$$\lim_{r\to 0} \frac{1}{\mu(C_r(x))} \int_{C_r(x)} |f(y) - f(x)|\,\mu(dy) = 0 \quad \textit{for } x \in N^c, \tag{1}$$

and in particular

$$\lim_{r\to 0}\frac{1}{\mu(C_r(x))}\int_{C_r(x)} f(y)\,\mu(dy)=f(x)\quad \textit{for } x\in N^c. \tag{2}$$

Proof. Let $f\in\mathcal{L}^1_{loc}(\mathbb{R}^n,\mathfrak{M}(\mu),\mu)$. Let $\mathbb{Q}$ be the set of all rational numbers. For each $q\in\mathbb{Q}$ define a function g_q on $\mathbb{R}^n$ by

$$g_q(y)=|f(y)-q|\quad\text{for } y\in\mathbb{R}^n.$$

Then $g_q\in\mathcal{L}^1_{loc}(\mathbb{R}^n,\mathfrak{M}(\mu),\mu)$ and $g_q\geq 0$ on $\mathbb{R}^n$. Thus by Proposition 9.5, there exists a null set N_q in $(\mathbb{R}^n,\mathfrak{B}_{\mathbb{R}^n},\mu)$ such that

$$\overline{g}_q\leq g_q\quad\text{on } N_q^c. \tag{3}$$

Now for $x\in N_q^c$ we have

$$\begin{aligned}
&\limsup_{r\to 0}\frac{1}{\mu(C_r(x))}\int_{C_r(x)}|f(y)-f(x)|\,\mu(dy)\\
\leq&\limsup_{r\to 0}\frac{1}{\mu(C_r(x))}\int_{C_r(x)}\{|f(y)-q|+|q-f(x)|\}\,\mu(dy)\\
=&\limsup_{r\to 0}\frac{1}{\mu(C_r(x))}\int_{C_r(x)}\{g_q(y)+|q-f(x)|\}\,\mu(dy)\\
\leq&\limsup_{r\to 0}\frac{1}{\mu(C_r(x))}\int_{C_r(x)}g_q(y)\,\mu(dy)+|q-f(x)|\\
\leq& g_q(x)+|q-f(x)|=2|q-f(x)|,
\end{aligned} \tag{4}$$

where the third inequality is by Definition 9.1 and (3). Let $N=\bigcup_{q\in\mathbb{Q}}N_q$. Since $\mathbb{Q}$ is a countable set, N is a null set in $(\mathbb{R}^n,\mathfrak{B}_{\mathbb{R}^n},\mu)$. Then for $x\in N^c$ we have for all $q\in\mathbb{Q}$

$$\limsup_{r\to 0}\frac{1}{\mu(C_r(x))}\int_{C_r(x)}|f(y)-f(x)|\,\mu(dy)\leq 2|q-f(x)|. \tag{5}$$

With $x\in N^c$ arbitrarily fixed, let $(q_\ell:\ell\in\mathbb{N})$ be a sequence in $\mathbb{Q}$ such that $\lim_{\ell\to\infty}q_\ell=f(x)$. Replacing q in (5) with q_ℓ and then letting $\ell\to\infty$, we have

$$\limsup_{r\to 0}\frac{1}{\mu(C_r(x))}\int_{C_r(x)}|f(y)-f(x)|\,\mu(dy)=0. \tag{6}$$

This implies

$$\liminf_{r\to 0}\frac{1}{\mu(C_r(x))}\int_{C_r(x)}|f(y)-f(x)|\,\mu(dy)=0,$$

and then

$$\lim_{r\to 0} \frac{1}{\mu(C_r(x))} \int_{C_r(x)} |f(y) - f(x)| \,\mu(dy) = 0,$$

for $x \in N^c$. This proves (1).

2. To derive (2) from (1), let us write $f(x) = \frac{1}{\mu(C_r(x))} \int_{C_r(x)} f(x)\,\mu(dy)$. Then

$$\frac{1}{\mu(C_r(x))} \int_{C_r(x)} f(y)\,\mu(dy) - f(x) = \frac{1}{\mu(C_r(x))} \int_{C_r(x)} \{f(y) - f(x)\}\,\mu(dy).$$

Then we have

$$(7) \qquad \left| \frac{1}{\mu(C_r(x))} \int_{C_r(x)} f(y)\,\mu(dy) - f(x) \right| \leq \frac{1}{\mu(C_r(x))} \int_{C_r(x)} |f(y) - f(x)|\,\mu(dy).$$

Letting $r \to 0$ and applying (1) to the right side of (7), we have

$$\lim_{r\to 0} \left| \frac{1}{\mu(C_r(x))} \int_{C_r(x)} f(y)\,\mu(dy) - f(x) \right| = 0,$$

for $x \in N^c$. This proves (2). ■

Comments on Theorem 9.6. Theorem 9.6 is similar to Theorem 8.12 in contents. The approaches are different. Whereas Theorem 8.12 is based on Theorem 7.8 for derivative of a Radon measure, Theorem 9.6 utilizes the limit superior of the averaging operators as defined in Definition 9.1.

[II] Differentiation of Integrals with Respect to Doubling Borel Regular Measures on a Metric Space

Lemma 9.7. *Let μ be a doubling Borel regular outer measure on a metric space (X, d) satisfying the following additional conditions:*
1° $\mu(A) = \inf\{\mu(V) : A \subset V, V$ is open$\}$ for every $A \subset X$.
2° $\mu(C_r(x))$ is a continuous function of $r \in (0, \infty)$ for each fixed $x \in X$. Consider the measure space $(X, \mathfrak{M}(\mu), \mu)$. For $f \in \mathcal{L}^1_{loc}(X, \mathfrak{M}(\mu), \mu)$ such that $f \geq 0$ on X, let

$$A_r f(x) = \frac{1}{\mu(C_r(x))} \int_{C_r(x)} f(y)\,\mu(dy) \quad \text{for } x \in X \text{ and } r \in (0, \infty).$$

Let $B \in \mathfrak{B}_X$ be a bounded set in X. For $t > 0$, let us define

$$(1) \qquad B'_t = \{x \in B : \liminf_{r\to 0} A_r f(x) \leq t\},$$

$$(2) \qquad B''_t = \{x \in B : \limsup_{r\to 0} A_r f(x) \geq t\}.$$

Then we have

$$\int_{B'_t} f\,d\mu \le t\mu(B'_t), \tag{3}$$

$$\int_{B''_t} f\,d\mu \ge t\mu(B''_t). \tag{4}$$

Proof. Note that by Theorem 8.10, $\liminf_{r\to 0} A_r f$ and $\limsup_{r\to 0} A_r f$ are $\mathfrak{B}_X$-measurable functions on X and thus $B'_t, B''_t \in \mathfrak{B}_X$.

1. Let us prove (3). Let $\varepsilon > 0$ be arbitrarily given. By $1°$, there exists an open set U in X such that $B'_t \subset U$ and $\mu(U) \le \mu(B'_t) + \varepsilon$. For $x \in B'_t$, (1) implies that there exists $r_x > 0$ such that

$$A_r f(x) \le t + \varepsilon \quad \text{for all } r \in (0, r_x). \tag{5}$$

Let us take r_x so small that $C_r(x) \subset U$ for all $r \in (0, r_x)$. Then the collection of closed balls $\mathcal{B} = \{C_r(x) : x \in B'_t, r \in (0, r_x)\}$ is a Vitali cover of the set B'_t by Definition 6.18 and this Vitali cover $\mathcal{B}$ is contained in the open set U. Then by Theorem 6.20 (Vitali Covering Theorem for a Doubling Borel Regular Outer Measure), there exists a countable disjoint subcollection of $\mathcal{B}$ denoted by $\{C_i : i \in \mathbb{N}\}$ such that

$$\mu\left(B'_t \setminus \bigcup_{i\in\mathbb{N}} C_i\right) = 0. \tag{6}$$

Note that since $C_i \in \mathcal{B}$ we have from (5)

$$\int_{C_i} f\,d\mu \le (t+\varepsilon)\mu(C_i). \tag{7}$$

Now $B'_t = \left(B'_t \cap \bigcup_{i\in\mathbb{N}} C_i\right) \cup \left(B'_t \setminus \bigcup_{i\in\mathbb{N}} C_i\right)$. Thus we have

$$\begin{aligned} \int_{B'_t} f\,d\mu &= \int_{B'_t\cap\bigcup_{i\in\mathbb{N}} C_i} f\,d\mu + \int_{B'_t\setminus\bigcup_{i\in\mathbb{N}} C_i} f\,d\mu \\ &= \int_{B'_t\cap\bigcup_{i\in\mathbb{N}} C_i} f\,d\mu \quad \text{by (6)} \\ &\le \int_{\bigcup_{i\in\mathbb{N}} C_i} f\,d\mu = \sum_{i\in\mathbb{N}} \int_{C_i} f\,d\mu \\ &\le (t+\varepsilon)\sum_{i\in\mathbb{N}} \mu(C_i) \quad \text{by (7)} \\ &= (t+\varepsilon)\mu\left(\bigcup_{i\in\mathbb{N}} C_i\right) \le (t+\varepsilon)\mu(U) \\ &\le (t+\varepsilon)\{\mu(B'_t)+\varepsilon\}. \end{aligned} \tag{8}$$

Since this holds for every $\varepsilon > 0$, letting $\varepsilon \to 0$ we have (3).

2. Let us prove (4). Let $\varepsilon > 0$ be arbitrarily given. By 1° we can select a decreasing sequence $(U_i : i \in \mathbb{N})$ of open sets in X containing B_t'' such that $\lim_{i\to\infty} \mu(U_i) = \mu(B_t'')$. Then by Observation 9.4, we have

$$\lim_{i\to\infty} \int_{U_i} f \, d\mu = \int_{B_t''} f \, d\mu.$$

Thus there exists an open set U in X such that $B_t'' \subset U$ and

$$(9) \qquad \int_U f \, d\mu \leq \int_{B_t''} f \, d\mu + \varepsilon.$$

For $x \in B_t''$, (2) implies that there exists $r_x > 0$ such that

$$(10) \qquad A_r f(x) \geq t + \varepsilon \quad \text{for all } r \in (0, r_x).$$

Let us take r_x so small that $C_r(x) \subset U$ for all $r \in (0, r_x)$. Then the collection of closed balls $\mathcal{B} = \{C_r(x) : x \in B_t'', r \in (0, r_x)\}$ is a Vitali cover of the set B_t'' by Definition 6.18 and this Vitali cover $\mathcal{B}$ is contained in the open set U. Then by Theorem 6.20 (Vitali Covering Theorem for a Doubling Borel Regular Outer Measure), there exists a countable disjoint subcollection of $\mathcal{B}$ denoted by $\{C_i : i \in \mathbb{N}\}$ such that

$$(11) \qquad \mu\left(B_t'' \setminus \bigcup_{i\in\mathbb{N}} C_i\right) = 0.$$

This implies by Observation 1.15 that

$$(12) \qquad \mu(B_t'') \leq \mu\left(\bigcup_{i\in\mathbb{N}} C_i\right).$$

Note also that since $C_i \in \mathcal{B}$ we have from (10) that

$$(13) \qquad \int_{C_i} f \, d\mu \geq (t + \varepsilon)\mu(C_i).$$

Thus starting with (9) we have

$$(14) \qquad \begin{aligned} \int_{B_t''} f \, d\mu + \varepsilon &\geq \int_U f \, d\mu \geq \int_{\bigcup_{i\in\mathbb{N}} C_i} f \, d\mu = \sum_{i\in\mathbb{N}} \int_{C_i} f \, d\mu \\ &\geq (t+\varepsilon) \sum_{i\in\mathbb{N}} \mu(C_i) \quad \text{by (13)} \\ &= (t+\varepsilon)\mu\left(\bigcup_{i\in\mathbb{N}} C_i\right) \\ &\geq (t+\varepsilon)\mu(B_t'') \quad \text{by (12)}. \end{aligned}$$

Since this holds for every $\varepsilon > 0$, letting $\varepsilon \to 0$ we have (4). ∎

Theorem 9.8. (Lebesgue Differentiation Theorem for Integrals with Respect to a Doubling Borel Regular Outer Measure on a Metric Space) *Let μ be a doubling Borel regular outer measure on a metric space (X, d) satisfying the following additional conditions:*
$1°$ $\mu(A) = \inf\{\mu(V) : A \subset V, V$ *is open*$\}$ *for every* $A \subset X$.
$2°$ $\mu(C_r(x))$ *is a continuous function of* $r \in (0, \infty)$ *for each fixed* $x \in X$.
Consider the measure space $(X, \mathfrak{M}(\mu), \mu)$. For $f \in \mathcal{L}^1_{loc}(X, \mathfrak{M}(\mu), \mu)$, let

$$A_r f(x) = \frac{1}{\mu(C_r(x))} \int_{C_r(x)} f(y)\, \mu(dy) \quad \text{for } x \in X \text{ and } r \in (0, \infty).$$

Then there exists a null set N in $(X, \mathfrak{B}_X, \mu)$ such that

$$\lim_{r \to 0} A_r f(x) = f(x) \quad \text{for } x \in N^c.$$

Proof. 1. Let $f \in \mathcal{L}^1_{loc}(X, \mathfrak{M}(\mu), \mu)$ and $f \geq 0$ on X. Let B be an arbitrary bounded set in X and $B \in \mathfrak{B}_X$. Let us show that there exists a null set N in $(X, \mathfrak{B}_X, \mu)$, $N \subset B$, such that

$$\lim_{r \to 0} A_r f(x) = f(x) \quad \text{for } x \in B \setminus N. \tag{1}$$

Note that by Theorem 8.10, $\liminf_{r \to 0} A_r f$ and $\limsup_{r \to 0} A_r f$ are $\mathfrak{B}_X$-measurable functions on X and thus B_t', $B_t'' \in \mathfrak{B}_X$. For $t > 0$, let us define

$$B_t' = \left\{x \in B : \liminf_{r \to 0} A_r f(x) \leq t\right\} \in \mathfrak{B}_X, \tag{2}$$

$$B_t'' = \left\{x \in B : \limsup_{r \to 0} A_r f(x) \geq t\right\} \in \mathfrak{B}_X. \tag{3}$$

Then by Lemma 9.7, we have

$$\int_{B_t'} f\, d\mu \leq t\mu(B_t'), \tag{4}$$

$$\int_{B_t''} f\, d\mu \geq t\mu(B_t''). \tag{5}$$

Let $\{q_i : i \in \mathbb{N}\}$ be the set of all positive rational numbers. For any two nonnegative extended real numbers α and β, we have $\alpha < \beta$ if and only if there exists $(i, j) \in \mathbb{N} \times \mathbb{N}$, $i \neq j$, such that $\alpha \leq q_i < q_j \leq \beta$. For $(i, j) \in \mathbb{N} \times \mathbb{N}$, $i \neq j$, let

$$B_{i,j} = \left\{x \in B : \liminf_{r \to 0} A_r f(x) \leq q_i < q_j \leq \limsup_{r \to 0} A_r f(x)\right\} \in \mathfrak{B}_X. \tag{6}$$

Let us show

$$(7) \qquad \mu(B_{i,j}) = 0 \quad \text{for every } (i,j) \in \mathbb{N} \times \mathbb{N}, i \neq j.$$

Note first that if $q_i \geq q_j$, then $B_{i,j} = \emptyset$ and thus $\mu(B_{i,j}) = 0$. Consider the case $q_i < q_j$. Let $B^* = \{x \in B : \liminf_{r \to 0} A_r f(x) \leq q_i\}$ and $B^{**} = \{x \in B : \limsup_{r \to 0} A_r f(x) \geq q_j\}$. Then we have

$$B_{i,j} = B^* \cap B^{**}.$$

By the $\mathfrak{B}_X$-measurability of $\liminf_{r \to 0} A_r f$ and $\limsup_{r \to 0} A_r f$, we have $B^*, B^{**} \in \mathfrak{B}_X$. Also as subsets of the bounded set B, B^* and B^{**} are bounded sets. In terms of B^* and B^{**}, the set $B_{i,j}$ can be written in two different ways as

$$(8) \qquad B_{i,j} = \{x \in B^{**} : \liminf_{r \to 0} A_r f(x) \leq q_i\},$$

$$(9) \qquad B_{i,j} = \{x \in B^* : \limsup_{r \to 0} A_r f(x) \geq q_j\}.$$

Applying (3) and (4) of Lemma 9.7 to (8) and (9) respectively, we have

$$\int_{B_{i,j}} f \, d\mu \leq q_i \mu(B_{i,j}),$$

$$\int_{B_{i,j}} f \, d\mu \geq q_j \mu(B_{i,j}).$$

Thus we have

$$q_j \mu(B_{i,j}) \leq \int_{B_{i,j}} f \, d\mu \leq q_i \mu(B_{i,j}).$$

Since $q_i < q_j$, we have $\mu(B_{i,j}) = 0$. This completes the proof of (7).

Now

$$\{x \in B : \liminf_{r \to 0} A_r f(x) < \limsup_{r \to 0} A_r f(x)\} = \bigcup_{(i,j) \in \mathbb{N} \times \mathbb{N}, i \neq j} B_{i,j}$$

and then by (7) we have

$$(10) \qquad \mu\{x \in B : \liminf_{r \to 0} A_r f(x) < \limsup_{r \to 0} A_r f(x)\} \leq \sum_{(i,j) \in \mathbb{N} \times \mathbb{N}, i \neq j} \mu(B_{i,j}) = 0.$$

Now $\liminf_{r \to 0} A_r f(x) \leq \limsup_{r \to 0} A_r f(x)$ for every $x \in X$. Thus (10) implies that there exists a null set N_0 in $(X, \mathfrak{B}_X, \mu)$, $N_0 \subset B$, such that $\liminf_{r \to 0} A_r f(x) = \limsup_{r \to 0} A_r f(x)$ for $x \in B \setminus N_0$, that is,

$$(11) \qquad \lim_{r \to 0} A_r f(x) \text{ exists for } x \in B \setminus N_0.$$

Let us show next that there exists a null set N_1 in $(X, \mathfrak{B}_X, \mu)$, $N_1 \subset B \setminus N_0$, such that

$$\lim_{r \to 0} A_r f(x) = f(x) \quad \text{for } x \in (B \setminus N_0) \setminus N_1.$$

For brevity let us write

$$g(x) := \lim_{r \to 0} A_r f(x) = \liminf_{r \to 0} A_r f(x) = \limsup_{r \to 0} A_r f(x) \quad \text{for } x \in B \setminus N_0. \tag{12}$$

To show that there exists a null set N_1 in $(X, \mathfrak{B}_X, \mu)$, $N_1 \subset B \setminus N_0$, such that $g(x) = f(x)$ for $x \in (B \setminus N_0) \setminus N_1$, it suffices to show that for every $E \mathfrak{B}_X$, $E \subset B \setminus N_0$ we have

$$\int_E g \, d\mu = \int_E f \, d\mu. \tag{13}$$

Let $E \in \mathfrak{B}_X$ and $E \subset B \setminus N_0$. For $\varepsilon > 0$ arbitrarily given, let

$$\begin{aligned} E_0 &= \{x \in E : 0 \leq g(x) < 1 + \varepsilon\}, \\ E_n &= \{x \in E : (1+\varepsilon)^n \leq g(x) < (1+\varepsilon)^{n+1}, \quad \text{for } n \in \mathbb{N}, \\ E_\infty &= \{x \in E : g(x) = \infty\}. \end{aligned}$$

Let us show that $\mu(E_\infty) = 0$. Now

$$\begin{aligned} E_\infty &= \{x \in E : g(x) = \infty\} = \left\{x \in E : \lim_{r \to 0} A_r f(x) = \infty\right\} \\ &= \left\{x \in E : \limsup_{r \to 0} A_r f(x) = \infty\right\} \\ &\subset \left\{x \in B : \limsup_{r \to 0} A_r f(x) = \infty\right\} \\ &\subset B_k'' \end{aligned}$$

as defined by (3). Thus by the monotonicity of the outer measure and by (5) we have

$$\mu(E_\infty) \leq \mu(B_k'') \leq \frac{1}{k} \int_{B_k''} f \, d\mu \leq \frac{1}{k} \int_B f \, d\mu.$$

Since $f \in \mathcal{L}^1_{loc}(X, \mathfrak{M}(\mu), \mu)$ and $f \geq 0$ on X and B is a bounded set we have $\int_B f \, d\mu \in [0, \infty)$. Thus letting $k \to \infty$, we have $\mu(E_\infty) = 0$. With $\mu(E_\infty) = 0$, we have

$$\int_E g \, d\mu = \int_{\bigcup_{n \in \mathbb{Z}_+} E_n} g \, d\mu = \sum_{n \in \mathbb{Z}_+} \int_{E_n} g \, d\mu. \tag{14}$$

Now for every $n \in \mathbb{Z}_+$ we have

$$(1+\varepsilon)^n \mu(E_n) \leq \int_{E_n} g \, d\mu \leq (1+\varepsilon)^{n+1} \mu(E_n). \tag{15}$$

By (14) and (15), we have

$$(16)\qquad \sum_{n\in\mathbb{Z}_+}(1+\varepsilon)^n\mu(E_n)\le\int_E g\,d\mu\le\sum_{n\in\mathbb{Z}_+}(1+\varepsilon)^{n+1}\mu(E_n).$$

Let us estimate $\mu(E_n)$. For this purpose let us write $E^*=\{x\in E_n : g(x)<(1+\varepsilon)^{n+1}\}$ and $E^{**}=\{x\in E_n : g(x)\ge(1+\varepsilon)^n\}$. Then $E^*, E^{**}\in\mathfrak{B}_X$, E^* and E^{**} are bounded sets and $E_n=E^*\cap E^{**}$. Recalling the definition of g by (12), we have

$$(17)\qquad E_n=\{x\in E^{**} : \liminf_{r\to 0} A_rf(x)<(1+\varepsilon)^{n+1}\}$$

$$(18)\qquad E_n=\{x\in E^{*} : \limsup_{r\to 0} A_rf(x)>(1+\varepsilon)^{n}\}$$

Applying (3) and (4) of Lemma 9.7 to (17) and (18) respectively, we have

$$(19)\qquad \int_{E_n} f\,d\mu\le(1+\varepsilon)^{n+1}\mu(E_n),$$

$$(20)\qquad \int_{E_n} f\,d\mu\ge(1+\varepsilon)^{n}\mu(E_n).$$

Substituting (19) in the first inequality in (16), we have

$$(21)\qquad \begin{aligned}\int_E g\,d\mu&\ge\sum_{n\in\mathbb{Z}_+}(1+\varepsilon)^n\frac{1}{(1+\varepsilon)^{n+1}}\int_{E_n} f\,d\mu\\&=\frac{1}{1+\varepsilon}\sum_{n\in\mathbb{Z}_+}\int_{E_n} f\,d\mu=\frac{1}{1+\varepsilon}\int_E f\,d\mu\end{aligned}$$

and substituting (20) in the second inequality in (16) we have

$$(22)\qquad \begin{aligned}\int_E g\,d\mu&\le\sum_{n\in\mathbb{Z}_+}(1+\varepsilon)^{n+1}\frac{1}{(1+\varepsilon)^{n}}\int_{E_n} f\,d\mu\\&=(1+\varepsilon)\sum_{n\in\mathbb{Z}_+}\int_{E_n} f\,d\mu=(1+\varepsilon)\int_E f\,d\mu.\end{aligned}$$

By (21) and (22) we have

$$(23)\qquad \frac{1}{1+\varepsilon}\int_E f\,d\mu\le\int_E g\,d\mu\le(1+\varepsilon)\int_E f\,d\mu.$$

Letting $\varepsilon\to 0$, we have (13). This completes the proof of (1).

2. Let $f\in\mathcal{L}^1_{loc}(X,\mathfrak{M}(\mu),\mu)$ and $f\ge 0$ on X. Let us show that there exists a null set N in $(X,\mathfrak{B}_X,\mu)$ such that

$$(24)\qquad \lim_{r\to 0} A_rf(x)=f(x)\quad\text{for } x\in N^c.$$

With $x_0 \in X$ arbitrarily chosen, let

$$B_1 = \{x \in X : d(x, x_0) \in [0, 1)\},$$

$$B_k = \{x \in X : d(x, x_0) \in [k-1, k)\} \text{ for } k \geq 2.$$

Then $\{B_k : k \in \mathbb{N}\}$ is a disjoint collection of bounded $\mathfrak{B}_X$-measurable sets such that $\bigcup_{k\in\mathbb{N}} B_k = X$. For each $k \in \mathbb{N}$, by our result in **1** there exists a null set N_k in $(X, \mathfrak{B}_X, \mu)$, $N_k \subset B_k$, such that

$$\lim_{r\to 0} A_r f(x) = f(x) \quad \text{for } x \in B_k \setminus N_k. \tag{25}$$

Let $N = \bigcup_{k\in\mathbb{N}} N_k$, a null set in $(X, \mathfrak{B}_X, \mu)$. Let $x \in N^c$. Since $X = \bigcup_{k\in\mathbb{N}} B_k$, we have $x \in B_k \setminus N \subset B_k \setminus N_k$ for some $k \in \mathbb{N}$. Then by (25) we have $\lim_{r\to 0} A_r f(x) = f(x)$ and this proves (24).

3. Let $f \in \mathcal{L}^1_{loc}(X, \mathfrak{M}(\mu), \mu)$. Let us write $f = f^+ - f^-$. By our result in **2**, there exist null sets N_1 and N_2 in $(X, \mathfrak{B}_X, \mu)$ such that $\lim_{r\to 0} A_r f^+(x) = f^+(x)$ for $x \in N_1^c$ and $\lim_{r\to 0} A_r f^-(x) = f^-(x)$ for $x \in N_2^c$. Let $N = N_1 \cup N_2$, a null set in $(X, \mathfrak{B}_X, \mu)$. Then for $x \in N^c$ we have

$$\begin{aligned}\lim_{r\to 0} A_r f(x) &= \lim_{r\to 0} A_r f^+(x) - \lim_{r\to 0} A_r f^-(x) \\ &= f^+(x) - f^-(x) = f(x). \quad \blacksquare\end{aligned}$$

§10 Hardy-Littlewood Maximal Functions

[I] Hardy-Littlewood Maximal Functions for Outer Measures

Definition 10.1. *Let μ and ν be Borel outer measures on a metric space (X,d). The Hardy-Littlewood maximal function for ν with respect to μ is defined by*

$$M_\mu\nu(x) = \begin{cases} \sup_{r\in(0,\infty)} \dfrac{\nu\big(C_r(x)\big)}{\mu\big(C_r(x)\big)} & \text{if } \mu\big(C_r(x)\big) > 0 \text{ for all } r \in (0,\infty), \\ \infty & \text{if } \mu\big(C_r(x)\big) = 0 \text{ for some } r \in (0,\infty). \end{cases}$$

Remark 10.2. If the Borel outer measure μ on X is such that $\operatorname{supp}\mu = X$, that is, μ is such that $\mu(O) > 0$ for every non-empty open set O in X (see Proposition 1.44), then Definition 10.1 reduces to

$$M_\mu\nu(x) = \sup_{r\in(0,\infty)} \frac{\nu\big(C_r(x)\big)}{\mu\big(C_r(x)\big)} \quad \text{for every } x \in X.$$

[I.1] Measurability of Hardy-Littlewood Maximal Function of an Outer Measure

Proposition 10.3. *Let μ and ν be boundedly finite Borel outer measures on a metric space (X,d). Assume that* $\operatorname{supp}\mu = X$. *Then for each $r \in (0,\infty)$, $\frac{\nu(C_r(x))}{\mu(C_r(x))}$ is a $\mathfrak{B}_X$-measurable function of $x \in X$.*

Proof. By Theorem 7.2(b), $\mu\big(C_r(x)\big)$ and $\nu\big(C_r(x)\big)$ are $\mathfrak{B}_X$-measurable functions of $x \in X$ for each $r \in (0,\infty)$. Since $\operatorname{supp}\mu = X$ we have $\mu\big(C_r(x)\big) > 0$ for every $x \in X$ and $r \in (0,\infty)$. Thus $\frac{\nu(C_r(x))}{\mu(C_r(x))}$ is defined for all $x \in X$ and $r \in (0,\infty)$. Then the $\mathfrak{B}_X$-measurability of $\mu\big(C_r(x)\big)$ and $\nu\big(C_r(x)\big)$ as functions of $x \in X$ implies that of their quotient $\frac{\nu(C_r(x))}{\mu(C_r(x))}$. ■

Theorem 10.4. *Let μ and ν be boundedly finite Borel outer measures on a metric space (X,d). Assume that* $\operatorname{supp}\mu = X$. *Suppose μ and ν satisfy the additional conditions that $\mu\big(C_r(x)\big)$ and $\nu\big(C_r(x)\big)$ are continuous functions of $r \in (0,\infty)$ for each fixed $x \in X$. Then the Hardy-Littlewood maximal function $M_\mu\nu(x)$ of ν with respect to μ is a $\mathfrak{B}_X$-measurable function of $x \in X$.*

Proof. The assumption $\operatorname{supp}\mu = X$ implies according to Remark 10.2 that we have

$$(1) \qquad M_\mu\nu(x) = \sup_{r\in(0,\infty)} \frac{\nu\big(C_r(x)\big)}{\mu\big(C_r(x)\big)} \quad \text{for every } x \in X.$$

By Proposition 10.3, for each $r \in (0,\infty)$, $\dfrac{\nu(C_r(x))}{\mu(C_r(x))}$ is a $\mathfrak{B}_X$-measurable function of $x \in X$. If $\mu(C_r(x))$ and $\nu(C_r(x))$ are continuous functions of $r \in (0,\infty)$ for each fixed $x \in X$, then $\dfrac{\nu(C_r(x))}{\mu(C_r(x))}$ is a continuous function of $r \in (0,\infty)$ for each $x \in X$. Then for an arbitrary countable dense subset $\{r_k : k \in \mathbb{N}\}$ of $(0,\infty)$, we have

$$\text{(2)} \qquad \sup_{r \in (0,\infty)} \frac{\nu(C_r(x))}{\mu(C_r(x))} = \sup_{k \in \mathbb{N}} \frac{\nu(C_{r_k}(x))}{\mu(C_{r_k}(x))} \quad \text{for } x \in X.$$

Since $\dfrac{\nu(C_{r_k}(x))}{\mu(C_{r_k}(x))}$ is a $\mathfrak{B}_X$-measurable function of $x \in X$ for every $k \in \mathbb{N}$, $\sup_{k \in \mathbb{N}} \dfrac{\nu(C_{r_k}(x))}{\mu(C_{r_k}(x))}$ is a $\mathfrak{B}_X$-measurable function of $x \in X$. Then by (1) and (2), $M_\mu\nu(x)$ is a $\mathfrak{B}_X$-measurable function of $x \in X$. ■

Next let us obtain some estimate of the Hardy-Littlewood maximal function $M_\mu\nu$ of a Borel outer measure ν with respect to a Borel outer measure μ on a metric space (X,d). Here measurability of the Hardy-Littlewood maximal function $M_\mu\nu$ is not required. See Theorem 10.8 below. Let us show first however that if $M_\mu\nu$ is $\mathfrak{B}_X$-measurable and μ-integrable on X then an estimate of $M_\mu\nu$ follows immediately.

Proposition 10.5. *Let μ and ν be Borel outer measures on a metric space (X,d). If the Hardy-Littlewood maximal function $M_\mu\nu$ is $\mathfrak{B}_X$-measurable and μ-integrable on X, then there exists a constant $C > 0$ such that*

$$\text{(1)} \qquad \mu\{x \in X : M_\mu\nu(x) > \lambda\} \le \frac{C}{\lambda}\nu(X) \ \textit{ for all } \lambda \in (0,\infty).$$

Proof. Assume that $M_\mu\nu$ is $\mathfrak{B}_X$-measurable and μ-integrable on X. Now if $\nu(X) = 0$ then $M_\mu\nu(x) = 0$ for every $x \in X$ and then $\mu\{x \in X : M_\mu\nu(x) > \lambda\} = \mu(\emptyset) = 0$ for every $\lambda \in (0,\infty)$ so that (1) holds trivially. If $\nu(X) = \infty$ then (1) holds trivially. Thus it remains to consider the case that $\nu(X) \in (0,\infty)$. For $\lambda \in (0,\infty)$, let us write

$$A_\lambda := \{x \in X : M_\mu\nu(x) > \lambda\}.$$

Now we have

$$\int_X M_\mu\nu(x)\,\mu(dx) \ge \int_{A_\lambda} M_\mu\nu(x)\,\mu(dx) \ge \lambda\mu(A_\lambda)$$

and then

$$\text{(2)} \qquad \mu(A_\lambda) \le \frac{1}{\lambda}\int_X M_\mu\nu(x)\,\mu(dx).$$

Let

$$(3) \qquad C := \frac{1}{\nu(X)} \int_X M_\mu \nu(x)\, \mu(dx).$$

Since $\nu(X) \in (0, \infty)$ and $\int_X M_\mu\nu(x)\,\mu(dx) < \infty$, we have $C \in (0, \infty)$. Now from (3) we have $\int_X M_\mu\nu(x)\,\mu(dx) = C\nu(X)$ and substituting this in (2) we have (1). ∎

Remark 10.6. In Proposition 10.5 above we assumed that the Hardy-Littlewood maximal function $M_\mu\nu$ is $\mathfrak{B}_X$-measurable and μ-integrable on X. We note here that even if a Hardy-Littlewood maximal function $M_\mu\nu$ is $\mathfrak{B}_X$-measurable on X, it still need not be μ-integrable on X. Proposition 10.7 below shows a $\mathfrak{B}_X$-measurable Hardy-Littlewood maximal function $M_\mu\nu$ which is not μ-integrable on X.

Proposition 10.7. *Consider the metric space $(X, d) = (\mathbb{R}, d_e)$ where d_e is the Euclidean metric on $\mathbb{R}$. Let $\mu = \mu_L^*$, the Lebesgue outer measure on $\mathbb{R}$, and let ν be a boundedly finite Borel outer measure on $\mathbb{R}$ satisfying the condition that $\nu(C_r(x))$ is a continuous function of $r \in (0, \infty)$ for each fixed $x \in \mathbb{R}$.*
(a) *The Hardy-Littlewood maximal function $M_{\mu_L^*}\nu$ is $\mathfrak{B}_{\mathbb{R}}$-measurable on $\mathbb{R}$.*
(b) *$M_{\mu_L^*}\nu$ is μ_L^*-integrable on $\mathbb{R}$ if and only if $\nu(\mathbb{R}) = 0$.*

Proof. 1. Note that μ_L^* is a Borel outer measure on $\mathbb{R}$ and $\mu_L^*(O) > 0$ for every non-empty open set O in $\mathbb{R}$. Thus by Remark 10.2 we have

$$(1) \qquad M_{\mu_L^*}\nu(x) = \sup_{r \in (0,\infty)} \frac{\nu(C_r(x))}{\mu_L^*(C_r(x))} \quad \text{for every } x \in \mathbb{R}.$$

Now both μ_L^* and ν are boundedly finite Borel outer measures on $\mathbb{R}$. Also $\mu_L^*(C_r(x))$ and $(C_r(x))$ are continuous functions of $r \in (0, \infty)$ for each fixed $x \in \mathbb{R}$. Then by Theorem 10.4, $\mu\{x \in X : M_\mu\nu(x)$ is a $\mathfrak{B}_{\mathbb{R}}$-measurable function of $x \in \mathbb{R}$.

2. If $\nu(\mathbb{R}) = 0$, then from (1) we have $M_{\mu_L^*}\nu(x) = 0$ for every $x \in \mathbb{R}$ and thus $\int_{\mathbb{R}} M_{\mu_L^*}\nu(x)\,\mu_L^*(dx) = 0$ so that $M_{\mu_L^*}\nu$ is μ_L^*-integrable on $\mathbb{R}$.

3. Let us show that if $\nu(\mathbb{R}) > 0$ then $M_{\mu_L^*}\nu$ is not μ_L^*-integrable on $\mathbb{R}$, that is, $\int_{\mathbb{R}} M_{\mu_L^*}\nu(x)\,\mu_L^*(dx) = \infty$.

Now if $\nu(\mathbb{R}) > 0$ then there exists $r_0 \in (0, \infty)$ such that $\nu(C_{r_0}(0)) > 0$. Indeed if $\nu(C_r(0)) = 0$ for every $r \in (0, \infty)$, then we have

$$\nu(\mathbb{R}) = \nu\left(\bigcup_{k \in \mathbb{N}} C_k(0)\right) = \lim_{n \to \infty} \nu(C_k(0)) = \lim_{k \to \infty} 0 = 0.$$

Let $\alpha := \nu(C_{r_0}(0)) > 0$. Let $x \in \mathbb{R}$ be arbitrarily chosen. Let $r_x = d_e(0, x) + r_0$. Then we have $C_{r_x}(x) \supset C_{r_0}(0)$ so that $\nu(C_{r_x}(x)) \geq \mu(C_{r_0}(0)) = \alpha$. We have also

$$\mu_L^*(C_{r_x}(x)) = 2r_x = 2\{d_e(0, x) + r_0\} = 2\{|x| + r_0\}.$$

Then by (1) we have

$$(2) \qquad M_{\mu_L^*}\nu(x) = \sup_{r\in(0,\infty)} \frac{\nu(C_r(x))}{\mu_L^*(C_r(x))} \geq \frac{\nu(C_{r_x}(x))}{\mu_L^*(C_{r_x}(x))} \geq \frac{\alpha}{2r_0 + 2|x|}.$$

Then we have

$$\int_{\mathbb{R}} M_{\mu_L^*}\nu(x)\,\mu_L^*(dx) \geq \frac{\alpha}{2}\int_{\mathbb{R}} \frac{1}{r_0 + |x|}\,\mu_L^*(dx) = \infty.$$

Thus $M_{\mu_L^*}\nu$ is not μ_L^*-integrable on $\mathbb{R}$. ■

[I.2] Estimate of Hardy-Littlewood Maximal Function of an Outer Measure

Next we have an estimate of the Hardy-Littlewood maximal function $M_\mu\nu$ of a Borel outer measure ν with respect to a doubling Borel regular outer measure μ on a metric space (X, d).

Theorem 10.8. *Let μ and ν be Borel outer measures on a metric space (X, d). Assume further that μ is a doubling Borel regular outer measure. Then there exists a constant $C > 0$ such that*

$$\mu\{x \in X : M_\mu\nu(x) > \lambda\} \leq \frac{C}{\lambda}\nu(X) \quad \text{for all } \lambda \in (0,\infty).$$

Indeed if γ is a doubling constant for μ then $C := \gamma^3$ will do.

Proof. 1. For each $k \in \mathbb{N}$, let us define

$$M_\mu^k\nu(x) = \begin{cases} \displaystyle\sup_{r\in(0,k)} \frac{\nu(C_r(x))}{\mu(C_r(x))} & \text{if } \mu(C_r(x)) > 0 \text{ for all } r \in (0,\infty), \\ \infty & \text{if } \mu(C_r(x)) = 0 \text{ for some } r \in (0,\infty). \end{cases}$$

For $\lambda \in (0,\infty)$, let

$$(1) \qquad A_\lambda^k = \{x \in X : M_\mu^k\nu(x) > \lambda\}.$$

Then for every $x \in A_\lambda^k$ there exists $r_x \in (0, k)$ such that $\nu(C_{r_x}(x)) > \lambda\mu(C_{r_x}(x))$. Consider the collection of closed balls $\mathcal{B} = \{C_{r_x}(x) : x \in A_\lambda^k\}$. The radii of the members of $\mathcal{B}$ are uniformly bounded by k. Thus by Theorem 6.6 ($5r$-covering Theorem in a Metric Space) there exists a disjoint subcollection $\mathcal{G}$ of $\mathcal{B}$ such that

$$(2) \qquad A_\lambda^k \subset \bigcup_{C\in\mathcal{B}} C \subset \bigcup_{C\in\mathcal{G}} \widetilde{C},$$

where $\widetilde{C}$ is a closed ball concentric with the closed ball C but with a radius 5 times that of C. Now every $C \in \mathcal{B}$ is a set of the type $C_{r_x}(x)$ with $x \in A_\lambda^k$ so that $\nu(C) = \nu(C_{r_x}(x)) > \lambda\mu(C_{r_x}(x)) > 0$. (Note that $\mu(C_{r_x}(x)) > 0$ since μ is a doubling Borel regular outer measure.) Now if $\nu(X) = \infty$ then the theorem is trivially true. Thus let us consider the case $\nu(X) < \infty$. Now $\mathcal{G}$ is a disjoint collection of closed balls C with $\nu(C) > 0$. Then the condition $\nu(X) < \infty$ implies that $\mathcal{G}$ is at most a countable collection. Let $\mathcal{G} = \{C_i : i \in \mathbb{N}\}$. Then from (2)

$$\mu(A_\lambda^k) \leq \mu\Big(\bigcup_{C \in \mathcal{G}} \widetilde{C}\Big) = \mu\Big(\bigcup_{i \in \mathbb{N}} \widetilde{C}_i\Big) \leq \sum_{i \in \mathbb{N}} \mu(\widetilde{C}_i). \tag{3}$$

Since μ is a doubling Borel regular outer measure on X, there exists a constant $\gamma \geq 1$ such that $\mu(C_{2r}(x)) \leq \gamma\mu(C_r(x))$ for all $x \in X$ and $r \in (0, \infty)$. Then

$$\begin{aligned}\mu(C_{5r}(x)) &\leq \gamma\mu(C_{\frac{5}{2}r}(x)) \leq \gamma^2\mu(C_{\frac{5}{4}r}(x)) \\ &\leq \gamma^2\mu(C_{2r}(x)) \leq \gamma^3\mu(C_r(x)).\end{aligned}$$

Applying this to $\widetilde{C}_i$ which is concentric with C and has a radius 5 times that of C, we have $\mu(\widetilde{C}_i) \leq \gamma^3\mu(C_i)$. Substituting this in (3), we have

$$\begin{aligned}\mu(A_\lambda^k) &\leq \gamma^3 \sum_{i \in \mathbb{N}} \mu(C_i) \leq \frac{\gamma^3}{\lambda} \sum_{i \in \mathbb{N}} \nu(C_i) \\ &= \frac{\gamma^3}{\lambda}\nu\Big(\bigcup_{i \in \mathbb{N}} C_i\Big) \leq \frac{\gamma^3}{\lambda}\nu(X).\end{aligned} \tag{4}$$

2. Let us observe that for $x \in X$ such that $\mu(C_r(x)) > 0$ for all $r \in (0, \infty)$, we have

$$M_\mu\nu(x) = \sup_{r \in (0,\infty)} \frac{\nu(C_r(x))}{\mu(C_r(x))} = \lim_{k \to \infty} \sup_{r \in (0,k)} \frac{\nu(C_r(x))}{\mu(C_r(x))} = \lim_{k \to \infty} M_\mu^k\nu(x).$$

On the other hand, for $x \in X$ such that $\mu(C_r(x)) = 0$ for some $r \in (0, \infty)$, we have $M_\mu\nu(x) = \infty = M_\mu^k\nu(x)$ for all $k \in \mathbb{N}$. Thus, as $k \to \infty$, we have

$$M_\mu^k\nu(x) \uparrow M_\mu\nu(x) \quad \text{for every } x \in X.$$

For $\lambda \in (0, \infty)$, let

$$A_\lambda = \{x \in X : M_\mu\nu(x) > \lambda\}. \tag{5}$$

Since $M_\mu^k\nu \leq M_\mu\nu$ on X we have $A_\lambda^k \subset A_\lambda$ for every $k \in \mathbb{N}$ so that $\bigcup_{k \in \mathbb{N}} A_\lambda^k \subset A_\lambda$. On the other hand, if $x \in A_\lambda$ then $M_\mu\nu(x) > \lambda$. Since $M_\mu^k\nu(x) \uparrow M_\mu\nu(x)$ as $k \to \infty$ there exists $k \in \mathbb{N}$ such that $M_\mu^k\nu(x) > \lambda$ and then $x \in A_\lambda^k$. Thus every $x \in A_\lambda$ is contained in A_λ^k for some $k \in \mathbb{N}$. Then we have $A_\lambda \subset \bigcup_{k \in \mathbb{N}} A_\lambda^k$. Therefore we have $A_\lambda = \bigcup_{k \in \mathbb{N}} A_\lambda^k$. Since $(M_\mu^k\nu : k \in \mathbb{N})$ is an increasing sequence of functions, $(A_\lambda^k : k \in \mathbb{N})$ is an increasing sequence of sets. Thus $\bigcup_{k \in \mathbb{N}} A_\lambda^k = \lim_{n \to \infty} A_\lambda^k$ and hence $A_\lambda = \lim_{n \to \infty} A_\lambda^k$. Then since μ is a doubling Borel regular outer measure and thus a Borel regular outer measure, $A_\lambda^k \uparrow A_\lambda$

implies $\mu(A_\lambda) = \lim_{k\to\infty} \mu(A_\lambda^k)$ by Theorem 1.36. Then by (4) we have $\mu(A_\lambda) \le \frac{\gamma^3}{\lambda}\nu(X)$, that is, recalling (5), we have

$$\mu\{x \in X : M_\mu \nu(x) > \lambda\} \le \frac{\gamma^3}{\lambda}\nu(X).$$

Let $C = \gamma^3$ and the proof is complete. ∎

[II] Hardy-Littlewood Maximal Functions for Functions

Definition 10.9. *Let μ be a Borel outer measure on a metric space (X, d). Consider the measure space $(X, \mathfrak{M}(\mu), \mu)$. Let $f \in \mathcal{L}^1_{loc}(X, \mathfrak{M}(\mu), \mu)$. The Hardy-Littlewood maximal function for f with respect to μ is defined by*

$$M_\mu f(x) = \begin{cases} \sup\limits_{r\in(0,\infty)} \dfrac{1}{\mu(C_r(x))} \displaystyle\int_{C_r(x)} |f|\, d\mu & \text{if } \mu(C_r(x)) > 0 \text{ for all } r \in (0,\infty), \\ \infty & \text{if } \mu(C_r(x)) = 0 \text{ for some } r \in (0,\infty). \end{cases}$$

Note that $M_\mu|f| = M_\mu f$.

Remark 10.10. If the Borel outer measure μ on X is such that $\operatorname{supp}\mu = X$, that is, μ is such that $\mu(O) > 0$ for every non-empty open set O in X, then Definition 10.9 reduces to

$$M_\mu f(x) = \sup_{r\in(0,\infty)} \frac{1}{\mu(C_r(x))} \int_{C_r(x)} |f|\, d\mu \quad \text{for every } x \in X.$$

[II.1] Measurability of Hardy-Littlewood Maximal Function of a Function

Proposition 10.11. *Let μ be boundedly finite Borel outer measure on a metric space (X, d). Assume that* $\operatorname{supp}\mu = X$. *Let $f \in \mathcal{L}^1_{loc}(X, \mathfrak{M}(\mu), \mu)$. Then for each $r \in (0, \infty)$, $\frac{1}{\mu(C_r(x))}\int_{C_r(x)} |f|\, d\mu$ is a $\mathfrak{B}_X$-measurable function of $x \in X$.*

Proof. By Theorem 7.2(b), $\mu(C_r(x))$ is a $\mathfrak{B}_X$-measurable function of $x \in X$ for each $r \in (0,\infty)$. By Proposition 8.1, $\int_{C_r(x)} |f|\, d\mu$ is a $\mathfrak{B}_X$-measurable function of $x \in X$ for each $r \in (0,\infty)$. Since $\operatorname{supp}\mu = X$ we have $\mu(C_r(x)) > 0$ for every $x \in X$ and $r \in (0,\infty)$. Thus $\frac{1}{\mu(C_r(x))}\int_{C_r(x)} |f|\, d\mu$ is defined for all $x \in X$ and $r \in (0,\infty)$. Then the $\mathfrak{B}_X$-measurability of $\mu(C_r(x))$ and $\int_{C_r(x)} |f|\, d\mu$ as functions of $x \in X$ implies that of their quotient $\frac{1}{\mu(C_r(x))}\int_{C_r(x)} |f|\, d\mu$. ∎

Theorem 10.12. *Let μ be a boundedly finite Borel outer measures on a metric space (X, d). Assume that* $\operatorname{supp}\mu = X$. *Suppose μ satisfies the additional conditions that $\mu(C_r(x))$ is* *a continuous functions of $r \in (0,\infty)$ for each fixed $x \in X$. Consider the measure space $(X, \mathfrak{M}(\mu), \mu)$. Let $f \in \mathcal{L}^1_{loc}(X, \mathfrak{M}(\mu), \mu)$. Then the Hardy-Littlewood maximal function $M_\mu f(x)$ of f with respect to μ is a $\mathfrak{B}_X$-measurable function of $x \in X$.*

Proof. The assumption $\operatorname{supp}\mu = X$ implies according to Remark 10.10 that we have

$$(1) \qquad M_\mu f(x) = \sup_{r \in (0,\infty)} \frac{1}{\mu\big(C_r(x)\big)} \int_{C_r(x)} |f| \, d\mu \quad \text{for every } x \in X.$$

By Proposition 10.11, for each $r \in (0,\infty)$, $\frac{1}{\mu(C_r(x))} \int_{C_r(x)} |f| \, d\mu$ is a $\mathfrak{B}_X$-measurable function of $x \in X$. If $\mu\big(C_r(x)\big)$ is a continuous functions of $r \in (0,\infty)$ for each fixed $x \in X$, then by (b) of Theorem 8.7, $\frac{1}{\mu(C_r(x))} \int_{C_r(x)} |f| \, d\mu$ is a continuous function of $r \in (0,\infty)$ for each $x \in X$. Then for an arbitrary countable dense subset $\{r_k : k \in \mathbb{N}\}$ of $(0,\infty)$, we have

$$(2) \qquad \sup_{r \in (0,\infty)} \frac{1}{\mu\big(C_r(x)\big)} \int_{C_r(x)} |f| \, d\mu = \sup_{k \in \mathbb{N}} \frac{1}{\mu\big(C_{r_k}(x)\big)} \int_{C_{r_k}(x)} |f| \, d\mu \quad \text{for } x \in X.$$

Since $\frac{1}{\mu(C_{r_k}(x))} \int_{C_{r_k}(x)} |f| \, d\mu$ is a $\mathfrak{B}_X$-measurable function of $x \in X$ for every $k \in \mathbb{N}$, $\sup_{k \in \mathbb{N}} \frac{1}{\mu(C_{r_k}(x))} \int_{C_{r_k}(x)} |f| \, d\mu$ is a $\mathfrak{B}_X$-measurable function of $x \in X$. Then by (1) and (2), $M_\mu f(x)$ is a $\mathfrak{B}_X$-measurable function of $x \in X$. ∎

[II.2] Estimates of Hardy-Littlewood Maximal Function of a Function

The following estimate of Hardy-Littlewood maximal function of a function does not require measurability of the Hardy-Littlewood maximal function.

Theorem 10.13. *Let μ be a doubling Borel regular outer measure on a metric space (X, d). Consider the measure space $(X, \mathfrak{M}(\mu), \mu)$. Let $f \in \mathcal{L}^1_{loc}(X, \mathfrak{M}(\mu), \mu)$. Then there exists a constant $C > 0$ such that*

$$\mu\{x \in X : M_\mu f(x) > \lambda\} \le \frac{C}{\lambda} \int_X |f| \, d\mu \quad \textit{for all } \lambda \in (0,\infty).$$

Indeed if γ is a doubling constant for μ then $C := \gamma^3$ will do.

Proof. If μ is a doubling Borel regular outer measure then μ is certainly a Borel outer measure. Let $f \in \mathcal{L}^1_{loc}(X, \mathfrak{M}(\mu), \mu)$. Let us define a set function ν on the σ-algebra $\mathfrak{M}(\mu)$ by setting

$$(1) \qquad \nu(E) = \int_E |f| \, d\mu \quad \text{for } E \in \mathfrak{M}(\mu).$$

Then ν is a measure on $\mathfrak{M}(\mu)$. Since $\mathfrak{B}_X \subset \mathfrak{M}(\mu)$, ν is actually a Borel measure. According to Theorem 10.8, if $C = \gamma^3$ where γ is a doubling constant for μ then

$$(2) \qquad \mu\{x \in X : M_\mu \nu(x) > \lambda\} \le \frac{C}{\lambda}\nu(X) \quad \text{for all } \lambda \in (0,\infty).$$

Now by (1) we have $\nu(C_r(x)) = \int_{C_r(x)} |f|\,d\mu$ and $\nu(X) = \int_X |f|\,d\mu$. By Definition 10.1

$$M_\mu \nu(x) = \sup_{r\in(0,\infty)} \frac{\nu(C_r(x))}{\mu(C_r(x))} = \sup_{r\in(0,\infty)} \frac{\int_{C_r(x)} |f|\,d\mu}{\mu(C_r(x))}$$

if $\mu(C_r(x)) > 0$ for all $r \in (0,\infty)$ and $M_\mu\nu(x) = \infty$ if $\mu(C_r(x)) = 0$ for some $r \in (0,\infty)$. By the expression above of $M_\mu\nu$ and by the definition of $M_\mu f$ in Definition 10.9, we have $M_\mu\nu(x) = M_\mu f(x)$ for $x \in X$. Then (2) can be written as

$$\mu\{x \in X : M_\mu f(x) > \lambda\} \le \frac{C}{\lambda}\int_X |f|\,d\mu \quad \text{for all } \lambda \in (0,\infty).$$

This completes the proof. ■

The following estimate requires the measurability of the Hardy-Littlewood maximal function as it involves integration of the maximal function.

Theorem 10.14. *Let μ be a doubling Borel regular outer measure on a metric space (X,d) satisfying the following condition:*
1° $\mu(C_r(x))$ is a continuous function of $r \in (0,\infty)$ for each fixed $x \in X$.
Consider the measure space $(X, \mathfrak{M}(\mu), \mu)$. Let $f \in L^p(X, \mathfrak{M}(\mu), \mu)$ for some $p \in (1,\infty)$. Then there exists $C_p > 0$ such that

$$\|M_\mu f\|_p \le C_p \|F\|_p.$$

Indeed if γ is a doubling constant of μ then $C_p := 2\gamma^{3/p}\big(\frac{p}{p-1}\big)^{1/p}$ will do.

Proof. Let $f \in L^p(X, \mathfrak{M}(\mu), \mu)$ where $p \in (1,\infty)$. Assume further that $f \ge 0$ on X. For $t \in (0,\infty)$ we have

$$(1) \qquad f = \Big(f - \frac{t}{2}\Big) + \frac{t}{2} \le \Big(f - \frac{t}{2}\Big)^+ + \frac{t}{2} \quad \text{on } X.$$

Thus we have

$$(2) \qquad \begin{aligned} M_\mu f(x) &= \sup_{r\in(0,\infty)} \frac{\int_{C_r(x)} f\,d\mu}{\mu(C_r(x))} \\ &\le \sup_{r\in(0,\infty)} \frac{\int_{C_r(x)} \{(f-\frac{t}{2})^+ + \frac{t}{2}\}\,d\mu}{\mu(C_r(x))} \\ &= M_\mu(f-\tfrac{t}{2})^+(x) + \tfrac{t}{2}. \end{aligned}$$

Now by (2) we have

$$\text{(3)} \qquad \{x \in X : M_\mu f(x) > t\} \subset \{x \in X : M_\mu (f - \tfrac{t}{2})^+(x) + \tfrac{t}{2} > t\}$$
$$= \{x \in X : M_\mu (f - \tfrac{t}{2})^+(x) > \tfrac{t}{2}\}.$$

According to Cavalieri's Theorem (see Theorem 23.68, [LRA]), if $(X, \mathfrak{A}, \mu)$ is a σ-finite measure space and f is a nonnegative real-valued $\mathfrak{A}$-measurable function on X and if $p \geq 1$ then we have

$$\int_X (f(x))^p \, \mu(dx) = p \int_{[0,\infty)} t^{p-1} \mu\{x \in X : f(x) > t\} \, \mu_L(dt),$$

where μ_L is the Lebesgue measure on $\mathbb{R}$.

Now a doubling Borel regular outer measure is a σ-finite outer measure by Observation 6.14. By Theorem 10.12, condition 1° implies that $M_\mu f$ is a $\mathfrak{B}_X$-measurable function on X. Thus by Cavalieri's Theorem and by (3) we have

$$\text{(4)} \qquad \int_X (M_\mu f(x))^p \, \mu(dx)$$
$$= p \int_{[0,\infty)} t^{p-1} \mu\{x \in X : M_\mu f(x) > t\} \, \mu_L(dt) p$$
$$\leq p \int_{[0,\infty)} t^{p-1} \mu\{x \in X : M_\mu (f - \tfrac{t}{2})^+(x) > \tfrac{t}{2}\} \, \mu_L(dt).$$

Letting $u = \frac{t}{2}$, we have

$$\int_{[0,\infty)} t^{p-1} \mu\{x \in X : M_\mu (f - \tfrac{t}{2})^+(x) > \tfrac{t}{2}\} \, \mu_L(dt)$$
$$= \int_{[0,\infty)} (2u)^{p-1} \mu\{x \in X : M_\mu (f - u)^+(x) > u\} \, 2\mu_L(du)$$
$$= 2^p \int_{[0,\infty)} u^{p-1} \mu\{x \in X : M_\mu (f - u)^+(x) > u\} \, \mu_L(du).$$

Substituting this in (4) we have

$$\text{(5)} \qquad \int_X (M_\mu f(x))^p \, \mu(dx)$$
$$\leq p \, 2^p \int_{[0,\infty)} u^{p-1} \mu\{x \in X : M_\mu (f - u)^+(x) > u\} \, \mu_L(du).$$

According to Theorem 10.13, we have for all $\lambda \in (0, \infty)$

$$\mu\{x \in X : M_\mu f(x) > \lambda\} \leq \frac{\gamma^3}{\lambda} \int_X |f| \, d\mu.$$

Then we have

$$\begin{aligned}
&\int_X \left(M_\mu f(x)\right)^p \mu(dx) \\
&\leq p\,2^p \int_{[0,\infty)} u^{p-1}\frac{\gamma^3}{u}\Big\{\int_X (f-u)^+(x)\,\mu(dx)\Big\}\,\mu_L(du) \\
&= \gamma^3 p\,2^p \int_{[0,\infty)} u^{p-2}\Big\{\int_{\{x\in X: f(x)>u\}} \{f(x)-u\}\,\mu(dx)\Big\}\,\mu_L(du) \\
&\leq \gamma^3 p\,2^p \int_{[0,\infty)} u^{p-2}\Big\{\int_X f(x)\mathbf{1}_{\{x\in X: f(x)>u\}}(x)\,\mu(dx)\Big\}\,\mu_L(du) \\
&= \gamma^3 p\,2^p \int_X f(x)\Big\{\int_{[0,f(x))} u^{p-2}\,\mu_L(du)\Big\}\,\mu(dx) \\
&= \gamma^3 p\,2^p \int_X f(x)\frac{1}{p-1}f(x)^{p-1}\,\mu(dx) \\
&= \gamma^3 2^p \frac{p}{p-1}\int_X f(x)^p\,\mu(dx) \\
&= \gamma^3 2^p \frac{p}{p-1}\|f\|_p^p.
\end{aligned}$$

Thus taking the p-th roots we have

$$\|M_\mu f\|_p \leq C_p\|f\|_p \quad \text{where } C_p := 2\gamma^{3/p}\Big(\frac{p}{p-1}\Big)^{1/p}.$$

This proves the theorem for the particular case that $f \geq 0$ on X.

Consider the general case that $f \in L^p(X, \mathfrak{M}(\mu), \mu)$. Consider $|f|$. Then by our result above we have $\big\|M_\mu|f|\big\|_p \leq C_p\big\||f|\big\|_p$. But $M_\mu f = M_\mu|f|$ and $\big\||f|\big\|_p = \|f\|_p$. Thus we have $\|M_\mu f\| \leq C_p\|f\|_p$. ■

Applying Theorem 10.13 we have an alternate proof for the Lebesgue differentiation theorem for integrals with respect to a doubling Borel regular outer measure on Metric Space. Compare this with Theorem 9.8.

Theorem 10.15. (Lebesgue Differentiation Theorem for Integrals with Respect to a Doubling Borel Regular Outer Measure on a Metric Space) *Let (X, d) be a metric space and let μ be a doubling Borel regular outer measure on X satisfying the following condition:*
1° for every $E \in \mathfrak{M}(\mu)$ and $\varepsilon > 0$ there exists a closed set $F \subset E$ such that $\mu(E \setminus F) < \varepsilon$.
Consider the measure space $(X, \mathfrak{M}(\mu), \mu)$. Let $f \in \mathcal{L}^1_{\text{loc}}(X, \mathfrak{M}(\mu), \mu)$. Then there exists a null set N in $(X, \mathfrak{B}_X, \mu)$ such that

$$\lim_{r\to 0}\frac{1}{\mu(C_r(x))}\int_{C_r(x)} |f(y) - f(x)|\,\mu(dy) \quad \textit{for } x \in N^c,$$

and in particular

$$\lim_{r\to 0}\frac{1}{\mu(C_r(x))}\int_{C_r(x)} f(y)\,\mu(dy) = f(x) \quad \textit{for } x\in N^c.$$

Proof. 1. Let $f\in\mathcal{L}^1_{loc}(X,\mathfrak{M}(\mu),\mu)$. Let us consider first the case that there exists an open ball in X, denoted by X_0, such that $f=0$ on X_0^c. Since μ is a doubling Borel regular outer measure on X we have $X_0\in\mathfrak{B}_X\subset\mathfrak{M}(\mu)$ and $\mu(X_0)\in(0,\infty)$.

Consider the measure space $(X_0,\mathfrak{M}(\mu)\cap X_0,\mu)$. Since $\mu(X_0)<\infty$, Theorem 14.9 is applicable to $(X_0,\mathfrak{M}(\mu)\cap X_0,\mu)$. Thus for every $i\in\mathbb{N}$ there exists a continuous function f_i on X_0 such that

$$(1)\qquad \int_{X_0}|f-f_i|\,d\mu<\frac{1}{2^i}.$$

Let us extend the domain of definition of f_i from X_0 to X by setting $f_i=0$ on X_0^c. Note that whereas f_i is continuous on X_0, f_i need not be continuous on X.

Now since $f\in\mathcal{L}^1_{loc}(X,\mathfrak{M}(\mu),\mu)$ and X_0 is a bounded set in X, f is μ-integrable on X_0. Then since $f=0$ on X_0^c, f is μ-integrable on X. Then since $f_i=0$ on X_0^c we have

$$\int_X|f_i|\,d\mu=\int_{X_0}|f_i|\,d\mu\le\int_{X_0}|f_i-f|\,d\mu+\int_{X_0}|f|\,d\mu<\frac{1}{2^i}+\int_{X_0}|f|\,d\mu<\infty.$$

Thus f_i is f is μ-integrable on X_0. Therefore $f,f_i\in L^1(X,\mathfrak{M}(\mu),\mu)$. For $i\in\mathbb{N}$, let

$$(2)\qquad g_i=f-f_i \quad \text{on X.}$$

Then from (1) we have

$$(3)\qquad \|g_i\|_1=\int_X|f-f_i|\,d\mu=\int_{X_0}|f-f_i|\,d\mu<\frac{1}{2^i}.$$

Let

$$(4)\qquad N_i=\left\{x\in X:|g_i(x)|>\left(\tfrac{2}{3}\right)^i\right\}\cup\left\{x\in X:M_\mu g_i(x)>\left(\tfrac{2}{3}\right)^i\right\}\in\mathfrak{M}(\mu).$$

By Markov's Inequality, for an arbitrary measure space $(X,\mathfrak{A},\mu)$, if $\varphi\in L^p(X,\mathfrak{A},\mu)$ where $p\ge 1$ then for every $t>0$ we have

$$\mu\{x\in X:|\varphi(x)|\ge t\}\le t^{-p}\|\varphi\|_p^p.$$

Applying Markov's Inequality and Theorem 10.13, we have

$$(5)\qquad \begin{aligned}\mu(N_i)&\le\mu\left\{x\in X:|g_i(x)|>\left(\tfrac{2}{3}\right)^i\right\}+\mu\left\{x\in X:M_\mu g_i(x)>\left(\tfrac{2}{3}\right)^i\right\}\\&\le\left(\tfrac{3}{2}\right)^i\|g_i\|_1+\left(\tfrac{3}{2}\right)^iC\|g_i\|_1=\left(\tfrac{3}{2}\right)^i(1+C)\|g_i\|_1\\&\le\left(\tfrac{3}{4}\right)^i(1+C).\end{aligned}$$

Let

$$N = \limsup_{i\to\infty} N_i = \bigcap_{k\geq 1}\bigcup_{i\geq k} N_i \in \mathfrak{M}(\mu). \tag{6}$$

Then we have

$$\mu\left(\bigcup_{i\geq k} N_i\right) \leq \sum_{i\geq k} \mu(N_i) \leq (1+C)\sum_{i\geq k}\left(\tfrac{3}{4}\right)^i = 4(1+C)\left(\tfrac{3}{4}\right)^k$$

and then

$$\mu(N) = \mu\left(\bigcap_{k\geq 1}\bigcup_{i\geq k} N_i\right) = \lim_{k\to\infty}\mu\left(\bigcup_{i\geq k} N_i\right) = 4(1+C)\lim_{k\to\infty}\left(\tfrac{3}{4}\right)^k = 0. \tag{7}$$

Thus N is a null set in $(X, \mathfrak{M}(\mu), \mu)$.

Let us show that

$$\lim_{r\to 0}\frac{1}{\mu(C_r(x))}\int_{C_r(x)} |f(y) - f(x)|\,\mu(dy) = 0 \quad \text{for } x \in X_0 \cap N^c. \tag{8}$$

Let $x \in X_0 \cap N^c$ be arbitrarily chosen and fixed. Now $N^c = \bigcup_{k\geq 1}\bigcap_{i\geq k} N_i^c$. Since $x \in N^c$, we have $x \in \bigcap_{i\geq k_0} N_i^c$ for some $k_0 \in \mathbb{N}$ and hence $x \in N_i^c$ for all $i \geq k_0$. Thus by (4)

$$|g_i(x)| \leq \left(\tfrac{2}{3}\right)^i \text{ and } M_\mu g_i(x) \leq \left(\tfrac{2}{3}\right)^i \quad \text{for } i \geq k_0. \tag{9}$$

Now since $f = f_i + g_i$, we have $f(y) - f(x) = \{f_i(y) + g_i(y)\} + \{f_i(x) + g_i(x)\}$ and hence $|f(y) - f(x)| \leq |f_i(y) - f_i(x)| + |g_i(y) - g_i(x)|$. Thus we have

$$\begin{aligned}
&\limsup_{r\to 0}\frac{1}{\mu(C_r(x))}\int_{C_r(x)} |f(y) - f(x)|\,\mu(dy) \\
&\leq \limsup_{r\to 0}\frac{1}{\mu(C_r(x))}\int_{C_r(x)} |f_i(y) - f_i(x)|\,\mu(dy) \\
&\quad + \limsup_{r\to 0}\frac{1}{\mu(C_r(x))}\int_{C_r(x)} |g_i(y) - g_i(x)|\,\mu(dy).
\end{aligned} \tag{10}$$

Now f_i is continuous on the open ball X_0 and $x \in X_0$. Thus for every $\varepsilon > 0$ there exists $\delta > 0$ such that $|f(y) - f(x)| < \varepsilon$ for $y \in C_r(x)$ when $r \in (0, \delta)$. Then we have

$$\frac{1}{\mu(C_r(x))}\int_{C_r(x)} |f_i(y) - f_i(x)|\,\mu(dy) \leq \varepsilon \quad \text{for all } r \in (0, \delta)$$

and then

$$\limsup_{r\to 0}\frac{1}{\mu(C_r(x))}\int_{C_r(x)} |f_i(y) - f_i(x)|\,\mu(dy) \leq \varepsilon.$$

Since this holds for an arbitrary $\varepsilon > 0$ we have

$$\limsup_{r \to 0} \frac{1}{\mu(C_r(x))} \int_{C_r(x)} |f_i(y) - f_i(x)| \, \mu(dy) = 0. \tag{11}$$

Estimating the second term on the right side of the inequality (10), we have

$$\begin{aligned}
&\frac{1}{\mu(C_r(x))} \int_{C_r(x)} |g_i(y) - g_i(x)| \, \mu(dy) \\
&\leq \frac{1}{\mu(C_r(x))} \int_{C_r(x)} |g_i(y)| \, \mu(dy) + \frac{1}{\mu(C_r(x))} \int_{C_r(x)} |g_i(x)| \, \mu(dy) \\
&\leq \sup_{r \in (0,\infty)} \frac{1}{\mu(C_r(x))} \int_{C_r(x)} |g_i(y)| \, \mu(dy) + |g_i(x)| \\
&= M_\mu g_i(x) + |g_i(x)| \\
&\leq \left(\tfrac{2}{3}\right)^i + \left(\tfrac{2}{3}\right)^i = 2\left(\tfrac{2}{3}\right)^i \quad \text{for } i \geq k_0
\end{aligned}$$

where the last inequality is by (9). Thus we have

$$\limsup_{r \to 0} \frac{1}{\mu(C_r(x))} \int_{C_r(x)} |g_i(y) - g_i(x)| \, \mu(dy) \leq 2\left(\tfrac{2}{3}\right)^i \quad \text{for } i \geq k_0. \tag{12}$$

Substituting (11) and (12) in (10) we have

$$\limsup_{r \to 0} \frac{1}{\mu(C_r(x))} \int_{C_r(x)} |f(y) - f(x)| \, \mu(dy) \leq 2\left(\tfrac{2}{3}\right)^i \quad \text{for } i \geq k_0.$$

Letting $i \to \infty$, we have (8). Now since μ is a doubling Borel regular outer measure on X, μ is a Borel regular outer measure on X. Then for every subset A of X there exists $B \in \mathfrak{B}_X$ such that $A \subset B$ and $\mu(A) = \mu(B)$. Thus for our $N \in \mathfrak{M}(\mu)$ there exists $N_0 \in \mathfrak{B}_X$ such that $N \subset N_0$ and $\mu(N_0) = \mu(N) = 0$. Note that $N_0^c \subset N^c$ and $X_0 \cap N_0^c \subset X_0 \cap N^c$. Thus by (8), there exists a null set N_0 in $(X, \mathfrak{B}_X, \mu)$ such that

$$\lim_{r \to 0} \frac{1}{\mu(C_r(x))} \int_{C_r(x)} |f(y) - f(x)| \, \mu(dy) = 0 \quad \text{for } x \in X_0 \cap N_0^c. \tag{13}$$

2. Now consider the general case that $f \in \mathcal{L}^1_{loc}(X, \mathfrak{M}(\mu), \mu)$. Let $x_0 \in X$ be arbitrarily chosen and consider the open balls $\{B_k(x_0) : k \in \mathbb{N}\}$. We have $X = \bigcup_{k \in \mathbb{N}} B_k(x_0)$. For each $k \in \mathbb{N}$ let us define a function f_k on X by

$$f_k(x) = \begin{cases} f(x) & \text{for } x \in B_k(x_0), \\ 0 & \text{for } x \in B_k(x_0)^c. \end{cases}$$

Then by (13) there exists a null set N_k in $(X, \mathfrak{B}_X, \mu)$ such that

$$\lim_{r \to 0} \frac{1}{\mu(C_r(x))} \int_{C_r(x)} |f_k(y) - f_k(x)| \, \mu(dy) = 0 \quad \text{for } x \in B_k(x_0) \cap N_k^c. \tag{14}$$

Let $N = \bigcup_{k \in \mathbb{N}} N_k$, a null set in $(X, \mathfrak{B}_X, \mu)$. Let $x \in N^c$. Since $X = \bigcup_{k \in \mathbb{N}} B_k(x_0)$, we have $x \in B_k(x_0) \cap N^c \subset B_k(x_0) \cap N_k^c$ for some $k \in \mathbb{N}$. For our $x \in N^c$ the equality (14) holds. Since x is in the open ball $B_k(x_0)$, we have $C_r(x) \subset B_k(x_0)$ for sufficiently small $r > 0$. Then we have $f_k(y) = f(y)$ for $y \in C_r(x) \subset B_k(x_0)$. Thus we have

$$\begin{aligned}&\frac{1}{\mu(C_r(x))} \int_{C_r(x)} |f(y) - f(x)| \, \mu(dy) \\ &= \frac{1}{\mu(C_r(x))} \int_{C_r(x)} |f_k(y) - f_k(x)| \, \mu(dy)\end{aligned}$$

for sufficient small $r > 0$. Then for our $x \in N^c$ we have

$$\begin{aligned}&\lim_{r \to 0} \frac{1}{\mu(C_r(x))} \int_{C_r(x)} |f(y) - f(x)| \, \mu(dy) \\ &= \lim_{r \to 0} \frac{1}{\mu(C_r(x))} \int_{C_r(x)} |f_k(y) - f_k(x)| \, \mu(dy) \\ &= 0.\end{aligned}$$

This completes the proof. ∎

[III] Estimates of Hardy-Littlewood Maximal Functions on $\mathbb{R}^n$

By means of the Besicovitch Covering Theorem on $\mathbb{R}^n$ (Theorem 6.24), which is independent of the outer measures on $\mathbb{R}^n$, we can derive estimates for Hardy-Littlewood maximal functions that are valid for all Borel outer measures on $\mathbb{R}^n$.

Theorem 10.16. *Let μ and ν be Borel outer measures on $\mathbb{R}^n$. Then for the Hardy-Littlewood maximal function $M_\mu \nu$ of ν with respect to μ there exists $C > 0$ such that*

$$\mu\{x \in \mathbb{R}^n : M_\mu \nu(x) > \lambda\} \leq \frac{C}{\lambda} \nu(\mathbb{R}^n) \quad \textit{for all } \lambda \in (0, \infty).$$

Indeed the constant $P(n)$ in the Besicovitch Covering Theorem on $\mathbb{R}^n$ (Theorem 6.24) will do as C.

Proof. For each $k \in \mathbb{N}$ let us define

$$M_\mu^k \nu(x) = \begin{cases} \displaystyle\sup_{r \in (0,k)} \frac{\nu(C_r(x))}{\mu(C_r(x))} & \text{if } \mu(C_r(x)) > 0 \text{ for all } r \in (0, \infty), \\ \infty & \text{if } \mu(C_r(x)) = 0 \text{ for some } r \in (0, \infty). \end{cases}$$

For $\lambda \in (0, \infty)$, let

$$A_\lambda^k = \{x \in C_k(0) : M_\lambda^k \nu(x) > \lambda\}. \tag{1}$$

If $x \in A_\lambda^k$ then there exists $r(x) \in (0, k)$ such that $\nu\big(C_{r(x)}(x)\big) > \lambda\mu\big(C_{r(x)}(x)\big)$. Consider the collection of closed balls $\mathcal{B} = \big\{C_{r(x)}(x) : x \in A_\lambda^k\big\}$. Now $A_\lambda^k \subset C_k(0)$ is a bounded set. Also $\sup\big\{r(x) : x \in A_\lambda^k\big\} \leq k < \infty$. Thus by Theorem 6.24 (Besicovitch Covering Theorem) there exists a countable subcollection of $\mathcal{B}$, say $\big\{C_{r(x_i)}(x_i) : i \in \mathbb{N}\big\}$ such that

$$\mathbf{1}_{A_\lambda^k} \leq \sum_{i\in\mathbb{N}} \mathbf{1}_{C_{r(x_i)}(x_i)} \leq P(n) \quad \text{on } \mathbb{R}^n, \tag{2}$$

where $P(n)$ is a positive integer. Then we have

$$\begin{aligned}
\mu(A_\lambda^k) &= \int_{\mathbb{R}^n} \mathbf{1}_{A_\lambda^k}\, d\mu \leq \int_{\mathbb{R}^n} \sum_{i\in\mathbb{N}} \mathbf{1}_{C_{r(x_i)}(x_i)}\, d\mu \\
&= \sum_{i\in\mathbb{N}} \int_{\mathbb{R}^n} \mathbf{1}_{C_{r(x_i)}(x_i)}\, d\mu = \sum_{i\in\mathbb{N}} \mu\big(C_{r(x_i)}(x_i)\big) \\
&\leq \frac{1}{\lambda} \sum_{i\in\mathbb{N}} \nu\big(C_{r(x_i)}(x_i)\big) = \frac{1}{\lambda} \sum_{i\in\mathbb{N}} \int_{\mathbb{R}^n} \mathbf{1}_{C_{r(x_i)}(x_i)}\, d\nu \\
&= \frac{1}{\lambda} \int_{\mathbb{R}^n} \sum_{i\in\mathbb{N}} \mathbf{1}_{C_{r(x_i)}(x_i)}\, d\nu \leq \frac{1}{\lambda} \int_{\mathbb{R}^n} P(n)\, d\nu \\
&= \frac{P(n)}{\lambda} \nu(\mathbb{R}^n).
\end{aligned} \tag{3}$$

2. Let us observe that for $x \in \mathbb{R}^n$ such that $\mu\big(C_r(x)\big) > 0$ for all $r \in (0, \infty)$ we have

$$M_\mu \nu(x) = \sup_{r\in(0,\infty)} \frac{\nu\big(C_r(x)\big)}{\mu\big(C_r(x)\big)} = \lim_{k\to\infty} \sup_{r\in(0,k)} \frac{\nu\big(C_r(x)\big)}{\mu\big(C_r(x)\big)} = \lim_{k\to\infty} M_\mu^k \nu(x). \tag{4}$$

On the other hand for $x \in \mathbb{R}^n$ such that $\mu\big(C_r(x)\big) = 0$ for some $r \in (0, \infty)$, we have $M_\mu\nu(x) = \infty = M_\mu^k\nu(x)$ for all $k \in \mathbb{N}$. Thus as $k \to \infty$ we have

$$M_\mu^k \nu(x) \uparrow M_\mu \nu(x) \quad \text{for all } x \in \mathbb{R}^n. \tag{5}$$

For $\lambda \in (0, \infty)$, let

$$A_\lambda = \big\{x \in \mathbb{R}^n : M_\mu\nu(x) > \lambda\big\}.$$

It is easily verified that $A_\lambda \supset A_\lambda^k$ for every $k \in \mathbb{N}$ so that $A_\lambda \supset \bigcup_{k\in\mathbb{N}} A_\lambda^k$ and on the other hand if $x \in A_\lambda$ then $x \in A_\lambda^k$ for some $k \in \mathbb{N}$ so that $A_\lambda \subset \bigcup_{\in\mathbb{N}} A_\lambda^k$ and hence $A_\lambda = \bigcup_{\in\mathbb{N}} A_\lambda^k = \lim_{k\to\infty} A_\lambda^k$. Note that since $\big(M_\mu^k : k \in \mathbb{N}\big)$ is an increasing sequence by (5), $\big(A_\lambda^k : k \in \mathbb{N}\big)$ is an increasing sequence. Then by (3) we have

$$\mu(A_\lambda) \leq \frac{P(n)}{\lambda} \nu(\mathbb{R}^n).$$

This completes the proof. ∎

Theorem 10.17. *Let μ be Borel outer measure on $\mathbb{R}^n$. Consider the measure space $(\mathbb{R}^n, \mathfrak{M}(\mu), \mu)$. Let $f \in \mathcal{L}^1_{loc}(\mathbb{R}^n, \mathfrak{M}(\mu), \mu)$. Then for the Hardy-Littlewood maximal function $M_\mu f$ of f with respect to μ there exists $C > 0$ such that*

$$\mu\{x \in \mathbb{R}^n : M_\mu f(x) > \lambda\} \leq \frac{C}{\lambda} \int_{\mathbb{R}^n} |f| \, d\mu \quad \text{for all } \lambda \in (0, \infty).$$

Indeed the constant $P(n)$ in the Besicovitch Covering Theorem on $\mathbb{R}^n$ (Theorem 6.24) will do as C.

Proof. Theorem 10.17 follows from Theorem 10.16 in the same way that Theorem 10.13 followed from Theorem 10.8.

Let $f \in \mathcal{L}^1_{loc}(\mathbb{R}^n, \mathfrak{M}(\mu), \mu)$. Let us define a set function ν on the σ-algebra $\mathfrak{M}(\mu)$ by setting

$$\nu(E) = \int_E |f| \, d\mu \quad \text{for } E \in \mathfrak{M}(\mu).$$

Then ν is a measure on $\mathfrak{M}(\mu)$ and since $\mathfrak{B}_{\mathbb{R}^n} \subset \mathfrak{M}(\mu)$, ν is a Borel measure. The rest of the proof follows by applying Theorem 10.16 in the same way that Theorem 10.13 was proved by applying Theorem 10.8. ■

Theorem 10.18. *Let μ be a Borel outer measure on $\mathbb{R}^n$ satisfying the following conditions:*
1° μ is σ-finite.
2° $\mu(C_r(x))$ is a continuous function of $r \in (0, \infty)$ for each fixed $x \in \mathbb{R}^n$.
Consider the measure space $(\mathbb{R}^n, \mathfrak{M}(\mu), \mu)$. Let $f \in L^p(\mathbb{R}^n, \mathfrak{M}(\mu), \mu)$ for some $p \in (1, \infty)$. Then for the Hardy-Littlewood maximal function $M_\mu f$ of f with respect to μ there exists $C_p > 0$ such that

$$\|M_\mu f\|_p \leq C_p \|f\|_p.$$

Proof. 1. Let $f \in L^p(\mathbb{R}^n, \mathfrak{M}(\mu), \mu)$ and $f \geq 0$ on $\mathbb{R}^n$. Let $t > 0$ and define

$$g(x) = \begin{cases} f(x) & \text{if } f(x) \geq \frac{t}{2}, \\ 0 & \text{if } f(x) < \frac{t}{2}. \end{cases} \tag{1}$$

Then $f \leq g + \frac{t}{2}$ on $\mathbb{R}^n$ and hence $M_\mu f \leq M_\mu g + \frac{t}{2}$ on $\mathbb{R}^n$. Therefore we have

$$\{x \in \mathbb{R}^n : M_\mu f(x) > t\} \subset \{x \in \mathbb{R}^n : M_\mu g(x) > \tfrac{t}{2}\}$$

and then

$$\begin{aligned} \mu\{x \in \mathbb{R}^n : M_\mu f(x) > t\} &\leq \mu\{x \in \mathbb{R}^n : M_\mu g(x) > \tfrac{t}{2}\} \\ &\leq \frac{2}{t} C \int_{\mathbb{R}^n} g \, d\mu \\ &= \frac{2}{t} C \int_{\{\mathbb{R}^n : f \geq \frac{t}{2}\}} f \, d\mu \end{aligned} \tag{2}$$

where the second inequality is by Theorem 10.17 and the equality is by the definition of g by (1).

According to Cavalieri's Theorem (see Theorem 23.68, [LRA]), if $(X, \mathfrak{A}, \mu)$ is a σ-finite measure space and f is a nonnegative real-valued $\mathfrak{A}$-measurable function on X and if $p \geq 1$ then we have

$$\int_X (f(x))^p \, \mu(dx) = p \int_{[0,\infty)} t^{p-1} \mu\{x \in X : f(x) > t\} \, \mu_L(dt),$$

where μ_L is the Lebesgue measure on $\mathbb{R}$.

By condition 1° our Borel outer measure μ is σ-finite. Also condition 2° implies that $M_\mu f$ is a $\mathfrak{B}_X$-measurable function on X according to Theorem 10.12. Thus by Cavalieri's Theorem we have

$$\begin{aligned}
\int_{\mathbb{R}^n} (M_\mu f(x))^p \, \mu(dx) &= p \int_{[0,\infty)} t^{p-1} \mu\{x \in \mathbb{R}^n : M_\mu f(x) > t\} \, \mu_L(dt) \\
&\leq 2C\, p \int_{[0,\infty)} t^{p-2} \left\{ \int_{\{x \in \mathbb{R}^n : 2f(x) \geq t\}} f(x) \, \mu(dx) \right\} \mu_L(dt) \\
&= 2C\, p \int_{\mathbb{R}^n} f(x) \left\{ \int_{[0, 2f(x))} t^{p-2} \, \mu_L(dt) \right\} \mu(dx) \\
&= 2C\, p \int_{\mathbb{R}^n} f(x) \frac{1}{p-1} (2f(x))^{p-1} \, \mu(dx) \\
&= 2^p C \frac{p}{p-1} \int_{\mathbb{R}^n} f(x)^p \, \mu(dx) \\
&= 2^p C \frac{p}{p-1} \|f\|_p^p,
\end{aligned}$$

where the inequality is by (2) and the second equality is by Fubini's Theorem. Then taking the p-th roots, we have

$$\|M_\mu f\|_p \leq C_p \|f\|_p \quad \text{where } C_p = 2\Big(C \frac{p}{p-1}\Big)^{1/p}. \tag{3}$$

Consider the general case that $f \in L^p(\mathbb{R}^n, \mathfrak{M}(\mu), \mu)$. Consider $|f|$. Then by our result above we have $\big\|M_\mu |f|\big\|_p \leq C_p \big\||f|\big\|_p$. But $M_\mu f = M_\mu |f|$ and $\big\||f|\big\|_p = \|f\|_p$. Thus we have $\|M_\mu f\| \leq C_p \|f\|_p$. ∎

§11 Density of Sets

[I] Density of Sets

Definition 11.1. **(μ-Density of a Set)** *Let μ be a Borel outer measure on a metric space (X, d) such that $\mu(O) > 0$ for every non-empty open set O in X. Let $A \subset X$ and $x \in X$. We say that the μ-density of A at x is equal to $\theta \in [0,1]$ if*

$$\Theta_\mu A(x) := \lim_{r \to 0} \frac{\mu(C_r(x) \cap A)}{\mu(C_r(x))} = \theta.$$

In this case we call x a point of μ-density θ for the set A.
(Note that the limit may not exist but if it exists then it is in $[0,1]$.)
In particular if $\lim_{r \to 0} \{\mu(C_r(x))\}^{-1} \mu(C_r(x) \cap A) = 1$ then we call x a point of μ-density 1 for A and if $\lim_{r \to 0} \{\mu(C_r(x))\}^{-1} \mu(C_r(x) \cap A) = 0$ then we call x a point of μ-density 0 for A.

Remark 11.2. As in Definition 11.1 above, let μ be a Borel outer measure on a metric space (X, d) such that $\mu(O) > 0$ for every non-empty open set O in X and let $A \subset X$. Consider the outer measure $\mu|_A$ on X, that is, $\mu|_A(E) := \mu(E \cap A)$ for every $E \subset X$. By Theorem 3.23, $\mu|_A$ is a Borel outer measure on X. Then

$$\Theta_\mu A(x) := \lim_{r \to 0} \frac{\mu(C_r(x) \cap A)}{\mu(C_r(x))} = \lim_{r \to 0} \frac{\mu|_A(C_r(x))}{\mu(C_r(x))} = D_\mu \mu|_A(x),$$

that is, the μ-density of a set $A \subset X$ at a point $x \in X$ is equal to the derivative of the outer measure $\mu|_A$ on X with respect to the outer measure μ at the point x.

Theorem 11.3. **(Lebesgue Density Theorem with Respect to a Radon Outer Measure on $\mathbb{R}^n$)** *Let μ be a Radon outer measure on $\mathbb{R}^n$ such that $\mu(O) > 0$ for every non-empty open set O in $\mathbb{R}^n$. Consider the measure space $(\mathbb{R}^n, \mathfrak{M}(\mu), \mu)$. Let $E \in \mathfrak{M}(\mu)$. Then there exists a null set N in the measure space $(\mathbb{R}^n, \mathfrak{B}_{\mathbb{R}^n}, \mu)$ such that*

$$\Theta_\mu E(x) = \lim_{r \to 0} \frac{\mu(C_r(x) \cap E)}{\mu(C_r(x))} = \begin{cases} 1 & \text{for } x \in E \cap N^c, \\ 0 & \text{for } x \in E^c \cap N^c. \end{cases}$$

Proof. If $E \in \mathfrak{M}(\mu)$ then $\mathbf{1}_E \in \mathcal{L}^1_{loc}(\mathbb{R}^n, \mathfrak{M}(\mu), \mu)$. Then by Theorem 8.12 there exists a null set N in $(\mathbb{R}^n, \mathfrak{M}(\mu), \mu)$ such that

$$\lim_{r \to 0} \frac{1}{\mu(C_r(x))} \int_{C_r(x)} \mathbf{1}_E \, d\mu = \mathbf{1}_E(x) \quad \text{for } x \in N^c,$$

that is,

$$\lim_{r \to 0} \frac{\mu(C_r(x) \cap E)}{\mu(C_r(x))} = \mathbf{1}_E(x) \quad \text{for } x \in N^c.$$

Now $\mathbf{1}_E(x) = 1$ for $x \in E$ and $\mathbf{1}_E(x) = 0$ for $x \in E^c$. Thus the Theorem follows. ∎

Theorem 11.4. (Lebesgue Density Theorem with Respect to a Doubling Borel Regular Outer Measure on a Metric Space) *Let (X, d) be a metric space and let μ be a doubling Borel regular outer measure on X satisfying the following condition:*
1° for every $E \in \mathfrak{M}(\mu)$ and $\varepsilon > 0$ there exists a closed set $F \subset E$ such that $\mu(E \setminus F) < \varepsilon$.
Consider the measure space $(X, \mathfrak{M}(\mu), \mu)$. Let $E \in \mathfrak{M}(\mu)$. Then there exists a null set N in the measure space $(X, \mathfrak{B}_X, \mu)$ such that

$$\Theta_\mu E(x) = \lim_{r \to 0} \frac{\mu(C_r(x) \cap E)}{\mu(C_r(x))} = \begin{cases} 1 & \text{for } x \in E \cap N^c, \\ 0 & \text{for } x \in E^c \cap N^c. \end{cases}$$

Proof. Let $E \in \mathfrak{M}(\mu)$. Then $\mathbf{1}_E \in \mathcal{L}^1_{loc}(X, \mathfrak{M}(\mu), \mu)$. Thus by Theorem 9.8, there exists a null set N in $(X, \mathfrak{B}_X, \mu)$ such that

$$\lim_{r \to 0} \frac{1}{\mu(C_r(x))} \int_{C_r(x)} \mathbf{1}_E(y)\, \mu(dy) = \mathbf{1}_E(x) \quad \text{for } x \in N^c,$$

that is,

$$\lim_{r \to 0} \frac{\mu(C_r(x) \cap E)}{\mu(C_r(x))} = \mathbf{1}_E(x) \quad \text{for } x \in N^c.$$

Now $\mathbf{1}_E(x) = 1$ for $x \in E$ and $\mathbf{1}_E(x) = 0$ for $x \in E^c$. Thus the theorem follows. ∎

[II] Set of Points of μ-density 1 of a Set

Lemma 11.5. *Let μ be a regular outer measure on a set X so that for every $A \subset X$ there exists $B \in \mathfrak{M}(\mu)$ such that $A \subset B$ and $\mu(A) = \mu(B)$. If $\mu(A) < \infty$ then we have $\mu(E \cap A) = \mu(E \cap B)$ for every $E \in \mathfrak{M}(\mu)$.*

Proof. Let $E \in \mathfrak{M}(\mu)$. Since $A \subset B$, we have $E \cap A \subset E \cap B$ and $\mu(E \cap A) \le \mu(E \cap B)$.

It remains to prove the reverse inequality. Now we have $B = (E \cap B) \cup (B \setminus E)$. Since $E \cap B$ and $B \setminus E$ are disjoint members of $\mathfrak{M}(\mu)$, we have $\mu(B) = \mu(E \cap B) + \mu(B \setminus E)$. Now $\mu(B \setminus E) \le \mu(B) = \mu(A) < \infty$. Thus subtracting $\mu(B \setminus E) < \infty$ from both sides of the last equality we have

$$\mu(E \cap B) = \mu(B) - \mu(B \setminus E) \le \mu(A) - \mu(A \setminus E), \tag{1}$$

where the inequality is from the fact that $\mu(B) = \mu(A)$ and the fact that $A \subset B$ implies $A \setminus E \subset B \setminus E$ and $\mu(A \setminus E) \le \mu(B \setminus E)$. Next writing $A = (A \setminus E) \cup (E \cap A)$, we have $\mu(A) \le \mu(A \setminus E) + \mu(E \cap A)$ by the subadditivity of the outer measure μ. Then $\mu(A) - \mu(A \setminus E) \le \mu(E \cap A)$. Substituting this in (2), we have $\mu(E \cap B) \le \mu(E \cap A)$. ∎

Proposition 11.6. (A Criterion for Measurability) *Let μ be an outer measure on a set X. Consider the measure space $(X, \mathfrak{M}(\mu), \mu)$. Let $A \subset X$ and $\mu(A) < \infty$. Suppose there exists $E \in \mathfrak{M}(\mu)$ such that*
1° $A \cup E \in \mathfrak{M}(\mu)$.
2° $\mu(A) = \mu(A \cup E)$.
3° $\mu(A \setminus E) = 0$.
Then we have
4° $A \in \mathfrak{M}(\mu) \Leftrightarrow \mu(E \setminus A) = 0$.

Proof. Let us observe that since μ is an outer measure on X, we have

$$C \subset X, \mu(C) = 0 \Rightarrow C \in \mathfrak{M}(\mu). \tag{1}$$

(See for instance Lemma 2.6, [LRA].)

1. Suppose $A \in \mathfrak{M}(\mu)$. Now $E \setminus A = (A \cup E) \setminus A$. Then since $A, A \cup E \in \mathfrak{M}(\mu)$, $A \subset A \cup E$ and $\mu(A) < \infty$, we have

$$\mu(E \setminus A) = \mu((A \cup E) \setminus A) = \mu(A \cup E) - \mu(A) = \mu(A) - \mu(A) = 0.$$

2. Conversely suppose $\mu(E \setminus A) = 0$. Now we have

$$A = (A \setminus E) \cup (A \cap E), \tag{2}$$

$$A \triangle E = (A \setminus E) \cup (E \setminus A). \tag{3}$$

By the subadditivity of the outer measure μ, we have $\mu(A \triangle E) \le \mu(A \setminus E) + \mu(E \setminus A) = 0$. Thus $\mu(A \triangle E) = 0$ and then $A \triangle E \in \mathfrak{M}(\mu)$ by (1). Now we have

$$A \cap E = (A \cup E) \setminus (A \triangle E). \tag{4}$$

Since $A \cup E, A \triangle E \in \mathfrak{M}(\mu)$, we have $A \cap E \in \mathfrak{M}(\mu)$. We have also $\mu(A \setminus E) = 0$, which implies $A \setminus E \in \mathfrak{M}(\mu)$ by (1). Now that $A \setminus E \in \mathfrak{M}(\mu)$ and $A \cap E \in \mathfrak{M}(\mu)$, the equality (2) implies that $A \in \mathfrak{M}(\mu)$. ∎

Theorem 11.7. *Let (X, d) be a metric space and let μ be a finite Borel regular outer measure on X satisfying the following conditions:*
1° *μ satisfies the Vitali condition.*
2° $\mu(A) = \inf \{\mu(V) : A \subset V, V \text{ is open}\}$ *for every $A \subset X$.*
3° $\mu(O) > 0$ *for every non-empty open set O in X.*
For an arbitrary $A \subset X$, let

$$P := \{x \in X : \Theta_\mu A(x) = 1\} = \left\{x \in X : \lim_{r \to 0} \frac{\mu(C_r(x) \cap A)}{\mu(C_r(x))} = 1\right\}. \tag{1}$$

Then we have
(a) $P \in \mathfrak{M}(\mu)$.

(b) $\mu(A \setminus P) = 0$.
(c) $A \cup P \in \mathfrak{M}(\mu)$ *and* $\mu(A) = \mu(A \cup P)$.
(d) $A \in \mathfrak{M}(\mu) \Leftrightarrow \mu(P \setminus A) = 0$.

Proof. 1. Let us prove (a). Let $A \subset X$. Since μ is a Borel regular outer measure and hence a regular outer measure on X, there exists $B \in \mathfrak{M}(\mu)$ such that $A \subset B$ and $\mu(A) = \mu(B)$. Since $\mu(A) \leq \mu(X) < \infty$ we have $\mu(E \cap A) = \mu(E \cap B)$ for every $E \in \mathfrak{M}(\mu)$ by Lemma 11.5. Thus we have

$$\mu\big(C_r(x) \cap A\big) = \mu\big(C_r(x) \cap B\big) \quad \text{for every } x \in X \text{ and } r > 0. \tag{2}$$

Consider an outer measure λ on X defined by $\lambda := \mu|_B$. By Theorem 3.25, λ is a Borel regular outer measure on X. Then by (2), we have

$$\mu\big(C_r(x) \cap A\big) = \mu\big(C_r(x) \cap B\big) = \lambda\big(C_r(x)\big) \quad \text{for every } x \in X \text{ and } r > 0.$$

Thus we have

$$\lim_{r \to 0} \frac{\mu\big(C_r(x) \cap A\big)}{\mu\big(C_r(x)\big)} = 1 \Leftrightarrow \lim_{r \to 0} \frac{\lambda\big(C_r(x)\big)}{\mu\big(C_r(x)\big)} = 1 \Leftrightarrow D_\mu \lambda(x) = 1, \tag{3}$$

by Definition 7.3 and Definition 7.4. By (1), Definition 11.1, and (3), we have

$$P = \left\{ x \in X : \lim_{r \to 0} \frac{\mu\big(C_r(x) \cap A\big)}{\mu\big(C_r(x)\big)} = 1 \right\} = \big\{ x \in X : D_\mu \lambda(x) = 1 \big\}. \tag{4}$$

Since μ satisfies conditions 1° and 2°, Theorem 7.8 is applicable to μ. Thus $D_\mu \lambda$ is a $\mathfrak{M}(\mu)$-measurable function on X. Then by (4) we have $P \in \mathfrak{M}(\mu)$. This proves (a).

2. To prove (b) and (c), we show first that

$$\mu(P \setminus B) = 0. \tag{5}$$

$$\mu(B \setminus P) = 0. \tag{6}$$

Now since $\lambda = \mu|_B$, we have $\lambda \ll \mu$. Then by Theorem 7.11, we have

$$\int_E D_\mu \lambda \, d\mu = \lambda(E) \quad \text{for every } E \in \mathfrak{M}(\mu). \tag{7}$$

Since $D_\mu \lambda = 1$ on the set P according to (4), we have by (7)

$$\mu(P) = \int_P 1 \, d\mu = \int_P D_\mu \lambda \, d\mu = \lambda(P) = \mu(P \cap B).$$

Since $P = (P \cap B) \cup (P \setminus B)$, we have by the last equality

$$\mu(P) = \mu(P \cap B) + \mu(P \setminus B) = \mu(P) + \mu(P \setminus B).$$

Subtracting $\mu(P) < \infty$ from both sides of the last equality, we have $\mu(P \setminus B) = 0$. This proves (5).

To prove (6), assume the contrary, that is, $\mu(B \setminus P) > 0$. Now by (4) we have $D_\mu \lambda < 1$ on $B \setminus P$. Thus we have

$$\mu(B \setminus P) = \int_{B \setminus P} 1 \, d\mu > \int_{B \setminus P} D_\mu \lambda \, d\mu = \lambda(B \setminus P)$$
$$= \mu((B \setminus P) \cap B) = \mu(B \setminus P),$$

where the first inequality is by the fact that on any measure space $(X, \mathfrak{A}, \mu)$ if f and g are extended real-valued $\mathfrak{A}$-measurable functions on a set $E \in \mathfrak{A}$ with $\mu(E) > 0$ and if $f < g$ on E then $\int_E f \, d\mu < \int_E g \, d\mu$ and the second equality is by (7). Thus we have $\mu(B \setminus P) > \mu(B \setminus P)$, a contradiction. This proves (6).

Now since $A \subset B$ we have $(A \setminus P) \subset (B \setminus P)$ and $\mu(A \setminus P) \le \mu(B \setminus P) = 0$ by (6). Thus $\mu(A \setminus P) = 0$. This proves (b).

3. To prove (c), note that since $A \subset B$ we have $A \cup P \subset B \cup P = B \cup (P \setminus B)$. Then

$$\mu(A \cup P) \le \mu(B \cup (P \setminus B)) \le \mu(B) + \mu(P \setminus B)$$
$$= \mu(A) + 0 = \mu(A) \quad \text{by (5).}$$

Since $\mu(A \cup P) \ge \mu(A)$, we have $\mu(A) = \mu(A \cup P)$. We have $A \cup P = P \cup (A \setminus P)$. Now $P \in \mathfrak{M}(\mu)$ by (a) and since $\mu(A \setminus P) = 0$ by (b) we have $A \setminus P \in \mathfrak{M}(\mu)$. Thus we have $A \cup P \in \mathfrak{M}(\mu)$. This completes the proof of (c).

4. (b) and (c) imply (d) by Proposition 11.6. ■

Theorem 11.8. *Let (X, d) be a metric space and let μ be a finite Borel regular outer measure on X satisfying the following conditions:*
$1°$ *μ satisfies the Vitali condition.*
$2°$ *$\mu(A) = \inf\{\mu(V) : A \subset V, V$ is open$\}$ for every $A \subset X$.*
$3°$ *$\mu(O) > 0$ for every non-empty open set O in X.*
For an arbitrary $A \subset X$, let

$$(1) \qquad Q := \{x \in X : \Theta_\mu A^c(x) = 0\} = \left\{ x \in X : \lim_{r \to 0} \frac{\mu(C_r(x) \cap A^c)}{\mu(C_r(x))} = 0 \right\}.$$

Then we have
(a) *$Q \in \mathfrak{M}(\mu)$.*
(b) *$\mu(Q \setminus A) = 0$.*
(c) *$A^c \cup Q^c \in \mathfrak{M}(\mu)$ and $\mu(A^c) = \mu(A^c \cup Q^c)$.*
(d) *$A \in \mathfrak{M}(\mu) \Leftrightarrow \mu(A \setminus Q) = 0$.*

Proof. 1. Let us prove (a). Let $A \subset X$. Since μ is a Borel regular outer measure and hence a regular outer measure on X, there exists $C \in \mathfrak{M}(\mu)$ such that $A^c \subset C$ and $\mu(A^c) = \mu(C)$.

Since $\mu(A^c) \le \mu(X) < \infty$ we have $\mu(E \cap A^c) = \mu(E \cap C)$ for every $E \in \mathfrak{M}(\mu)$ by Lemma 11.5. Thus we have

(2) $$\mu\big(C_r(x) \cap A^c\big) = \mu\big(C_r(x) \cap C\big) \quad \text{for every } x \in X \text{ and } r > 0.$$

Consider an outer measure ν on X defined by $\nu := \mu|_C$. By Theorem 3.25, ν is a Borel regular outer measure on X. Then by (2), we have

$$\mu\big(C_r(x) \cap A^c\big) = \mu\big(C_r(x) \cap C\big) = \nu\big(C_r(x)\big) \quad \text{for every } x \in X \text{ and } r > 0.$$

Thus we have

(3) $$\lim_{r \to 0} \frac{\mu\big(C_r(x) \cap A^c\big)}{\mu\big(C_r(x)\big)} = 0 \Leftrightarrow \lim_{r \to 0} \frac{\nu\big(C_r(x)\big)}{\mu\big(C_r(x)\big)} = 0 \Leftrightarrow D_\mu \nu(x) = 0,$$

by Definition 7.3 and Definition 7.4. By (1), Definition 11.1, and (3), we have

(4) $$Q = \left\{ x \in X : \lim_{r \to 0} \frac{\mu\big(C_r(x) \cap A^c\big)}{\mu\big(C_r(x)\big)} = 0 \right\} = \big\{ x \in X : D_\mu \nu(x) = 0 \big\}.$$

Since μ satisfies conditions 1° and 2°, Theorem 7.8 is applicable to μ. Thus $D_\mu \nu$ is a $\mathfrak{M}(\mu)$-measurable function on X. Then by (4) we have $Q \in \mathfrak{M}(\mu)$. This proves (a).

2. To prove (b) and (c), we show first that

(5) $$\mu(Q \cap C) = 0.$$

(6) $$\mu\big((C \cup Q)^c\big) = 0.$$

Now since $\nu = \mu|_B$, we have $\nu \ll \mu$. Then by Theorem 7.11, we have

(7) $$\int_E D_\mu \nu \, d\mu = \nu(E) \quad \text{for every } E \in \mathfrak{M}(\mu).$$

Since $D_\mu \nu = 0$ on the set Q according to (4), we have by (7)

$$\mu(Q) = \int_Q 1 \, d\mu = \int_Q \{1 - D_\mu \nu\} \, d\mu = \mu(Q) - \nu(Q) = \mu(Q) - \mu(Q \cap C).$$

Subtracting $\mu(Q) \le \mu(X) < \infty$ from both sides of the last equality we have $\mu(Q \cap C) = 0$. This proves (5).

To prove (6), assume the contrary, that is, $\mu\big((C \cup Q)^c\big) > 0$. Then we have

(8) $$\mu\big((C \cup Q)^c\big) = \int_{(C \cup Q)^c} 1 \, d\mu > \int_{(C \cup Q)^c} \{1 - D_\mu \nu\} \, d\mu,$$

since $D_\mu \nu > 0$ on $Q^c \supset (C \cup Q)^c$ so that $1 > 1 - D_\mu \nu$ on $(C \cup Q)^c$ and then $\mu((C \cup Q)^c) > 0$ implies the strict inequality above. On the other hand we have

$$\text{(9)} \qquad \int_{(C\cup Q)^c} \{1 - D_\mu \nu\}\, d\mu = \mu((C \cup Q)^c) - \nu((C \cup Q)^c)$$

$$= \mu((C \cup Q)^c) - \mu((C \cup Q)^c \cap C)$$

$$= \mu((C \cup Q)^c) - \mu(\emptyset)$$

$$= \mu((C \cup Q)^c).$$

Substituting (9) in (8) we have $\mu((C \cup Q)^c) > \mu((C \cup Q)^c)$ which is a contradiction. This proves (6). Now $Q \setminus A = Q \cap A^c \subset Q \cap C$ so that $\mu(Q \setminus A) \leq \mu(Q \cap C) = 0$ by (6). This proves (b).

3. Let us prove (c). Now we have $A^c \subset C$ and then

$$A^c \cup Q^c \subset C \cup Q^c = C \cup (Q^c \setminus C) = C \cup (Q^c \cap C^c) = C \cup (C \cup Q)^c.$$

Thus recalling $\mu(C) = \mu(A^c)$ and applying (6) we have

$$\mu(A^c \cup Q^c) \leq \mu(C) + \mu((C \cup Q)^c) = \mu(A^c).$$

But $A^c \subset A^c \cup Q^c$ and $\mu(A^c) \leq \mu(A^c \cup Q^c)$. Therefore we have $\mu(A^c) = \mu(A^c \cup Q^c)$.

Let us show that $A^c \cup Q^c \in \mathfrak{M}(\mu)$. Let us write

$$\text{(10)} \qquad A^c \cup Q^c = (A^c \setminus Q^c) \cup Q^c.$$

Now we have $A^c \setminus Q^c \subset C \setminus Q^c = C \cap Q$ and $\mu(A^c \setminus Q^c) \leq \mu(C \cap Q) = 0$ by (5). Thus $\mu(A^c \setminus Q^c) = 0$ and this implies $A^c \setminus Q^c \in \mathfrak{M}(\mu)$. On the other hand $Q \in \mathfrak{M}(\mu)$ by (a) and thus $Q^c \in \mathfrak{M}(\mu)$. Now that $A^c \setminus Q^c \in \mathfrak{M}(\mu)$ and $Q^c \in \mathfrak{M}(\mu)$, we have $A^c \cup Q^c \in \mathfrak{M}(\mu)$ by (10). This completes the proof of (c).

4. (b) and (c) imply (d) by Proposition 11.6. ∎

§12 Approximate Limit and Approximate Continuity

[I] Approximate Limit

Let (X, d) and (Y, ρ) be metric spaces. Let f be a mapping of a set $D \subset X$ into Y and let $a \in \overline{D}$. We say that $\lim_{x \to a} f(x) = y \in Y$ if for every $\varepsilon > 0$ there exists $\delta > 0$ such that

$$f\big(B_\delta(a) \cap \{a\}^c \cap D\big) \subset B_\varepsilon(y),$$

or equivalently $\lim_{x \to a} f(x) = y \in Y$ if for every $\varepsilon > 0$ there exists $\delta > 0$ such that

$$f\big(C_\delta(a) \cap \{a\}^c \cap D\big) \subset B_\varepsilon(y).$$

Note that $C_\delta(a) \cap \{a\}^c = C_\delta(a) \setminus \{a\} = \big\{x \in X : 0 < d(a, x) \le \delta\big\}$.
It is easily verified that $\lim_{x \to a} f(x) = y$ if and only if for every sequence $(x_n : n \in \mathbb{N})$ in $D \cap \{a\}^c$ such that $\lim_{n \to \infty} x_n = a$ we have $\lim_{n \to \infty} f(x_n) = y$.

Definition 12.1. *Let (X, d) and (Y, ρ) be metric spaces. Let μ be a Borel outer measure on X satisfying the following condition:*
$1°$ $\mu\big(B_r(x)\big) \in (0, \infty)$ *for every $x \in X$ and $r > 0$.*
Let $D \subset X$ and $a \in \overline{D}$. Let f be a mapping of D into Y. We say that $y \in Y$ is an approximate limit of f at a and we write

$$\text{(1)} \qquad \operatorname{ap} \lim_{x \to a} f(x) = y$$

if we have

$$\text{(2)} \qquad \lim_{r \to 0} \frac{\mu\big(C_r(a) \cap \{D \setminus f^{-1}(W)\}\big)}{\mu\big(C_r(a)\big)} = 0 \quad \textit{for every } \varepsilon > 0.$$

(Condition (2) says that a is a point of μ-density 0 for the set $D \setminus f^{-1}(W)$ for every $\varepsilon > 0$.)

Observation 12.2. In the setting of Definition 12.1, consider the case that the set $D \subset X$ is in $\mathfrak{M}(\mu)$ and the mapping f of D into Y is $\mathfrak{M}(\mu)/\mathfrak{B}_Y$-measurable. Regarding condition (2) in Definition 12.1, note that μ-density of D may not exist at $a \in \overline{D}$, that is, $\lim_{r \to 0} \big\{\mu\big(C_r(a)\big)\big\}^{-1} \mu\big(C_r(a) \cap D\big)$ may not exist. But if μ-density of D exist at $a \in \overline{D}$ and is equal to $\theta \in [0, 1]$, then condition (2) in Definition 12.1 is equivalent to the following condition

$$\text{(3)} \qquad \lim_{r \to 0} \frac{\mu\big(C_r(a) \cap \{D \cap f^{-1}(W)\}\big)}{\mu\big(C_r(a)\big)} = \theta \quad \text{for every } \varepsilon > 0.$$

Proof. Suppose μ-density of D exist at $a \in \overline{D}$ and is equal to $\theta \in [0,1]$, that is, we have

$$\lim_{r\to 0} \frac{\mu(C_r(a) \cap D)}{\mu(C_r(a))} = \theta.$$

Now for every $\varepsilon > 0$, the two sets $D \setminus f^{-1}(W)$ and $D \cap f^{-1}(W)$ are disjoint and their union is equal to D. Then $C_r(a) \cap \{D \setminus f^{-1}(W)\}$ and $C_r(a) \cap \{D \cap f^{-1}(W)\}$ are disjoint and their union is equal to $C_r(a) \cap D$. Since $D \in \mathfrak{M}(\mu)$ and f is a $\mathfrak{M}(\mu)/\mathfrak{B}_Y$-measurable mapping of D into Y, we have $f^{-1}(W) \in \mathfrak{M}(\mu)$. Then $C_r(a) \cap \{D \setminus f^{-1}(W)\} \in \mathfrak{M}(\mu)$ and $C_r(a) \cap \{D \cap f^{-1}(W)\} \in \mathfrak{M}(\mu)$ also. Thus we have

$$\mu(C_r(a) \cap \{D \setminus f^{-1}(W)\}) + \mu(C_r(a) \cap \{D \cap f^{-1}(W)\}) = \mu(C_r(a) \cap D).$$

Then we have

$$\lim_{r\to 0} \frac{\mu(C_r(a) \cap \{D \setminus f^{-1}(W)\})}{\mu(C_r(a))} + \lim_{r\to 0} \frac{\mu(C_r(a) \cap \{D \cap f^{-1}(W)\})}{\mu(C_r(a))}$$
$$= \lim_{r\to 0} \frac{\mu(C_r(a) \cap D)}{\mu(C_r(a))} = \theta.$$

Thus (2) holds if and only if (3) holds, that is, (2) and (3) are equivalent. ■

Proposition 12.3. *Let (X, d) and (Y, ρ) be metric spaces. Let μ be a Borel outer measure on X satisfying the following condition:*
$1°$ $\mu(B_r(x)) \in (0, \infty)$ *for every $x \in X$ and $r > 0$.*
Let $D \subset X$ and $a \in \overline{D}$. Let f be a mapping of D into Y. Let $y \in Y$. Then the following two conditions are equivalent:

(1) $$\lim_{r\to 0} \frac{\mu(C_r(a) \cap \{D \setminus f^{-1}(W)\})}{\mu(C_r(a))} = 0 \quad \textit{for every open set } W \textit{ containing } y.$$

(2) $$\lim_{r\to 0} \frac{\mu(C_r(a) \cap \{D \setminus f^{-1}(B_\varepsilon(y))\})}{\mu(C_r(a))} = 0 \quad \textit{for every } \varepsilon > 0.$$

Proof. Clearly condition (1) contains condition (2). It remains to show that condition (2) implies condition (1). Thus assume (2). Now if W is an open set in Y containing y then we have $W \supset B_\varepsilon(y)$ for some $\varepsilon > 0$ and this implies $f^{-1}(W) \supset f^{-1}(B_\varepsilon(y))$ and then $D \setminus f^{-1}(W) \subset D \setminus f^{-1}(B_\varepsilon(y))$ so that $C_r(a) \cap \{D \setminus f^{-1}(W)\} \subset C_r(a) \cap \{D \setminus f^{-1}(B_\varepsilon(y))\}$ and finally

$$\lim_{r\to 0} \frac{\mu(C_r(a) \cap \{D \setminus f^{-1}(W)\})}{\mu(C_r(a))} \leq \lim_{r\to 0} \frac{\mu(C_r(a) \cap \{D \setminus f^{-1}(B_\varepsilon(y))\})}{\mu(C_r(a))} = 0$$

by (2). This shows that (2) implies (1). ∎

Example 12.4. Consider a mapping of $(\mathbb{R}, \mathfrak{M}_L, \mu_L)$ into $\mathbb{R}$ defined as follows. Let Q be the set of rational numbers and P be the set of irrational numbers. Let $\alpha, \beta \in \mathbb{R}$ and $\alpha \neq \beta$. Define a mapping f of $\mathbb{R}$ into $\mathbb{R}$ by setting

$$f(x) = \begin{cases} \alpha & \text{for } x \in Q, \\ \beta & \text{for } x \in P. \end{cases}$$

Then for every $a \in \mathbb{R}$ we have $\text{ap} \lim_{x \to a} f(x) = \beta$.

Proof. We are to show that for every $\varepsilon > 0$ we have

$$\text{(1)} \qquad \lim_{r \to 0} \frac{\mu_L\big(C_r(a) \cap \{\mathbb{R} \setminus f^{-1}(B_\varepsilon(\beta))\}\big)}{\mu_L\big(C_r(a)\big)} = 0.$$

Let $\eta = d(\alpha, \beta) > 0$. Let $\varepsilon > 0$. Then we have

$$f^{-1}\big(B_\varepsilon(\beta)\big) = \begin{cases} P & \text{if } \varepsilon < \eta \\ bbr & \text{if } \varepsilon \geq \eta \end{cases}$$

and then

$$\mathbb{R} \setminus f^{-1}\big(B_\varepsilon(\beta)\big) = \begin{cases} Q & \text{if } \varepsilon < \eta \\ \emptyset & \text{if } \varepsilon \geq \eta. \end{cases}$$

Thus we have

$$C_r(a) \cap \{\mathbb{R} \setminus f^{-1}\big(B_\varepsilon(\beta)\big)\} = \begin{cases} C_r(a) \cap Q & \text{if } \varepsilon < \eta \\ \emptyset & \text{if } \varepsilon \geq \eta \end{cases}$$

and then

$$\mu_L\big(C_r(a) \cap \{\mathbb{R} \setminus f^{-1}\big(B_\varepsilon(\beta)\big)\}\big) = \begin{cases} \mu_L\big(C_r(a) \cap Q\big) \leq \mu_L(Q) = 0 & \text{if } \varepsilon < \eta \\ \mu_L(\emptyset) = 0 & \text{if } \varepsilon \geq \eta. \end{cases}$$

Therefore we have

$$\lim_{r \to 0} \frac{\mu_L\big(C_r(a) \cap \{\mathbb{R} \setminus f^{-1}\big(B_\varepsilon(\beta)\big)\}\big)}{\mu_L\big(C_r(a)\big)} = \lim_{r \to 0} \frac{0}{2r} = 0.$$

This completes the verification of (1). ∎

Example 12.5. Consider a mapping f of $(\mathbb{R}, \mathfrak{M}_L, \mu_L)$ into $\mathbb{R}$ defined by

$$f(x) = \begin{cases} \alpha & \text{for } x \in (-\infty, 0) \\ \beta & \text{for } x \in [0, \infty) \end{cases}$$

where $\alpha, \beta \in \mathbb{R}$ and $\alpha \neq \beta$. Then ap $\lim_{x\to 0} f(x)$ does not exist.

Proof. Let $\eta = d(\alpha, \beta) > 0$. To show that ap $\lim_{x\to 0} f(x)$ does not exist, that is, there does not exist $\gamma \in \mathbb{R}$ such that ap $\lim_{x\to 0} f(x) = \gamma$, we show that for any $\gamma \in \mathbb{R}$ we have for our $\eta > 0$ we have

$$(1) \qquad \liminf_{r\to 0} \frac{\mu_L\big(C_r(0) \cap \{\mathbb{R} \setminus f^{-1}(B_\eta(\gamma))\}\big)}{\mu_L\big(C_r(0)\big)} \geq \frac{1}{2} \neq 0.$$

Now for any $\gamma \in \mathbb{R}$, $B_\eta(\gamma)$ can not contain both α and β. Thus either $\alpha \notin B_\eta(\gamma)$ or $\beta \notin B_\eta(\gamma)$ or $\alpha, \beta \notin B_\eta(\gamma)$. Let us consider the case $\beta \notin B_\eta(\gamma)$. Then we have

$$\begin{aligned}
&f^{-1}(B_\eta(\gamma)) \cap [0, \infty) = \emptyset,\\
&\mathbb{R} \setminus f^{-1}(B_\eta(\gamma)) \supset [0, \infty),\\
&C_r(0) \cap \{\mathbb{R} \setminus f^{-1}(B_\eta(\gamma))\} \supset [0, r),\\
&\mu_L\big(C_r(0) \cap \{\mathbb{R} \setminus f^{-1}(B_\eta(\gamma))\}\big) \geq \mu_L\big([0, r)\big) = r,
\end{aligned}$$

and hence

$$(2) \qquad \frac{\mu_L\big(C_r(0) \cap \{\mathbb{R} \setminus f^{-1}(B_\eta(\gamma))\}\big)}{\mu_L\big(C_r(0)\big)} \geq \frac{r}{2r} = \frac{1}{2}.$$

Then (1) follows from (2). ∎

Proposition 12.6. *Let (X, d) and (Y, ρ) be metric spaces. Let μ be a Borel outer measure on X satisfying the following condition:*
1° $\mu\big(B_r(x)\big) \in (0, \infty)$ for every $x \in X$ and $r > 0$.
Consider the measure space $(X, \mathfrak{M}(\mu), \mu)$. Let $D \in \mathfrak{M}(\mu)$ and $a \in \overline{D}$. Let f be a $\mathfrak{M}(\mu)/\mathfrak{B}_Y$-measurable mapping of D into Y.
If μ-density of D exist at $a \in \overline{D}$ and is equal to 0, then every $y \in Y$ is an approximate limit of f at a, that is, ap $\lim_{x\to a} f(x) = y$.

Proof. If μ-density of D exist at $a \in \overline{D}$ and is equal to 0, then we have

$$(1) \qquad \lim_{r\to 0} \frac{\mu\big(C_r(a) \cap D\big)}{\mu\big(C_r(a)\big)} = 0.$$

Let $y \in Y$ be arbitrarily chosen. Then for every $\varepsilon > 0$ we have

$$(2) \qquad \lim_{r\to 0} \frac{\mu\big(C_r(a) \cap \{D \setminus f^{-1}(B_\varepsilon(y))\}\big)}{\mu\big(C_r(a)\big)} \leq \lim_{r\to 0} \frac{\mu\big(C_r(a) \cap D\big)}{\mu\big(C_r(a)\big)} = 0.$$

This shows that ap $\lim_{x \to a} f(x) = y$. ∎

Remark 12.7. Regarding the condition that μ-density of D exist at $a \in \overline{D}$ and is equal to 0 in Proposition 12.6, let us point out that this can never happen if $D = X$. Indeed for any $x \in X$ we have

$$\lim_{r \to 0} \frac{\mu(C_r(a) \cap X)}{\mu(C_r(a))} = \lim_{r \to 0} \frac{\mu(C_r(a))}{\mu(C_r(a))} = 1.$$

Thus the μ-density of X at any $a \in X$ always exist and is equal to 1 and hence never equal to 0.

Theorem 12.8. *Let (X, d) and (Y, ρ) be metric spaces. Let μ be a Borel outer measure on X satisfying the following condition:*
1° $\mu(B_r(x)) \in (0, \infty)$ for every $x \in X$ and $r > 0$.
Consider the measure space $(X, \mathfrak{M}(\mu), \mu)$. Let $D \in \mathfrak{M}(\mu)$ and $a \in \overline{D}$. Let f be a $\mathfrak{M}(\mu)/\mathfrak{B}_Y$-measurable mapping of D into Y.
Suppose μ-density of D exist at $a \in \overline{D}$ and is equal to $\theta > 0$. Then if ap $\lim_{x \to a} f(x)$ *exists it is unique.*

Proof. If μ-density of D exist at $a \in \overline{D}$ and is equal to $\theta > 0$, then we have

$$(1) \qquad \lim_{r \to 0} \frac{\mu(C_r(a) \cap D)}{\mu(C_r(a))} = \theta > 0.$$

Suppose ap $\lim_{x \to a} f(x) = y \in Y$ and ap $\lim_{x \to a} f(x) = z \in Y$ also.

Let us show that $y = z$. Suppose $y \neq z$. Then $c := \rho(y, z) > 0$. We claim that

$$(2) \qquad D \subset \{D \setminus f^{-1}(B_{c/3}(y))\} \cup \{D \setminus f^{-1}(B_{c/3}(z))\}.$$

To prove (2), assume the contrary. Then there exists $x \in D$ such that $x \notin \{D \setminus f^{-1}(B_{c/3}(y))\}$ and $x \notin \{D \setminus f^{-1}(B_{c/3}(z))\}$. Then $x \in D$ implies that $x \in f^{-1}(B_{c/3}(y))$ and also $x \in f^{-1}(B_{c/3}(z))$ so that $x \in f^{-1}(B_{c/3}(y)) \cap f^{-1}(B_{c/3}(z))$. But $B_{c/3}(y) \cap B_{c/3}(z) = \emptyset$ implies that $f^{-1}(B_{c/3}(y)) \cap f^{-1}(B_{c/3}(z)) = \emptyset$. This is a contradiction.

Now by (2) we have

$$C_r(a) \cap D \subset \left[C_r(a) \cap \{D \setminus f^{-1}(B_{c/3}(y))\}\right] \cup \left[C_r(a) \cap \{D \setminus f^{-1}(B_{c/3}(z))\}\right].$$

Then

$$\begin{aligned} 0 < \theta = \lim_{r \to 0} \frac{\mu(C_r(a) \cap D)}{\mu(C_r(a))} &\le \lim_{r \to 0} \frac{\mu(C_r(a) \cap \{D \setminus f^{-1}(B_{c/3}(y))\})}{\mu(C_r(a))} \\ &\quad + \lim_{r \to 0} \frac{\mu(C_r(a) \cap \{D \setminus f^{-1}(B_{c/3}(z))\})}{\mu(C_r(a))} \\ &= 0 \end{aligned}$$

since ap $\lim\limits_{x \to a} f(x) = y \in Y$ and ap $\lim\limits_{x \to a} f(x) = z \in Y$. This is a contradiction. Therefore we have $y = z$. ∎

Theorem 12.9. *Let (X, d) and (Y, ρ) be metric spaces. Let μ be a Borel outer measure on X satisfying the following condition:*
1° $\mu(B_r(x)) \in (0, \infty)$ for every $x \in X$ and $r > 0$.
Consider the measure space $(X, \mathfrak{M}(\mu), \mu)$. Let $D \in \mathfrak{M}(\mu)$ and let f be a $\mathfrak{M}(\mu)/\mathfrak{B}_Y$-measurable mapping of D into Y. Let $a \in \overline{D}$ be such that $\mu(\{a\}) = 0$ and let $y \in Y$. Suppose for every $\varepsilon > 0$ there exist $\delta > 0$ and a null set N in $(X, \mathfrak{M}(\mu), \mu)$ such that

$$(1) \qquad f\big(C_r(a) \cap \{a\}^c \cap D \cap N^c\big) \subset B_\varepsilon(y) \quad \text{for all } r \in (0, \delta].$$

Then we have ap $\lim\limits_{x \to a} f(x) = y$.

Proof. We are to show that under the assumption (1), we have for every $\varepsilon > 0$

$$(2) \qquad \lim_{r \to 0} \frac{\mu\big(C_r(a) \cap \{D \setminus f^{-1}(B_\varepsilon(y))\}\big)}{\mu(C_r(a))} = 0.$$

Now by (1), for every $\varepsilon > 0$ there exist $\delta > 0$ and a null set N in $(X, \mathfrak{M}(\mu), \mu)$ such that

$$C_r(a) \cap \{a\}^c \cap D \cap N^c \subset f^{-1}(B_\varepsilon(y))$$

for all $r \in (0, \delta]$, and hence

$$(3) \qquad \{C_r(a) \cap \{a\}^c \cap D \cap N^c\} \cap \{D \setminus f^{-1}(B_\varepsilon(y))\} = \emptyset$$

for all $r \in (0, \delta]$.

We have

$$\begin{aligned}
(4) \qquad & C_r(a) \cap \{D \setminus f^{-1}(B_\varepsilon(y))\} \\
&= \Big[\{C_r(a) \cap D \cap N^c\} \cup \{C_r(a) \cap (D \cap N^c)^c\}\Big] \cap \{D \setminus f^{-1}(B_\varepsilon(y))\} \\
&\subset \Big[\{C_r(a) \cap D \cap N^c\} \cap \{D \setminus f^{-1}(B_\varepsilon(y))\}\Big] \cup N
\end{aligned}$$

and

$$\begin{aligned}
(5) \quad & \{C_r(a) \cap D \cap N^c\} \cap \{D \setminus f^{-1}(B_\varepsilon(y))\} \\
&= \Big[\{C_r(a) \cap \{a\} \cap D \cap N^c\} \cup \{C_r(a) \cap \{a\}^c \cap D \cap N^c\}\Big] \cap \{D \setminus f^{-1}(B_\varepsilon(y))\} \\
&= \{\{a\} \cap D \cap N^c\} \cap \{D \setminus f^{-1}(B_\varepsilon(y))\} \quad \text{by (3)} \\
&= \emptyset \text{ or } \{a\} \\
&\subset \{a\}.
\end{aligned}$$

From (4) and (5) we have for all $r \in (0, \delta]$

$$C_r(a) \cap \{D \setminus f^{-1}(B_\varepsilon(y))\} \subset \{a\} \cup N \tag{6}$$

and hence for all $r \in (0, \delta]$ we have

$$\mu(C_r(a) \cap \{D \setminus f^{-1}(B_\varepsilon(y))\}) \leq \mu(\{a\} \cup N) = 0. \tag{7}$$

Then (2) follows from (7). ∎

Corollary 12.10. *Let (X, d) and (Y, ρ) be metric spaces. Let μ be a Borel outer measure on X satisfying the following condition:*
1° $\mu(B_r(x)) \in (0, \infty)$ *for every $x \in X$ and $r > 0$.*
Consider the measure space $(X, \mathfrak{M}(\mu), \mu)$. Let $D \in \mathfrak{M}(\mu)$ and let f be a $\mathfrak{M}(\mu)/\mathfrak{B}_Y$-measurable mapping of D into Y. Let $a \in \overline{D}$ be such that $\mu(\{a\}) = 0$ and let $y \in Y$. If $\lim_{x \to a} f(x) = y$ *then* ap $\lim_{x \to a} f(x) = y$.

Proof. If $\lim_{x \to a} f(x) = y$, the for every $\varepsilon > 0$ there exists $\delta > 0$ such that

$$f(C_r(a) \cap \{a\}^c \cap D) \subset B_\varepsilon(y) \quad \text{for all } r \in (0, \delta].$$

This is the condition (1) in Theorem 12.9 with $N = \emptyset$. Thus ap $\lim_{x \to a} f(x) = y$ by Theorem 12.9. ∎

[II] Approximate Continuity

Definition 12.11. *Let (X, d) and (Y, ρ) be metric spaces. Let μ be a Borel outer measure on X satisfying the following condition:*
$1°$ $\mu(B_r(x)) \in (0, \infty)$ *for every* $x \in X$ *and* $r > 0$.
Let $D \subset X$ and $a \in \overline{D}$. Let f be a mapping of D into Y. We say that f is approximately continuous at $a \in D$ if we have:
$1°$ $\operatorname{ap}\lim_{x \to a} f(x)$ *exists.*
$2°$ $\operatorname{ap}\lim_{x \to a} f(x) = f(a)$.
Thus f is approximately continuous at $a \in D$ if and only if for every open set W in Y containing $f(a)$ we have

$$(1) \qquad \lim_{r \to 0} \frac{\mu(C_r(a) \cap \{D \setminus f^{-1}(W)\})}{\mu(C_r(a))} = 0,$$

or equivalently, if and only if for every $\varepsilon > 0$ we have

$$(2) \qquad \lim_{r \to 0} \frac{\mu(C_r(a) \cap \{D \setminus f^{-1}(B_\varepsilon(f(a)))\})}{\mu(C_r(a))} = 0.$$

Theorem 12.12. *Let (X, d) be a metric space and (Y, ρ) be a separable metric space. Let μ be a doubling Borel regular outer measure on X satisfying the following condition:*
$1°$ *for every $E \in \mathfrak{M}(\mu)$ and $\varepsilon > 0$ there exists a closed set $F \subset E$ such that $\mu(E \setminus F) < \varepsilon$. Consider the measure space $(X, \mathfrak{M}(\mu), \mu)$. Let f be a mapping of X into Y. If f is $\mathfrak{M}(\mu)/\mathfrak{B}_Y$-measurable on X, then there exists a null set N in the measure space $(X, \mathfrak{B}_X, \mu)$ such that f is approximately continuous at every $x \in N^c$.*

Proof. Suppose f is $\mathfrak{M}(\mu)/\mathfrak{B}_Y$-measurable on X. Since μ is a doubling Borel regular outer measure on X, the measure space $(X, \mathfrak{M}(\mu), \mu)$ is σ-finite by Observation 6.14. Then by Theorem 12.13 there exists a countable disjoint collection of bounded closed sets, $\{K_i : i \in \mathbb{N}\}$, such that

$$(1) \qquad \mu\left(X \setminus \bigcup_{i \in \mathbb{N}} K_i\right) = 0,$$

$$(2) \qquad f|_{K_i} \text{ is continuous for } i \in \mathbb{N}.$$

Since μ satisfies condition $1°$, Theorem 11.4 is applicable and thus for each $i \in \mathbb{N}$ there exists a null set N_i in $(X, \mathfrak{B}_X, \mu)$ such that

$$(3) \qquad \lim_{r \to 0} \frac{\mu(C_r(x) \cap K_i^c)}{\mu(C_r(x))} = 0 \quad \text{for } x \in K_i \cap N_i^c.$$

Let us define

(4) $$N_\infty = \bigcup_{i\in\mathbb{N}} N_i,$$

(5) $$A = \bigcup_{i\in\mathbb{N}} (K_i \cap N_i^c),$$

(6) $$N = X \setminus A.$$

Then N_∞ is a null set in $(X, \mathfrak{B}_X, \mu)$ and $A \in \mathfrak{B}_X$. Let us show that N is a null set in $(X, \mathfrak{B}_X), \mu)$. Now we have

$$\begin{aligned} X \setminus A = X \setminus \bigcup_{i\in\mathbb{N}}(K_i \setminus N_i) &\subset X \setminus \bigcup_{i\in\mathbb{N}}(K_i \setminus N_\infty) \\ &= X \setminus \left\{ \left(\bigcup_{i\in\mathbb{N}} K_i \right) \setminus N_\infty \right\} \\ &\subset \left\{ X \setminus \left(\bigcup_{i\in\mathbb{N}} K_i \right) \right\} \cup N_\infty, \end{aligned}$$

and

$$\mu(X \setminus A) \le \mu \left\{ X \setminus \left(\bigcup_{i\in\mathbb{N}} K_i \right) \right\} + \mu(N_\infty) = 0,$$

by (1). This shows that N is a null set in $(X, \mathfrak{B}_X, \mu)$.

Now that N is a null set in $(X, \mathfrak{B}_X, \mu)$, it remains to show that f is approximately continuous at every $x \in N^c$. Thus we are to show that for every $x \in N^c$ we have

(7) $$\lim_{r\to 0} \frac{\mu\big(C_r(x) \cap \{X \setminus f^{-1}(B_\varepsilon(f(x)))\}\big)}{\mu\big(C_r(x)\big)} = 0 \quad \text{for every } \varepsilon > 0.$$

Now let $x \in N^c = A = \bigcup_{i\in\mathbb{N}}(K_i \cap N_i^c)$. Then $x \in K_i \cap N_i^c$ for some $i \in \mathbb{N}$. Now since $f|_{K_i}$ is continuous, for every $\varepsilon > 0$ there exists $\delta > 0$ such that $\rho\big(f(x), f(x')\big) < \varepsilon$ for $x' \in K_i \cap C_\delta(x)$ and thus $\rho\big(f(x), f(x')\big) < \varepsilon$ for $x' \in K_i \cap C_r(x)$ for $r \in (0, \delta]$. Then we have $C_r(x) \cap K_i \subset f^{-1}\big(B_\varepsilon(f(x))\big)$ for $r \in (0, \delta]$ and then

$$\{C_r(x) \cap K_i\} \cap \{X \setminus f^{-1}(B_\varepsilon(f(x)))\} = \emptyset \quad \text{for } r \in (0, \delta]$$

and consequently

$$C_r(x) \cap \{X \setminus f^{-1}(B_\varepsilon(f(x)))\} \subset C_r(x) \setminus K_i \quad \text{for } r \in (0, \delta]\,.$$

Then

$$\lim_{r\to 0} \frac{\mu\big(C_r(x) \cap \{X \setminus f^{-1}(B_\varepsilon(f(x)))\}\big)}{\mu\big(C_r(x)\big)} \le \lim_{r\to 0} \frac{\mu\big(C_r(x) \setminus K_i\big)}{\mu\big(C_r(x)\big)} = 0$$

by (3). This proves (7). ∎

Theorem 12.13. *Let (X,d) be a metric space and (Y,ρ) be a separable metric space. Let μ be a doubling Borel regular outer measure on X satisfying the following condition:*
$1°$ $\mu(A)=\inf\{\mu(V): A\subset V, V$ *is open*$\}$ *for every* $A\subset X$.
Consider the measure space $(X,\mathfrak{M}(\mu),\mu)$. Let f be a mapping of X into Y. If f is approximately continuous at $(\mathfrak{M}(\mu),\mu)$-a.e. $x\in X$, then f is $\mathfrak{M}(\mu)/\mathfrak{B}_Y$-measurable.

Proof. 1. Let us prove the theorem for the particular case that $\mu(X)<\infty$ and f is approximately continuous at every $x\in X$. Now f is a $\mathfrak{M}(\mu)/\mathfrak{B}_Y$-measurable mapping if and only if $f^{-1}(\mathfrak{B}_Y)\subset\mathfrak{M}(\mu)$. Let $\mathfrak{O}_Y$ be the collection of all open sets in Y. Then we have $\mathfrak{B}_Y=\sigma(\mathfrak{O}_Y)$. By Theorem 1.14, [LRA], $f^{-1}(\mathfrak{B}_Y)=f^{-1}(\sigma(\mathfrak{O}_Y))=\sigma\big(f^{-1}(\mathfrak{O}_Y)\big)$. Thus if we show that $f^{-1}(\mathfrak{O}_Y)\subset\mathfrak{M}(\mu)$, then $\sigma\big(f^{-1}(\mathfrak{O}_Y)\big)\subset\sigma\big(\mathfrak{M}(\mu)\big)=\mathfrak{M}(\mu)$ so that $f^{-1}(\mathfrak{B}_Y)=\sigma\big(f^{-1}(\mathfrak{O}_Y)\big)\subset\mathfrak{M}(\mu)$. Thus to show that f is $\mathfrak{M}(\mu)/\mathfrak{B}_Y$-measurable, it suffices to show that $f^{-1}(\mathfrak{B}_Y)\subset\mathfrak{M}(\mu)$, that is, for every open set W in Y we have

$$f^{-1}(W)\in\mathfrak{M}(\mu). \tag{1}$$

Now since μ is a doubling Borel regular outer measure, μ satisfies the Vitali condition by Theorem 6.20. By assumption, μ satisfies condition $1°$ and $\mu(X)<\infty$. Thus according to Theorem 11.8, for any $A\subset X$ we have

$$A\in\mathfrak{M}(\mu)\Leftrightarrow\mu(A\setminus Q_A)=0,$$

where the set Q_A is defined by

$$Q_A=\left\{x\in X:\lim_{r\to0}\frac{\mu(C_r(x)\cap A^c}{\mu(C_r(x))}=0\right\}.$$

Thus to show that $f^{-1}(W)\in\mathfrak{M}(\mu)$, we show that if we define

$$Q=\left\{x\in X:\lim_{r\to0}\frac{\mu\big(C_r(x)\cap\big(f^{-1}(W)\big)^c}{\mu\big(C_r(x)\big)}=0\right\},$$

then

$$\mu\big(f^{-1}(W)\setminus Q\big)=0.$$

Now since f is approximately continuous at every $x\in X$, for every $x\in f^{-1}(W)$ the open set W is a neighborhood of $f(x)$ and thus by (2) of Definition 12.11 we have

$$\lim_{r\to0}\frac{\mu\big(C_r(x)\cap\big(f^{-1}(W)\big)^c}{\mu\big(C_r(x)\big)}=0.$$

This shows that $x\in Q$. Then by the arbitrariness of $x\in f^{-1}(W)$, we have $f^{-1}(W)\subset Q$. Then $f^{-1}(W)\setminus Q=\emptyset$ and $\mu\big(f^{-1}(W)\setminus Q\big)=\mu(\emptyset)=0$. Thus we have $f^{-1}(W)\in\mathfrak{M}(\mu)$. This proves that f is $\mathfrak{M}(\mu)/\mathfrak{B}_Y$-measurable.

2. Let us remove the assumption that $\mu(X) < \infty$ but retain the assumption that f is approximately continuous at every $x \in X$. Let $x_0 \in X$ be arbitrarily chosen and for $i \in \mathbb{N}$ let

$$X_i = \{x \in X : d(x_0, x) < i\}.$$

Then $X = \bigcup_{i \in \mathbb{N}} X_i$ and since X_i is a bounded set in X and μ is a doubling Borel regular outer measure we have $\mu(X_i) < \infty$. Now $X_i \in \mathfrak{B}_X \subset \mathfrak{M}(\mu)$. Consider the measure space $(X_i, \mathfrak{M}(\mu) \cap X_i, \mu)$. Then μ is a doubling Borel regular outer measure on X_i satisfying condition 1°. Thus by our result in **1**, $f|_{X_i}$ is a $\mathfrak{M}(\mu) \cap X_i/\mathfrak{B}_Y$-measurable mapping of X_i into Y. Since this holds for every $i \in \mathbb{N}$, f is a $\mathfrak{M}(\mu)/\mathfrak{B}_Y$-measurable mapping of X into Y.

3. Suppose f is approximately continuous at $(\mathfrak{M}(\mu), \mu)$-a.e. $x \in X$. Then there exists a null set N in the measure space $(X, \mathfrak{M}(\mu), \mu)$ such that f is approximately continuous at every $x \in N^c$. Let $X_0 = N^c$. Then f is approximately continuous at every $x \in X_0$ so that by our result in **2** f is a $\mathfrak{M}(\mu)/\mathfrak{B}_Y$-measurable mapping of X_0 into Y. Since $N$is a null set in the complete measure space $(X, \mathfrak{M}(\mu), \mu)$, any mapping of N into Y is $\mathfrak{M}(\mu)/\mathfrak{B}_Y$-measurable. In particular f is a $\mathfrak{M}(\mu)/\mathfrak{B}_Y$-measurable mapping of N into Y. Now that f is a $\mathfrak{M}(\mu)/\mathfrak{B}_Y$-measurable mapping of $X_0 = N^c$ into Y and f is a $\mathfrak{M}(\mu)/\mathfrak{B}_Y$-measurable mapping of N into Y, f is a $\mathfrak{M}(\mu)/\mathfrak{B}_Y$-measurable mapping of X into Y. ∎

§13 Lipschitz Mappings

[I] Lipschitz Mappings of Metric Spaces

Definition 13.1. (Lipschitz Mapping and Lipschitz Coefficient) *Let (X, d_X) and (Y, d_Y) be metric spaces. A mapping f of $A \subset X$ into Y is called a Lipschitz mapping if there exists $C \geq 0$ such that*

$$d_Y\big(f(x'), f(x'')\big) \leq C d_X(x', x'') \quad \textit{for } x', x'' \in A.$$

We call C a Lipschitz coefficient of the Lipschitz mapping f.
We write $\Gamma(f)$ for the set of all Lipschitz coefficients of a Lipschitz mapping f.

Remark 13.2. Regarding Definition 13.1, note that if $x', x'' \in A$ and $x' = x''$ then $d_X(x', x'') = 0$ and $f(x') = f(x'')$ so that $d_Y\big(f(x'), f(x'')\big) = 0$ also. Then for any $C \geq 0$ we have $d_Y\big(f(x'), f(x'')\big) = 0 = C \cdot 0 = C d_X(x', x'')$. Thus f is a Lipschitz mapping if and only if there exists $C \geq 0$ such that

$$d_Y\big(f(x'), f(x'')\big) \leq C d_X(x', x'') \quad \text{for } x', x'' \in A, x' \neq x''.$$

Observation 13.3. **(a)** Note that if $C \in \Gamma(f)$ then $[C, \infty) \subset \Gamma(f)$.
(b) Since $\Gamma(f) \subset [0, \infty)$ we have $\inf \Gamma(f) \geq 0$. In Proposition 13.4 below, we show that $\inf \Gamma(f) \in \Gamma(f)$, that is, $\Gamma(f)$ has a minimum.

Proposition 13.4. *Let (X, d_X) and (Y, d_Y) be metric spaces. Let f be a Lipschitz mapping of $A \subset X$ into Y and let $\Gamma(f)$ be the set of all Lipschitz coefficients of f. Then*

$$\gamma := \inf \Gamma(f) \in \Gamma(f),$$

that is, $\Gamma(f)$ has a minimum.

Proof. Let $\eta > 0$. Since $\gamma = \inf \Gamma(f)$, there exists $C \in \Gamma(f)$ such that $C \leq \gamma + \eta$. As we noted in Observation 13.3, if $C \in \Gamma(f)$ then $[C, \infty) \subset \Gamma(f)$. Then since $C \leq \gamma + \eta$, we have

$$\gamma + \eta \in \Gamma(f). \tag{1}$$

To show that $\gamma \in \Gamma(f)$, we show that

$$d_Y\big(f(x'), f(x'')\big) \leq \gamma d_X(x', x'') \quad \text{for } x', x'' \in A. \tag{2}$$

To prove (2), assume the contrary. Then there exist $x_1, x_2 \in A$ such that

$$d_Y\big(f(x_1), f(x_2)\big) > \gamma d_X(x_1, x_2). \tag{3}$$

For such x_1 and x_2, if $x_1 = x_2$ then we would have $f(x_1) = f(x_2)$ and then

$$0 = d_Y\big(f(x_1), f(x_2)\big) > \gamma d_X(x_1, x_2) = \gamma \cdot 0 = 0$$

so that we have $0 > 0$, a contradiction. Therefore we have $x_1 \neq x_2$ and $d(x_1, x_2) > 0$. Now (3) implies that there exists $\varepsilon > 0$ such that

$$\begin{aligned} (4) \qquad d_Y\big(f(x_1), f(x_2)\big) &= \gamma d_X(x_1, x_2) + \varepsilon \\ &= \gamma d_X(x_1, x_2) + \frac{\varepsilon}{d_X(x_1, x_2)} d_X(x_1, x_2) \\ &= \left\{\gamma + \frac{\varepsilon}{d_X(x_1, x_2)}\right\} d_X(x_1, x_2) \\ &> \left\{\gamma + \frac{1}{2}\frac{\varepsilon}{d_X(x_1, x_2)}\right\} d_X(x_1, x_2). \end{aligned}$$

Since $\gamma = \inf \Gamma(f)$ and $\frac{1}{2}\frac{\varepsilon}{d_X(x_1,x_2)} > 0$, there exists $C \in \Gamma(f)$ such that $C \leq \gamma + \frac{1}{2}\frac{\varepsilon}{d_X(x_1,x_2)}$. Now $C \in \Gamma(f)$ implies $[C, \infty) \subset \Gamma(f)$. Thus $\gamma + \frac{1}{2}\frac{\varepsilon}{d_X(x_1,x_2)} \in \Gamma(f)$. Therefore we have for all $x', x'' \in A$

$$(5) \qquad d_Y\big(f(x'), f(x'')\big) \leq \left\{\gamma + \frac{1}{2}\frac{\varepsilon}{d_X(x_1, x_2)}\right\} d_X(x', x'').$$

This contradicts (4). Thus (3) is impossible and (2) is valid. ■

Definition 13.5. (Lipschitz Constant of a Lipschitz Mapping) *Let (X, d_X) and (Y, d_Y) be metric spaces. Let f be a Lipschitz mapping of $A \subset X$ into Y and let $\Gamma(f)$ be the set of all Lipschitz coefficients of f. Let*

$$\mathrm{Lip}(f) = \min \Gamma(f).$$

We call $\mathrm{Lip}(f)$ *the Lipschitz constant of the Lipschitz mapping f. We have then*

$$\Gamma(f) = [\mathrm{Lip}(f), \infty).$$

Definition 13.6. *Let (X, d_X) and (Y, d_Y) be metric spaces. Let $A \subset X$ and let f be a mapping of A into Y. We define*

$$\Lambda(f) = \sup\left\{\frac{d_Y\big(f(x'), f(x'')\big)}{d_X(x', x'')} : x', x'' \in A, x' \neq x''\right\} \in [0, \infty].$$

Theorem 13.7. *(X, d_X) and (Y, d_Y) be metric spaces. Let $A \subset X$ and let f be a mapping of A into Y.*

(a) *If f is a Lipschitz mapping then $\Lambda(f) < \infty$.*
(b) *If $\Lambda(f) < \infty$ then f is a Lipschitz mapping and moreover $\Lambda(f) = \mathrm{Lip}(f)$.*
(c) *If f is a Lipschitz mapping then $\Lambda(f) = \mathrm{Lip}(f)$.*

Proof. 1. Let us prove (a). Let f be a Lipschitz mapping. To show $\Lambda(f) < \infty$, assume the contrary, that is, $\Lambda(f) = \infty$. Then we have

$$\sup\left\{\frac{d_Y\big(f(x'), f(x'')\big)}{d_X(x', x'')} : x', x'' \in A, x' \neq x''\right\} = \infty.$$

Thus for every $C \geq 0$ there exist $x', x'' \in A$, $x' \neq x''$ such that

$$\frac{d_Y\big(f(x'), f(x'')\big)}{d_X(x', x'')} > C, \quad \text{that is,} \quad d_Y\big(f(x'), f(x'')\big) > C d_X(x', x'').$$

This implies that f is not a Lipschitz mapping, contradicting the assumption that f is a Lipschitz mapping. Therefore we must have $\Lambda(f) < \infty$.

2. Let us prove (b). Suppose $\Lambda(f) < \infty$. Let us show that f is a Lipschitz mapping. Let $x', x'' \in A$ and $x' \neq x''$. Then $d_X(x', x'') > 0$ and

$$\begin{aligned} d_Y\big(f(x'), f(x'')\big) &= \frac{d_Y\big(f(x'), f(x'')\big)}{d_X(x', x'')} d_X(x', x'') \\ &\leq \sup\left\{\frac{d_Y\big(f(x'), f(x'')\big)}{d_X(x', x'')} : x', x'' \in A, x' \neq x''\right\} d_X(x', x'') \\ &= \Lambda(f) d_X(x', x''). \end{aligned}$$

Since $\Lambda(f) < \infty$, this shows that f is a Lipschitz mapping and $\Lambda(f)$ is a Lipschitz coefficient of f.

To show that $\Lambda(f) = \mathrm{Lip}(f)$, we show that $\Lambda(f)$ is the minimum of the set of all Lipschitz coefficients of f. To show this we show that for any $\varepsilon > 0$, $\Lambda(f) - \varepsilon$ is not a Lipschitz coefficient of f. By Definition 13.6 for $\Lambda(f)$ there exist $x', x'' \in A$, $x' \neq x''$, such that

$$\Lambda(f) - \varepsilon < \frac{d_Y\big(f(x'), f(x'')\big)}{d_X(x', x'')}.$$

Then we have

$$d_Y\big(f(x'), f(x'')\big) > \{\Lambda(f) - \varepsilon\} d_X(x', x'').$$

This shows that $\Lambda(f) - \varepsilon$ is not a Lipschitz coefficient of f. Therefore $\Lambda(f)$ is the minimum of the set of all Lipschitz coefficients of f, that is, $\Lambda(f) = \mathrm{Lip}(f)$.

3. (c) follows from (a) and (b) immediately. Let f be a Lipschitz mapping. Then by (a) we have $\Lambda(f) < \infty$. According to (b), $\Lambda(f) < \infty$ implies $\Lambda(f) = \mathrm{Lip}(f)$. ∎

Proposition 13.8. *Consider metric spaces (X, d_X) and $(\mathbb{R}, d_e)$ where d_e is the Euclidean metric on $\mathbb{R}$. Assume that X has at least two elements and let $x_0 \in X$ be arbitrarily chosen. Let f be a mapping of X into $\mathbb{R}$ defined by*

$$f(x) = d_X(x, x_0) \quad \textit{for } x \in X.$$

Then f is a Lipschitz mapping of X into $\mathbb{R}$ and $\text{Lip}(f) = 1$.

Proof. 1. Let us show that f is a Lipschitz mapping of X into $\mathbb{R}$ and 1 is a Lipschitz coefficient of f. Thus let us show that

$$(1) \qquad d_e\big(f(x'), f(x'')\big) \le 1 \cdot d_X(x', x'') \quad \text{for } x', x'' \in X.$$

Now we have

$$\begin{aligned} d_e\big(f(x'), f(x'')\big) &= |f(x') - f(x'')| \\ &= |d_X(x', x_0) - d_X(x'', x_0)| \le d_X(x', x'') \end{aligned}$$

where the inequality is a consequence of the triangle inequality of the metric d_X. This proves (1). Thus f is a Lipschitz mapping of X into $\mathbb{R}$ and 1 is a Lipschitz coefficient of f. Then since $\text{Lip}(f)$ is the minimum of all Lipschitz coefficients of f we have $\text{Lip}(f) \le 1$.

2. Let us show that $\text{Lip}(f) = 1$. Since 1 is a Lipschitz coefficient of f as we showed above, to show that $\text{Lip}(f) = 1$ we show that 1 is the minimum of all Lipschitz coefficients of f. Suppose for some $\delta > 0$, $1 - \delta$ is a Lipschitz coefficient of f. Then we have

$$(2) \qquad d_e\big(f(x'), f(x'')\big) \le (1 - \delta) d_X(x', x'') \quad \text{for } x', x'' \in X.$$

In particular for $x, x_0 \in X$ such that $x \ne x_0$, we have

$$(3) \qquad d_e\big(f(x), f(x_0)\big) \le (1 - \delta) d_X(x, x_0).$$

Now

$$(4) \qquad \begin{aligned} d_e\big(f(x), f(x_0)\big) &= |f(x) - f(x_0)| \\ &= |d_X(x, x_0) - d_X(x_0, x_0)| = d_X(x, x_0). \end{aligned}$$

By (3) and (4) we have

$$(5) \qquad d_X(x, x_0) \le (1 - \delta) d_X(x, x_0).$$

This contradicts the fact that $d_X(x, x_0) > 0$ and $1 - \delta < 1$. Thus 1 is the minimum of all Lipschitz coefficients of f. ∎

Theorem 13.9. (Transitivity of Lipschitz Mapping) *Let (X, d_X), (Y, d_Y) and (Z, d_Z) be metric spaces. Let $A \subset X$ and $B \subset Y$. Let f be a Lipschitz mapping of A into B and g*

be a Lipschitz mapping of B into Z. Then $g \circ f$ is a Lipschitz mapping of A into Z and $\mathrm{Lip}(g \circ f) \leq \mathrm{Lip}(g)\,\mathrm{Lip}(f)$.

Proof. Since f is a Lipschitz mapping of A into B we have

$$(1) \qquad d_Y\big(f(x'), f(x'')\big) \leq \mathrm{Lip}(f) d_X(x', x'') \quad \text{for } x', x'' \in A$$

and since g is a Lipschitz mapping of B into Z we have

$$(2) \qquad d_Z\big(g(y'), g(y'')\big) \leq \mathrm{Lip}(g) d_Y(y', y'') \quad \text{for } y', y'' \in B.$$

Then since f maps A into B, (2) implies

$$(3) \qquad d_Z\big(g(f(x')), g(f(x''))\big) \leq \mathrm{Lip}(g) d_Y\big(f(x'), f(x'')\big) \quad \text{for } x', x'' \in A.$$

Substituting (1) in (3), we have

$$(4) \qquad d_Z\big((g \circ f)(x'), (g \circ f)(x'')\big) \leq \mathrm{Lip}(g)\,\mathrm{Lip}(f) d_X(x', x'') \quad \text{for } x', x'' \in A.$$

This shows that $g \circ f$ is a Lipschitz mapping of A into Z and $\mathrm{Lip}(g)\,\mathrm{Lip}(f)$ is a Lipschitz coefficient of $g \circ f$. Then since $\mathrm{Lip}(g \circ f)$, the Lipschitz constant of $g \circ f$, is the minimum of all Lipschitz coefficients of $g \circ f$, we have $\mathrm{Lip}(g \circ f) \leq \mathrm{Lip}(g)\,\mathrm{Lip}(f)$. ∎

[II] Extension of Lipschitz Mappings

Definition 13.10. *Let (X, d_X) and (Y, d_Y) be metric spaces and let $A \subset B \subset X$. Let f be a Lipschitz mapping of A into Y. If there exists a Lipschitz mapping g of B into Y such that $f = g$ on A then we call g a Lipschitz extension of f to B.*

Proposition 13.11. *Let (X, d_X) and (Y, d_Y) be metric spaces and let $A \subset B \subset X$. Let f be a Lipschitz mapping of A into Y. If g is a Lipschitz extension of f to B then* $\mathrm{Lip}(f) \leq \mathrm{Lip}(g)$.

Proof. Let $\Gamma(f)$ and $\Gamma(g)$ be the sets of all Lipschitz coefficients of f and g respectively. Since $f = g$ on $A \subset B$, if $C \in \Gamma(g)$ then $C \in \Gamma(f)$. Thus $\Gamma(g) \subset \Gamma(f)$. According to Observation 13.3, we have $\Gamma(f) = [\mathrm{Lip}(f), \infty)$ and $\Gamma(g) = [\mathrm{Lip}(g), \infty)$. Then $[\mathrm{Lip}(g), \infty) \subset [\mathrm{Lip}(f), \infty)$ implies that $\mathrm{Lip}(f) \leq \mathrm{Lip}(g)$. ∎

Theorem 13.12. (McShane's Extension Theorem) *Consider a metric space (X, d_X) and the metric space $(\mathbb{R}, d_e)$ where d_e is the Euclidean metric on $\mathbb{R}$. Let $A \subset X$ and let f be a Lipschitz mapping of A into $\mathbb{R}$. Then there exists a Lipschitz extension $\widetilde{f}$ of f to X with* $\mathrm{Lip}\,(\widetilde{f}) = \mathrm{Lip}(f)$. *Indeed we have the following constructions:*
(a) *Let us define a mapping g of X into $\mathbb{R}$ by setting*

$$g(x) = \inf_{a \in A} \{f(a) + \mathrm{Lip}(f) d_X(a, x)\} \quad \text{for } x \in X. \tag{1}$$

Then g is a Lipschitz mapping of X into $\mathbb{R}$ such that $g = f$ on A and $\mathrm{Lip}(g) = \mathrm{Lip}(f)$.
(b) *Let us define a mapping h of X into $\mathbb{R}$ by setting*

$$h(x) = \sup_{a \in A} \{f(a) - \mathrm{Lip}(f) d_X(a, x)\} \quad \text{for } x \in X. \tag{2}$$

Then h is a Lipschitz mapping of X into $\mathbb{R}$ such that $h = f$ on A and $\mathrm{Lip}(h) = \mathrm{Lip}(f)$.

Proof. 1. Let us prove (a). Let g be a mapping of X into $\mathbb{R}$ defined by (1). Let us show that $g = f$ on A. Let $b \in A$. Then by (1) we have

$$\begin{aligned} g(b) &= \inf_{a \in A} \{f(a) + \mathrm{Lip}(f) d_X(a, b)\} \\ &\leq f(b) + \mathrm{Lip}(f) d_X(b, b) \\ &= f(b). \end{aligned} \tag{3}$$

On the other hand, for every $a \in A$ we have $d_e\big(f(a), f(b)\big) \leq \mathrm{Lip}(f) d_X(a, b)$ so that

$$\begin{aligned} &f(a) + \mathrm{Lip}(f) d_X(a, b) \geq f(a) + d_e\big(f(a), f(b)\big) \\ =&f(a) + |f(b) - f(a)| \geq f(a) + f(b) - f(a) \\ =&f(b). \end{aligned} \tag{4}$$

By (1) and (4), we have

$$g(b) = \inf_{a \in A} \{f(a) + \mathrm{Lip}(f) d_X(a, b)\} \geq f(b). \tag{5}$$

By (3) and (5), we have $g(b) = f(b)$ for every $b \in A$. Thus $g = f$ on A.

Let us show that g is a Lipschitz mapping of X into $\mathbb{R}$. Let $x, y \in X$. Then by (1)

$$\begin{aligned} g(x) &= \inf_{a \in A} \{f(a) + \mathrm{Lip}(f) d_X(a, x)\} \\ &\leq \inf_{a \in A} \{f(a) + \mathrm{Lip}(f)\big(d_X(a, y) + d_X(y, x)\big)\} \\ &= g(y) + \mathrm{Lip}(f) d_X(y, x), \end{aligned}$$

and then

$$g(x) - g(y) \leq \mathrm{Lip}(f) d_X(x, y). \tag{6}$$

Interchanging the roles of x and y in the argument above, we have

$$g(y) - g(x) \leq \mathrm{Lip}(f) d_X(x, y). \tag{7}$$

By (6) and (7), we have

$$|g(x) - g(y)| \leq \mathrm{Lip}(f) d_X(x, y) \quad \text{for } x, y \in X.$$

Then we have

$$d_e\big(g(x), g(y)\big) \leq \mathrm{Lip}(f) d_X(x, y) \quad \text{for } x, y \in X. \tag{8}$$

This shows that g is a Lipschitz mapping of X into $\mathbb{R}$ and $\mathrm{Lip}(f)$ is a Lipschitz coefficient of g.

Let us show that $\mathrm{Lip}(g) = \mathrm{Lip}(f)$. We showed above that $\mathrm{Lip}(f)$ is a Lipschitz coefficient of g. Since $\mathrm{Lip}(g)$ is the minimum of all Lipschitz coefficients of g, we have $\mathrm{Lip}(g) \leq \mathrm{Lip}(f)$. Now

$$d_e\big(g(x), g(y)\big) \leq \mathrm{Lip}(g) d_X(x, y) \quad \text{for } x, y \in X. \tag{9}$$

Since $g = f$ on A, (9) implies

$$d_e\big(f(x), f(y)\big) \leq \mathrm{Lip}(g) d_X(x, y) \quad \text{for } x, y \in A. \tag{10}$$

Then $\mathrm{Lip}(g)$ is a Lipschitz coefficient of f and hence $\mathrm{Lip}(g) \geq \mathrm{Lip}(f)$. Therefore we have $\mathrm{Lip}(g) = \mathrm{Lip}(f)$.

2. (b) is proved by a similar argument. Let us show that $h = f$ on A. Let $b \in A$. Then by (2) we have

$$\text{(11)} \qquad \begin{aligned} h(b) &= \sup_{a \in A} \{f(a) - \text{Lip}(f) d_X(a,b)\} \\ &\geq f(b) - \text{Lip}(f) d_X(b,b) \\ &= f(b). \end{aligned}$$

On the other hand, for every $a \in A$ we have $d_e\big(f(a), f(b)\big) \leq \text{Lip}(f) d_X(a,b)$ and thus

$$\text{(12)} \qquad \begin{aligned} f(a) - \text{Lip}(f) d_X(a,b) &\leq f(a) - d_e\big(f(a), f(b)\big) \\ = f(a) - |f(a) - f(b)| &\leq f(a) - \{f(a) - f(b)\} \\ = f(b). \end{aligned}$$

By (2) and (12), we have

$$\text{(13)} \qquad h(b) = \sup_{a \in A} \{f(a) - \text{Lip}(f) d_X(a,b)\} \leq f(b).$$

By (11) and (13), we have $h(b) = f(b)$ for every $b \in A$. Thus $h = f$ on A.

The fact that h is a Lipschitz mapping of X into $\mathbb{R}$ and $\text{Lip}(h) = \text{Lip}(f)$ is proved in the same way as for g. ∎

Theorem 13.13. *Consider a metric space (X, d_X) and the metric space $(\mathbb{R}, d_e)$ where d_e is the Euclidean metric on $\mathbb{R}$. Let $A \subset X$ and let f be a Lipschitz mapping of A into $\mathbb{R}$. Let $\widetilde{f}$ be a Lipschitz extension of f to X with* $\text{Lip}\big(\widetilde{f}\big) = \text{Lip}(f)$. *Then $h \leq \widetilde{f} \leq g$ where g and h are Lipschitz extensions of f to X with* $\text{Lip}(f) = \text{Lip}(g) = \text{Lip}(h)$ *as defined by (1) and (2) of Theorem 13.12.*

Proof. 1. Let us prove $\widetilde{f} \leq g$. Let $x \in X$. By the definition of g by (1) of Theorem 13.12, we have

$$\text{(1)} \qquad \begin{aligned} g(x) - \widetilde{f}(x) &= \inf_{a \in A} \{f(a) + \text{Lip}(f) d_X(a,x)\} - \widetilde{f}(x) \\ &= \inf_{a \in A} \{f(a) - \widetilde{f}(x) + \text{Lip}(f) d_X(a,x)\} \\ &= \inf_{a \in A} \{\widetilde{f}(a) - \widetilde{f}(x) + \text{Lip}\big(\widetilde{f}\big) d_X(a,x)\} \end{aligned}$$

where the last equality is by the fact that $\widetilde{f} = f$ on A and $\text{Lip}\big(\widetilde{f}\big) = \text{Lip}(f)$.

In (1) above, $x \in X$ is arbitrarily given and fixed and the infimum is over $a \in A$. Let us show that for every $a \in A$ we have

$$\text{(2)} \qquad \{\widetilde{f}(a) - \widetilde{f}(x) + \text{Lip}\big(\widetilde{f}\big) d_X(a,x)\} \geq 0.$$

Now for $a \in A$, there are two possibilities: either $\widetilde{f}(a) - \widetilde{f}(x) \geq 0$ or $\widetilde{f}(a) - \widetilde{f}(x) < 0$. If $\widetilde{f}(a) - \widetilde{f}(x) \geq 0$ then since $\mathrm{Lip}\,(\widetilde{f}) d_X(a,x) \geq 0$, (2) is valid. On the other hand if $\widetilde{f}(a) - \widetilde{f}(x) < 0$, then

$$\widetilde{f}(a) - \widetilde{f}(x) = -|\widetilde{f}(a) - \widetilde{f}(x)| = -d_e(\widetilde{f}(a), \widetilde{f}(x)). \tag{3}$$

But we have

$$0 \leq d_e(\widetilde{f}(a), \widetilde{f}(x)) \leq \mathrm{Lip}\,(\widetilde{f}) d_X(a,x). \tag{4}$$

Then by (3) and (4) we have

$$\widetilde{f}(a) - \widetilde{f}(x) + \mathrm{Lip}\,(\widetilde{f}) d_X(a,x) = -d_e(\widetilde{f}(a), \widetilde{f}(x)) + \mathrm{Lip}\,(\widetilde{f}) d_X(a,x) \geq 0$$

and (2) is valid. This completes the proof of (2). Now (2) implies

$$\inf_{a\in A} \{\widetilde{f}(a) - \widetilde{f}(x) + \mathrm{Lip}\,(\widetilde{f}) d_X(a,x)\} \geq 0. \tag{5}$$

By (1) and (5) we have $g(x) - \widetilde{f}(x) \geq 0$. This shows that $\widetilde{f} \leq g$ on X.

2. Let us prove $h \leq \widetilde{f}$. By the definition of h by (2) of Theorem 13.12, we have

$$\begin{aligned} h(x) - \widetilde{f}(x) &= \sup_{a\in A} \{f(a) - \mathrm{Lip}(f) d_X(a,x)\} - \widetilde{f}(x) \\ &= \sup_{a\in A} \{f(a) - \widetilde{f}(x) - \mathrm{Lip}(f) d_X(a,x)\} \\ &= \sup_{a\in A} \{\widetilde{f}(a) - \widetilde{f}(x) - \mathrm{Lip}\,(\widetilde{f}) d_X(a,x)\} \end{aligned} \tag{6}$$

where the last equality is by the fact that $\widetilde{f} = f$ on A and $\mathrm{Lip}\,(\widetilde{f}) = \mathrm{Lip}(f)$.

Let us show that for every $a \in A$ we have

$$\widetilde{f}(a) - \widetilde{f}(x) - \mathrm{Lip}\,(\widetilde{f}) d_X(a,x) \leq 0. \tag{7}$$

Now for $a \in A$, there are two possibilities: either $\widetilde{f}(a) - \widetilde{f}(x) \geq 0$ or $\widetilde{f}(a) - \widetilde{f}(x) < 0$. If $\widetilde{f}(a) - \widetilde{f}(x) \geq 0$, then

$$\widetilde{f}(a) - \widetilde{f}(x) = |\widetilde{f}(a) - \widetilde{f}(x)| = d_e(\widetilde{f}(a), \widetilde{f}(x)) \leq \mathrm{Lip}\,(\widetilde{f}) d_X(a,x)$$

and (7) is valid. If $\widetilde{f}(a) - \widetilde{f}(x) < 0$, then since $\mathrm{Lip}\,(\widetilde{f}) d_X(a,x) \geq 0$, (7) is valid. Now (7) implies

$$\sup_{a\in A} \{\widetilde{f}(a) - \widetilde{f}(x) - \mathrm{Lip}\,(\widetilde{f}) d_X(a,x)\} \leq 0. \tag{8}$$

By (6) and (8) we have $h(x) - \widetilde{f}(x) \leq 0$. This shows that $h \leq \widetilde{f}$ on X. ∎

We consider next Lipschitz extension of a Lipschitz mapping of $A \subset X$ into $\mathbb{R}^m$. Theorem 13.12 (McShane) was for the particular case $m = 1$.

Lemma 13.14. *Consider the metric spaces $(\mathbb{R}^m, d_e^m)$ and $(\mathbb{R}, d_e)$ where d_e^m and d_e are the Euclidean metrics on $\mathbb{R}^m$ and $\mathbb{R}$ respectively. For $i = 1, \dots, n$, let π_i be the projection of $\mathbb{R}^m = \mathbb{R}_1 \times \cdots \times \mathbb{R}_m$ onto $\mathbb{R}_i$, that is, $\pi_i(x) = \pi(x_1, \dots, x_m) = x_i$. Then π_i is a Lipschitz mapping of $\mathbb{R}^m$ into $\mathbb{R}$ with $\mathrm{Lip}(\pi_i) = 1$.*

Proof. 1. For $x', x'' \in \mathbb{R}^m$ we have

$$d_e\big(\pi_i(x'), \pi_i(x'')\big) = |\pi_i(x') - \pi_i(x'')| = |x_i' - x_i''|$$
$$\leq |x' - x''| = d_e^m(x', x'').$$

This shows that π_i is a Lipschitz mapping of $\mathbb{R}^m$ into $\mathbb{R}$ with 1 as a Lipschitz coefficient.

To show that $\mathrm{Lip}(\pi_i) = 1$, we show that for any $\varepsilon > 0$, $1 - \varepsilon$ cannot be a Lipschitz coefficient for π_i. Assume that $1-\varepsilon$ is a Lipschitz coefficient for π_i. Consider two points in $\mathbb{R}^m$ given by $x' = (0, \dots, 0, x_i', 0, \cdots, 0)$ and $x'' = (0, \dots, 0, x_i'', 0, \cdots, 0)$ where $x_i' \neq x_i''$. Since we assume that $1 - \varepsilon$ is a Lipschitz coefficient for π_i, we have for every $x', x'' \in \mathbb{R}^m$ and in particular for our x' and x''

$$d_e\big(\pi_i(x'), \pi_i(x'')\big) \leq \{1 - \varepsilon\} d_e^m(x', x''). \tag{1}$$

But

$$d_e\big(\pi_i(x'), \pi_i(x'')\big) = d_e(x_i', x_i'') = |x_i' - x_i''|$$

and

$$d_e^m(x', x'') = |x' - x''| = |x_i' - x_i''|$$

so that (1) reduces to

$$|x_i' - x_i''| \leq \{1 - \varepsilon\}|x_i' - x_i''|,$$

which is impossible since $|x_i' - x_i''| > 0$. ∎

Proposition 13.15. *Consider a metric space (X, d_X) and the metric space $(\mathbb{R}^m, d_e^m)$ where d_e^m is the Euclidean metrics on $\mathbb{R}^m$. Let f be a Lipschitz mapping of a set $A \subset X$ into $\mathbb{R}^m$. Let us write $f = (f_1, \dots, f_m)$ where $f_i = \pi_i \circ f$ and π_i is the projection of $\mathbb{R}^m = \mathbb{R}_1 \times \cdots \times \mathbb{R}_m$ onto $\mathbb{R}_i$. Then f_i is a Lipschitz mapping of A into $\mathbb{R}$ and $\mathrm{Lip}(f_i) \leq \mathrm{Lip}(f)$.*

Proof. Now f is a Lipschitz mapping of $A \subset X$ into $\mathbb{R}^m$ and according to Lemma 13.14, π_i is a Lipschitz mapping of $\mathbb{R}^m$ into $\mathbb{R}$ with $\mathrm{Lip}(\pi_i) = 1$. Then by Theorem 13.9 (Transitivity), $f_i = \pi_i \circ f$ is a Lipschitz mapping of A into $\mathbb{R}$ with $\mathrm{Lip}(f_i) \leq \mathrm{Lip}(\pi_i)\,\mathrm{Lip}(f) = \mathrm{Lip}(f)$. ∎

Theorem 13.16. *Consider a metric space (X, d_X) and the metric space $(\mathbb{R}^m, d_e^m)$ where d_e^m is the Euclidean metrics on $\mathbb{R}^m$. Let $A \subset X$ and let f be a Lipschitz mapping of A into $\mathbb{R}^m$. Then there exists a Lipschitz extension $\widetilde{f}$ of f to X with $\mathrm{Lip}\big(\widetilde{f}\big) \leq \sqrt{m}\,\mathrm{Lip}(f)$.*

Proof. Let us write $f = (f_1, \dots, f_m)$ where $f_i = \pi_i \circ f$ and π_i is the projection of $\mathbb{R}^m = \mathbb{R}_1 \times \cdots \times \mathbb{R}_m$ onto $\mathbb{R}_i$. By Proposition 13.15, f_i is a Lipschitz mapping of A into $\mathbb{R}$ and $\mathrm{Lip}(f_i) \le \mathrm{Lip}(f)$.

By Theorem 13.12 (McShane), there exists a Lipschitz extension $\widetilde{f_i}$ of f_i to X with $\mathrm{Lip}\left(\widetilde{f_i}\right) = \mathrm{Lip}(f_i)$. Let us define a mapping of X into $\mathbb{R}^m$ by

$$\widetilde{f} = \left(\widetilde{f_1}, \dots, \widetilde{f_m}\right).$$

Now since $\widetilde{f_i}$ is a Lipschitz extension of f_i to X, we have $\widetilde{f_i} = f_i$ on A. Then $\widetilde{f} = f$ on A.

Let us show that $\widetilde{f}$ is a Lipschitz mapping of X into $\mathbb{R}^m$. Now we have

$$\begin{aligned}\left\{d_e^m\left(\widetilde{f}(x'), \widetilde{f}(x'')\right)\right\}^2 &= \left|\widetilde{f}(x') - \widetilde{f}(x'')\right|^2 = \sum_{i=1}^m \left|\widetilde{f_i}(x') - \widetilde{f_i}(x'')\right|^2 \\ &= \sum_{i=1}^m \left\{d_e\left(\widetilde{f_i}(x'), \widetilde{f_i}(x'')\right)\right\}^2 \le \sum_{i=1}^m \left\{\mathrm{Lip}\left(\widetilde{f_i}\right) d_X(x', x'')\right\}^2 \\ &\le m\left\{\mathrm{Lip}(f) d_X(x', x'')\right\}^2\end{aligned}$$

where the last equality is by $\mathrm{Lip}\left(\widetilde{f_i}\right) = \mathrm{Lip}(f_i) \le \mathrm{Lip}(f)$. Thus we have

$$d_e^m\left(\widetilde{f}(x'), \widetilde{f}(x'')\right) \le \sqrt{m}\,\mathrm{Lip}(f) d_X(x', x'').$$

This shows that $\widetilde{f}$ is a Lipschitz mapping of X into $\mathbb{R}^m$ with $\sqrt{m}\,\mathrm{Lip}(f)$ as a Lipschitz coefficient. Then $\mathrm{Lip}\left(\widetilde{f}\right) \le \sqrt{m}\,\mathrm{Lip}(f)$. ∎

[III] Lipschitz Condition and Differentiability

Let f be a real-valued function on an interval I in $\mathbb{R}$. If f satisfies the Lipschitz condition

$$|f(x') - f(x'')| \leq C|x' - x''| \quad \text{for } x', x'' \in I$$

where $C \geq 0$, then f is absolutely continuous on I and consequently f is differentiable at μ_L-a.e. $x \in I$.

Now let f be a real-valued function on $\mathbb{R}^n$. Suppose f satisfies the Lipschitz condition

$$|f(x') - f(x'')| \leq C|x' - x''| \quad \text{for } x', x'' \in \mathbb{R}^n$$

where $C \geq 0$. We want to show that f is differentiable at μ_L^n-a.e. $x \in \mathbb{R}^n$. The first problem we encounter here is that the definition of differentiability of a function on $\mathbb{R}^n$ is a generalization of the definition of differentiability of a function on $\mathbb{R}$ and the second problem is that absolute continuity of a function on $\mathbb{R}^n$ is undefined.

[III.1] Differentiability of Functions on $\mathbb{R}^n$

Definition 13.17. (Differentiability and Differential of a Function on $\mathbb{R}^n$) *Let f be a mapping of an open set $\Omega \subset \mathbb{R}^n$ into $\mathbb{R}$. We say that f is differentiable at $x \in \Omega$ if there exists a linear mapping L of $\mathbb{R}^n$ into $\mathbb{R}$ such that*

$$\lim_{y \to x, y \in \Omega} \frac{|f(y) - f(x) - L(y - x)|}{|x - y|} = 0.$$

We call the linear mapping L the differential of f at $x \in \Omega$ and we write

$$df(x, \cdot) = L(\cdot).$$

(Note that the condition above can be rewritten as

$$\lim_{z \to 0} \frac{|f(x + z) - f(x) - L(z)|}{|z|} = 0,$$

where $z \in \mathbb{R}^n$ such that $x + z \in \Omega$.)

Theorem 13.18. (Uniqueness of the Differential) *Let f be a mapping of an open set $\Omega \subset \mathbb{R}^n$ into $\mathbb{R}$. Let $x \in \Omega$. If the differential of f at x, $df(x, \cdot)$, exists then it is unique.*

Proof. By Definition 13.17 the differential $df(x, \cdot)$ is a linear mapping L of $\mathbb{R}^n$ into $\mathbb{R}$ such that

$$\lim_{z \to 0} \frac{|f(x + z) - f(x) - L(z)|}{|z|} = 0,$$

where $z \in \mathbb{R}^n$ such that $x + z \in \Omega$. Suppose there exist two linear mappings L_1 and L_2 of $\mathbb{R}^n$ into $\mathbb{R}$ such that

$$\text{(2)} \qquad \lim_{z \to 0} \frac{|f(x + z) - f(x) - L_j(z)|}{|z|} = 0 \quad \text{for } j \in \{1, 2\}.$$

Let us show that $L_1 = L_2$. Now (1) is equivalent to

(3) $$\lim_{z\to\mathbf{0}} \frac{f(x+z) - f(x) - L_j(z)}{|z|} = 0 \quad \text{for } j \in \{1, 2\}.$$

Subtracting (2) for $j = 2$ from (2) for $j = 1$, we have

(4) $$\lim_{z\to\mathbf{0}} \frac{L_1(z) - L_2(z)}{|z|} = 0.$$

Since L_1 is a linear mapping of $\mathbb{R}^n$ into $\mathbb{R}$, L_1 can be represented as

(5) $$L_1(z) = a_1 z_1 + \cdots + a_n z_n \quad \text{for } z = (z_1, \ldots, z_n) \in \mathbb{R}^n,$$

where $\{a_1, \ldots, a_n\} \subset \mathbb{R}$ and similarly L_2 can be represented as

(6) $$L_2(z) = b_1 z_1 + \cdots + b_n z_n \quad \text{for } z = (z_1, \ldots, z_n) \in \mathbb{R}^n,$$

where $\{b_1, \ldots, b_n\} \subset \mathbb{R}$. Let $z = (z_1, 0, \ldots, 0)$ where $z_1 \in \mathbb{R}$. Then from (3) and by (4) and (5) we have

$$0 = \lim_{z\to\mathbf{0}} \frac{L_1(z) - L_2(z)}{|z|} = \lim_{z_1\to 0} \frac{a_1 z_1 - b_1 z_1}{|z_1|} = \lim_{z_1\to 0} \pm(a_1 - b_1) = \pm(a_1 - b_1).$$

Thus $a_1 = b_1$. We show similarly that $a_2 = b_2$, ..., $a_n = b_n$. Then we have $L_1 = L_2$. ∎

Definition 13.19. (Directional Derivative) *Let f be a mapping of an open set $\Omega \subset \mathbb{R}^n$ into $\mathbb{R}$. Consider $S(\mathbf{0}, 1) = \{x \in \mathbb{R}^n : d(x, \mathbf{0}) = 1\} = \{x \in \mathbb{R}^n : |x| = 1\}$. Let $v \in S(\mathbf{0}.1)$. Let $x \in \Omega$ and let $t \in \mathbb{R}$. Let $t \in \mathbb{R}$ be so small that $x + tv \in \Omega$. If*

$$\lim_{t\to 0} \frac{f(x+tv) - f(x)}{t} \in \mathbb{R}$$

then we call this limit the directional derivative of f at x in the direction of v and we write

$$D_v f(x) = \lim_{t\to 0} \frac{f(x+tv) - f(x)}{t}.$$

Remark 13.20. (Directional Derivatives and Partial derivatives) Let f be a mapping of an open set $\Omega \subset \mathbb{R}^n$ into $\mathbb{R}$. Let $x \in \Omega$. If the partial derivatives of f at x, that is, $\frac{\partial f}{\partial x_1}(x)$, ..., $\frac{\partial f}{\partial x_n}(x)$, exist, then

$$D_{v_1} f(x) = \frac{\partial f}{\partial x_1}(x), \ldots, D_{v_n} f(x) = \frac{\partial f}{\partial x_n}(x)$$

where $v_1 = (1, 0, \ldots, 0) \in S(\mathbf{0}, 1)$, ..., $v_n = (0, \ldots, 0, 1) \in S(\mathbf{0}, 1)$. Thus partial derivatives are particular cases of directional derivatives.

Proposition 13.21. *Let f be a mapping of an open set $\Omega \subset \mathbb{R}^n$ into $\mathbb{R}$. Let $x \in \Omega$ and $v \in S(\mathbf{0}, 1)$. If the directional derivative $D_v f(x)$ exists then so does the directional derivative $D_{-v} f(x)$ and moreover we have*

$$D_{-v} f(x) = -D_v f(x).$$

Proof. By Definition 13.19 we have

$$\begin{aligned} D_{-v} f(x) &= \lim_{t \to 0} \frac{f(x + t(-v)) - f(x)}{t} = \lim_{t \to 0} \frac{f(x - tv) - f(x)}{t} \\ &= \lim_{-s \to 0} \frac{f(x + sv) - f(x)}{-s} = -\lim_{s \to 0} \frac{f(x + sv) - f(x)}{s} \\ &= -D_v f(x). \quad \blacksquare \end{aligned}$$

Definition 13.22. (Gradient) *Let f be a mapping of an open set $\Omega \subset \mathbb{R}^n$ into $\mathbb{R}$. If the partial derivatives $\frac{\partial f}{\partial x_1}(x)$, ..., $\frac{\partial f}{\partial x_n}(x)$ exist at $x \in \Omega$ then we call the vector $\left(\frac{\partial f}{\partial x_1}(x), \ldots, \frac{\partial f}{\partial x_n}(x)\right) \in \mathbb{R}^n$ the gradient of f at x. Thus we define*

$$\operatorname{grad} f(x) = \left(\frac{\partial f}{\partial x_1}(x), \ldots, \frac{\partial f}{\partial x_n}(x) \right).$$

Proposition 13.23. *Let f be a mapping of an open set $\Omega \subset \mathbb{R}^n$ into $\mathbb{R}$. Suppose f is differentiable at $x \in \Omega$ so that the differential $df(x, \cdot)$ exists. Then for every $v \in S(\mathbf{0}, 1)$ the directional derivative $D_v f(x)$ exists and moreover it is given by*

$$D_v f(x) = df(x, v) = \langle \operatorname{grad} f(x), v \rangle. \tag{1}$$

Moreover the $1 \times n$ matrix A of the linear mapping $df(x, \cdot)$ of $\mathbb{R}^n$ into $\mathbb{R}$ is given by

$$A = \left[\frac{\partial f}{\partial x_1}(x), \ldots, \frac{\partial f}{\partial x_n}(x) \right]. \tag{2}$$

Proof. By Definition 13.19 we have

$$D_v f(x) = \lim_{t \to 0} \frac{f(x + tv) - f(x)}{t}. \tag{3}$$

If f is differentiable at $x \in \Omega$ then by Definition 13.17 there exists a linear mapping $df(x, \cdot)$ of $\mathbb{R}^n$ into $\mathbb{R}$ such that

(4) $$\lim_{z \to \mathbf{0}} \frac{f(x+z) - f(x) - df(x,z)}{|z|} = 0$$

Let $v \in S(\mathbf{0}, 1)$ be arbitrarily chosen. Let $z = tv$ for $t \in \mathbb{R}$. Then $z \to \mathbf{0}$ if and only if $t \to 0$. Thus from (4) we have

$$\lim_{t \to 0} \frac{f(x+tv) - f(x) - df(x,tv)}{|t|} = 0,$$

which is equivalent to

(5) $$\lim_{t \to 0} \frac{f(x+tv) - f(x) - df(x,tv)}{t} = 0.$$

By the linearity of $df(x, \cdot)$ we have $df(x, tv) = tdf(x, v)$. Substituting this in (5)

$$\lim_{t \to 0} \frac{f(x+tv) - f(x)}{t} = df(x, v).$$

This shows that $D_v f(x)$ exists and moreover

(6) $$D_v f(x) = df(x, v).$$

Since $df(x, \cdot)$ is a linear mapping of $\mathbb{R}^n$ into $\mathbb{R}$, $df(x, \cdot)$ can be represented as

(7) $$df(x, z) = a_1 z_1 + \cdots + a_n z_n \quad \text{for } z = (z_1, \dots, z_n) \in \mathbb{R}^n,$$

where $\{a_1, \dots, a_n\} \subset \mathbb{R}$. Let $v_1 = (1, 0, \dots, 0) \in S(\mathbf{0}, 1)$, ..., $v_n = (0, \dots, 0, 1) \in S(\mathbf{0}, 1)$. Then by (6) and (7) we have for $i \in \{1, \dots, n\}$

(8) $$D_{v_i} f(x) = df(x, v_i) = a_i.$$

Now the partial derivative $D_{v_i} f(x) = \frac{\partial f}{\partial x_i}(x)$. Thus $a_i = \frac{\partial f}{\partial x_i}(x)$ by (8). Substituting this in (7) we have by Definition 13.22

(9) $$df(x, z) = \frac{\partial f}{\partial x_1}(x) z_1 + \cdots + \frac{\partial f}{\partial x_n}(x) z_n = \langle \operatorname{grad} f(x), z \rangle .$$

With (6) and (9) we have (1).

Let A be the $1 \times n$ matrix of the linear mapping $df(x, cdot)$ of $\mathbb{R}^n$ into $\mathbb{R}$. Then $df(x, z) = Az$ for $z \in \mathbb{R}^n$. Then the first equality in (9) shows that A is given by (2). ■

Theorem 13.24. (Existence of Differential) *Let f be a mapping of an open set $\Omega \subset \mathbb{R}^n$ into $\mathbb{R}$. Suppose $f \in C^1(\Omega)$. Then f is differentiable at every $x \in \Omega$ and moreover the differential of f at x is given by*

(1) $$df(x, \cdot) = \langle \operatorname{grad} f(x), \cdot \rangle ,$$

that is,

$$(2)\qquad df(x,z) = \frac{\partial f}{\partial x_1}(x)z_1 + \cdots + \frac{\partial f}{\partial x_n}(x)z_n \quad \textit{for } z = (z_1, \ldots, z_n) \in \mathbb{R}^n.$$

Proof. Let $x \in \Omega$. Consider a linear mapping of $\mathbb{R}^n$ into $\mathbb{R}$ defined by

$$(3)\qquad L(z) = \frac{\partial f}{\partial x_1}(x)z_1 + \cdots + \frac{\partial f}{\partial x_n}(x)z_n \quad \text{for } z = (z_1, \ldots, z_n) \in \mathbb{R}^n.$$

Let us show that f is differentiable at x and the differential of f at x, $df(x,\cdot) = L(\cdot)$, that is,

$$(4)\qquad \lim_{z \to \mathbf{0}} \frac{f(x+z) - f(x) - L(z)}{|z|} = 0.$$

For simplicity let us prove (4) for the case $n = 2$. Let $x \in \Omega$. Now since $f \in C^1(\Omega)$, $\frac{\partial f}{\partial x_1}(x)$ and $\frac{\partial f}{\partial x_2}(x)$ are continuous functions of $x \in \Omega$. Thus for every $\varepsilon > 0$ there exists $\delta > 0$ such that

$$(5)\qquad \left|\frac{\partial f}{\partial x_1}(x') - \frac{\partial f}{\partial x_1}(x)\right| < \frac{\varepsilon}{2} \quad \text{for } x' \in \Omega \text{ such that } |x' - x| < \delta,$$

$$(6)\qquad \left|\frac{\partial f}{\partial x_2}(x'') - \frac{\partial f}{\partial x_2}(x)\right| < \frac{\varepsilon}{2} \quad \text{for } x'' \in \Omega \text{ such that } |x' - x| < \delta.$$

Since $x \in \Omega$ and Ω is an open set in $\mathbb{R}^2$, there exists an open ball $B(x,\delta) \subset \Omega$. Let $z \in \mathbb{R}^2$ be such that $|z| < \delta$ and $z \neq \mathbf{0}$. Then $x + z \in \Omega$. Then by the Mean Value Theorem we have

$$\begin{aligned}(7)\qquad f(x+z) - f(x) &= f(x_1 + z_1, x_2 + z_2) - f(x_1, x_2)\\ &= f(x_1 + z_1, x_2 + z_2) - f(x_1, x_2 + z_2) + f(x_1, x_2 + z_2) - f(x_1, x_2)\\ &= \frac{\partial f}{\partial x_1}(x_1 + \vartheta_1 z_1, x_2 + z_2)z_1 + \frac{\partial f}{\partial x_2}(x_1, x_2 + \vartheta_2 z_2)z_2,\\ &= \frac{\partial f}{\partial x_1}(x')z_1 + \frac{\partial f}{\partial x_2}(x'')z_2\end{aligned}$$

where $\vartheta_1, \vartheta_2 \in (0,1)$ and we set

$$(8)\qquad x' := (x_1 + \vartheta_1 z_1, x_2 + z_2) \quad \text{and} \quad x'' := (x_1, x_2 + \vartheta_2 z_2).$$

For x' and x'' defined above we have

$$(9)\qquad |x' - x| < |z| < \delta \quad \text{and} \quad |x'' - x| < |z| < \delta.$$

Now from (7) and (3) defining the linear mapping L, we have

$$\begin{aligned}
&|f(x+z) - f(x) - L(z)| \\
&= \left| \left\{ \frac{\partial f}{\partial x_1}(x')z_1 + \frac{\partial f}{\partial x_2}(x'')z_2 \right\} - \left\{ \frac{\partial f}{\partial x_1}(x)z_1 + \frac{\partial f}{\partial x_2}(x)z_2 \right\} \right| \\
&\leq \left| \frac{\partial f}{\partial x_1}(x') - \frac{\partial f}{\partial x_1}(x) \right| |z_1| + \left| \frac{\partial f}{\partial x_2}(x'') - \frac{\partial f}{\partial x_2}(x) \right| |z_2| \\
&< \left\{ \frac{\varepsilon}{2} + \frac{\varepsilon}{2} \right\} |z| = \varepsilon |z|
\end{aligned}$$

where the second inequality is by (5), (6) and (9). Thus we have

$$\frac{|f(x+z) - f(x) - L(z)|}{|z|} < \varepsilon \quad \text{for } |z| < \delta.$$

This proves (4). ■

[III.2] Differentiability of Lipschitz Functions on $\mathbb{R}^n$

Lemma 13.25. (Existence of Directional Derivative) *Let f be a Lipschitz mapping of an open set $\Omega \subset \mathbb{R}^n$ into $\mathbb{R}$. Then for every $v \in S(\mathbf{0}, 1)$, the directional derivative $D_v f(x)$ exists for μ_L^n-a.e. $x \in \Omega$ and $D_v f(x)$ is a $\mathfrak{M}_L^n$-measurable function on Ω.*

Proof. 1. Let $v \in S(\mathbf{0}, 1)$ and $x \in \Omega$. By Definition 13.10 we have

$$(1) \qquad D_v f(x) = \lim_{t \to 0} \frac{f(x+tv) - f(x)}{t},$$

provided that the limit exists in $\mathbb{R}$. Let us define

$$\overline{D}_v f(x) = \limsup_{t \to 0} \frac{f(x+tv) - f(x)}{t},$$
$$\underline{D}_v f(x) = \liminf_{t \to 0} \frac{f(x+tv) - f(x)}{t}.$$

Then $\underline{D}_v f(x)$ and $\overline{D}_v f(x)$ always exist, $\underline{D}_v f(x) \leq \overline{D}_v f(x)$, and $D_v f(x)$ exists if and only if $\underline{D}_v f(x) = \overline{D}_v f(x)$ and when this is the case we have $D_v f(x) = \underline{D}_v f(x) = \overline{D}_v f(x)$.

Let

$$(2) \qquad A_v = \{x \in \Omega : \ D_v f(x) \text{ does not exist }\} = \{x \in \Omega : \underline{D}_v f(x) < \overline{D}_v f(x)\}.$$

To prove our Lemma, we show that $\overline{D}_v f(x)$ and $\underline{D}_v f(x)$ are $\mathfrak{M}_L^n$-measurable functions on Ω and $\mu_L^n(A) = 0$.

Let us prove that $\overline{D}_v f(x)$ for $x \in \Omega$ is a $\mathfrak{M}_L^n$-measurable functions on Ω. Now

$$\overline{D}_v f(x) = \limsup_{t \to 0} \frac{f(x+tv) - f(x)}{t} = \lim_{k \to \infty} \sup_{0<|t|<\frac{1}{k}} \frac{f(x+tv) - f(x)}{t}.$$

With $x \in \Omega$ and $v \in S(\mathbf{0}, 1)$ fixed, consider the function

$$\varphi(t) = f(x+tv) \quad \text{for } t \in \mathbb{R}.$$

Since f is a Lipschitz mapping of Ω into $\mathbb{R}$, we have

$$|\varphi(t') - \varphi(t'')| = |f(x+t'v) - f(x+t''v)| \leq \text{Lip}(f)|(x+t'v) - (x+t''v)|$$
$$= \text{Lip}(f)|(t'-t'')v| = \text{Lip}(f)|t'-t''| \quad \text{for } t', t'' \in \mathbb{R}.$$

Thus $\varphi(t)$ is a continuous function of t. Then $\frac{1}{t}\{f(x+tv) - f(x)\}$ is a continuous function of $t \in \mathbb{R} \setminus \{0\}$. Thus if we let Q be the set of all rational numbers then

$$\sup_{0<|t|<\frac{1}{k}} \frac{f(x+tv) - f(x)}{t} = \sup_{0<|t|<\frac{1}{k}, t \in Q} \frac{f(x+tv) - f(x)}{t}.$$

Therefore we have

$$\overline{D}_v f(x) = \lim_{k\to\infty} \sup_{0<|t|<\frac{1}{k}, t\in Q} \frac{f(x+tv)-f(x)}{t}.$$

Since f is a Lipschitz mapping of Ω into $\mathbb{R}$, f is a continuous mapping of Ω into $\mathbb{R}$ and hence a $\mathfrak{B}_{\mathbb{R}^n}$-measurable mapping of Ω into $\mathbb{R}$. Then $\frac{1}{t}\{f(x+tv)-f(x)\}$ is a $\mathfrak{B}_{\mathbb{R}^n}$-measurable mapping of Ω into $\mathbb{R}$ for each fixed $t \in \mathbb{R} \setminus \{0\}$. Then $\overline{D}_v f(x)$ as given above is a $\mathfrak{B}_{\mathbb{R}^n}$-measurable and then a $\mathfrak{M}_L^n$-measurable mapping of Ω into $\mathbb{R}$.

Similarly $\underline{D}_v f(x)$ is a $\mathfrak{M}_L^n$-measurable mapping of Ω into $\mathbb{R}$. This implies that A_v is a $\mathfrak{M}_L^n$-measurable set. Since $D_v f(x)$ is defined to be $D_v f(x) = \underline{D}_v f(x) = \overline{D}_v f(x)$ for $x \in \Omega$ for which $\underline{D}_v f(x) = \overline{D}_v f(x)$, $D_v f(x)$ is a $\mathfrak{M}_L^n$-measurable mapping of Ω into $\mathbb{R}$. Thus it remains to show that $\mu_L^n(A_v) = 0$.

2. Let $v_1 = (1, 0, \ldots, 0) \in S(\mathbf{0}, 1)$. Let us show that for this particular case we have

$$\mu_L^n(A_{v_1}) = 0. \tag{3}$$

Note that $D_{v_1} f(x) = \frac{\partial f}{\partial x_1}(x)$ if it exists. Now we have by Fubini's Theorem

$$\begin{aligned} \mu_L^n(A_{v_1}) &= \int_{\mathbb{R}^n} \mathbf{1}_{A_{v_1}}\, d\mu_L^n \\ &= \int_{\mathbb{R}_2\times\cdots\times\mathbb{R}_n} \left[\int_{\mathbb{R}_1} \mathbf{1}_{A_{v_1}}(x_1, x_2, \ldots, x_n)\, \mu_L(dx_1)\right] \mu_L^{n-1}(d(x_2, \ldots, x_n)). \end{aligned} \tag{4}$$

With $(x_2, \ldots, x_n) \in \mathbb{R}_2 \times \cdots \times \mathbb{R}_n$ fixed, $f(\cdot, x_2, \ldots, x_n)$ is a function on $\mathbb{R}_1$. Since f is a Lipschitz mapping of Ω into $\mathbb{R}$ we have

$$\begin{aligned} &|f(x_1', x_2, \ldots, x_n) - f(x_1'', x_2, \ldots, x_n)| \\ &\leq \text{Lip}(f)|(x_1', x_2, \ldots, x_n) - (x_1'', x_2, \ldots, x_n)| \\ &= \text{Lip}(f)|x_1' - x_1''|. \end{aligned}$$

Thus $f(\cdot, x_2, \ldots, x_n)$ is a Lipschitz function on $\mathbb{R}_1$. Then $f(\cdot, x_2, \ldots, x_n)$ is an absolutely continuous function and thus $f(\cdot, x_2, \ldots, x_n)$ is differentiable at μ_L-a.e. $x_1 \in \mathbb{R}_1$, that is, $\frac{\partial f}{\partial x_1}(x_1, x_2, \ldots, x_n) = D_{v_1} f(x_1, x_2, \ldots, x_n)$ exists for μ_L-a.e. $x_1 \in \mathbb{R}_1$. Thus $(x_1, x_2, \ldots, x_n) \notin A_{v_1}$ for μ_L-a.e. $x_1 \in \mathbb{R}_1$. Then we have $\mathbf{1}_{A_{v_1}}(x_1, x_2, \ldots, x_n) = 0$ for μ_L-a.e. $x_1 \in \mathbb{R}_1$. Then

$$\int_{\mathbb{R}_1} \mathbf{1}_{A_{v_1}}(x_1, x_2, \ldots, x_n)\, \mu_L(dx_1) = 0.$$

Substituting this in (4), we have (3).

3. To complete the proof of the Lemma, it remains to show that for an arbitrary $v \in S(\mathbf{0}, 1)$ we have

(5) $$\mu_L^n(A_v) = 0.$$

Consider the Lebesgue measure space $(\mathbb{R}^n, \mathfrak{M}_L^n, \mu_L^n)$. Let T be a non-singular linear transformation of $\mathbb{R}^n$ into $\mathbb{R}^n$. Then $T(E) \in \mathfrak{M}_L^n$ for every $E \in \mathfrak{M}_L^n$ and moreover we have

$$\mu_L^n(T(E)) = |\det M_T| \mu_L^n(E) \quad \text{for } E \in \mathfrak{M}_L^n,$$

where M_T is the matrix of the linear transformation T. (See Theorem 24.31, [LRA].) In particular when T is a rotation then $\det M_T = 0$ and thus

(6) $$\mu_L^n(T(E)) = \mu_L^n(E) \quad \text{for } E \in \mathfrak{M}_L^n.$$

Thus the Lebesgue measure space $(\mathbb{R}^n, \mathfrak{M}_L^n, \mu_L^n)$ is invariant under rotation. Now let $v \in S(\mathbf{0}, 1)$ be arbitrarily chosen. Let T be a rotation of $\mathbb{R}^n$ such that $T(v) = v_1$. Then we have $T(A_v) = A_{v_1}$. Thus by (6) and (3) we have

$$\mu_L^n(A_v) = \mu_L^n(T(A_v)) = \mu_L^n(A_{v_1}) = 0.$$

This proves (5). ■

Lemma 13.26. (Existence of Gradient) *Let f be a Lipschitz mapping of an open set $\Omega \subset \mathbb{R}^n$ into $\mathbb{R}$. Then the gradient of f at $x \in \Omega$, $\operatorname{grad} f(x) = \left(\frac{\partial f}{\partial x_1}(x), \dots, \frac{\partial f}{\partial x_n}(x)\right)$ exists for μ_L^n-a.e. $x \in \Omega$.*

Proof. Partial derivatives are particular cases of directional derivatives. Thus by Lemma 13.25 for each $i = 1, \dots, n$, $\frac{\partial f}{\partial x_i}(x)$ exists for μ_L^n-a.e. $x \in \Omega$. Thus for each $i = 1, \dots, n$, there exists $A_i \in \mathfrak{M}_L^n$ such that $A_i \subset \Omega$ and $\mu_L^n(\Omega \setminus A_i) = 0$ and $\frac{\partial f}{\partial x_i}(x)$ exists for $x \in A_i$. Let $A = \bigcap_{i=1}^n A_i$. Then $A \in \mathfrak{M}_L^n$, $A \subset \Omega$ and

$$\Omega \setminus A = \Omega \cap A^c = \Omega \cap \left(\bigcap_{i=1}^n A_i\right)^c = \Omega \cap \left(\bigcup_{i=1}^n A_i^c\right) = \bigcup_{i=1}^n (\Omega \cap A_i^c) = \bigcup_{i=1}^n (\Omega \setminus A_i)$$

and then

$$\mu_L^n(\Omega \setminus A) = \mu_L^n\left(\bigcup_{i=1}^n (\Omega \setminus A_i)\right) \le \sum_{i=1}^n \mu_L^n(\Omega \setminus A_i) = 0.$$

For $x \in A$, $\frac{\partial f}{\partial x_1}(x)$, ..., $\frac{\partial f}{\partial x_n}(x)$ all exist and thus $\left(\frac{\partial f}{\partial x_1}(x), \dots, \frac{\partial f}{\partial x_n}(x)\right)$ exists. This shows that $\left(\frac{\partial f}{\partial x_1}(x), \dots, \frac{\partial f}{\partial x_n}(x)\right)$ exists for μ_L^n-a.e. $x \in \Omega$. ■

Lemma 13.27. (Integration by Parts) *Let f be a Lipschitz mapping of $\mathbb{R}^n$ into $\mathbb{R}$. Let $\zeta \in C_c^1(\Omega)$. Then for each $i \in \{1, \dots, n\}$, we have*

(1) $$\int_{\mathbb{R}^n} f(x) \frac{\partial \zeta}{\partial x_i}(x)\, \mu_L^n(dx) = -\int_{\mathbb{R}^n} \frac{\partial f}{\partial x_i}(x)\, \zeta(x)\, \mu_L^n(dx).$$

Proof. For simplicity of notations required, let us prove (1) for the case $i = 1$. Thus we are to prove that for every $\zeta \in C_c^1(\mathbb{R}^n)$ we have

$$\int_{\mathbb{R}^n} f(x)\,\frac{\partial \zeta}{\partial x_1}(x)\,\mu_L^n(dx) = -\int_{\mathbb{R}^n} \frac{\partial f}{\partial x_1}(x)\,\zeta(x)\,\mu_L^n(dx). \tag{2}$$

Since f is a Lipschitz mapping of $\mathbb{R}^n$ into $\mathbb{R}$, f is continuous on $\mathbb{R}^n$ and hence f is a $\mathfrak{B}_{\mathbb{R}^n}$-measurable mapping of $\mathbb{R}^n$ into $\mathbb{R}$ and a fortiori a $\mathfrak{M}_L^n$-measurable mapping of $\mathbb{R}^n$ into $\mathbb{R}$. Also by Lemma 13.25, $\frac{\partial \zeta}{\partial x_1}(x)$ exists for μ_L^n-a.e. $x \in \mathbb{R}^n$ and is a $\mathfrak{M}_L^n$-measurable mapping of $\mathbb{R}^n$ into $\mathbb{R}$. Since $\zeta \in C_c^1(\mathbb{R}^n)$, $\frac{\partial \zeta}{\partial x_1}(x)$ is a continuous function on $\mathbb{R}^n$ and hence it is a $\mathfrak{B}_{\mathbb{R}^n}$-measurable and a fortiori a $\mathfrak{M}_L^n$-measurable mapping of $\mathbb{R}^n$ into $\mathbb{R}$.

Now since f is a Lipschitz mapping of $\mathbb{R}^n$ into $\mathbb{R}$, we have

$$\begin{aligned}
&|f(x_1', x_2, \dots, x_n) - f(x_1'', x_2, \dots, x_n)| \\
&\le \operatorname{Lip}(f)|(x_1', x_2, \dots, x_n) - (x_1'', x_2, \dots, x_n)| \\
&= \operatorname{Lip}(f)|x_1' - x_1''| \quad \text{for } x_1', x_1'' \in \mathbb{R}_1.
\end{aligned}$$

Thus for fixed $(x_2, \dots, x_n) \in \mathbb{R}_2 \times \cdots \times \mathbb{R}_n$, the function $f(\cdot, x_2, \dots, x_n)$ on $\mathbb{R}_1$ is a Lipschitz function and hence it is an absolutely continuous function on $\mathbb{R}_1$ and consequently its derivative $\frac{\partial f}{\partial x_1}(x_1, x_2, \dots, x_n)$ exists for μ_L-a.e. $x_1 \in \mathbb{R}_1$.

Since $\frac{\partial \zeta}{\partial x_1}$ is a continuous function on $\mathbb{R}^n$, for fixed $(x_2, \dots, x_n) \in \mathbb{R}_2 \times \cdots \times \mathbb{R}_n$, the function $\frac{\partial \zeta}{\partial x_1}(\cdot, x_2, \dots, x_n)$ is a continuous function on $\mathbb{R}_1$ and this implies that $\zeta(\cdot, x_2, \dots, x_n)$ is an absolutely continuous function on $\mathbb{R}_1$.

Since $\zeta \in C_c^1(\mathbb{R}^n)$, there exists a closed ball $C(\mathbf{0}, a)$, where $a > 0$, in $\mathbb{R}^n$ such that $\zeta(x) = 0$ for $x \in C(\mathbf{0}, a)^c$ and consequently $\frac{\partial \zeta}{\partial x_1}(x) = 0$ for $x \in C(\mathbf{0}, a)^c$. Then for fixed $(x_2, \dots, x_n) \in \mathbb{R}_2 \times \cdots \times \mathbb{R}_n$, $\zeta(\cdot, x_2, \dots, x_n) = 0$ on $(-\infty, -a) \cup (a, \infty)$ and $\frac{\partial \zeta}{\partial x_1}(\cdot, x_2, \dots, x_n) = 0$ on $(-\infty, -a) \cup (a, \infty)$. Thus we have

$$\int_{\mathbb{R}} f(x_1, \cdot)\frac{\partial \zeta}{\partial x_1}(x_1, \cdot)\,\mu_L(dx_1) = \int_{[-a,a]} f(x_1, \cdot)\frac{\partial \zeta}{\partial x_1}(x_1, \cdot)\,\mu_L(dx_1), \tag{3}$$

$$\int_{\mathbb{R}} \frac{\partial f}{\partial x_1}(x_1, \cdot)\zeta(x_1, \cdot)\,\mu_L(dx_1) = \int_{[-a,a]} \frac{\partial f}{\partial x_1}(x_1, \cdot)\zeta(x_1, \cdot)\,\mu_L(dx_1). \tag{4}$$

Since $f(\cdot, x_2, \dots, x_n)$ and $\zeta(\cdot, x_2, \dots, x_n)$ are absolutely continuous functions on $[-a, a]$, we have by Theorem 13.29 (Integration by Parts), [LRA]

$$\begin{aligned}
&\int_{[-a,a]} f(x_1, \cdot)\frac{\partial \zeta}{\partial x_1}(x_1, \cdot)\,\mu_L(dx_1) + \int_{[-a,a]} \frac{\partial f}{\partial x_1}(x_1, \cdot)\zeta(x_1, \cdot)\,\mu_L(dx_1) \\
&= f(a, \cdot)\zeta(a, \cdot) - f(-a, \cdot)\zeta(-a, \cdot) = 0
\end{aligned}$$

since $\zeta(-a,\cdot)=\zeta(a,\cdot)=0$. Thus we have

$$\text{(5)}\qquad \int_{[-a,a]} f(x_1,\cdot)\frac{\partial\zeta}{\partial x_1}(x_1,\cdot)\,\mu_L(dx_1) = -\int_{[-a,a]} \frac{\partial f}{\partial x_1}(x_1,\cdot)\zeta(x_1,\cdot)\,\mu_L(dx_1).$$

Combining (3), (4) and (5), we have

$$\text{(6)}\qquad \int_{\mathbb{R}} f(x_1,\cdot)\frac{\partial\zeta}{\partial x_1}(x_1,\cdot)\,\mu_L(dx_1) = -\int_{\mathbb{R}} \frac{\partial f}{\partial x_1}(x_1,\cdot)\zeta(x_1,\cdot)\,\mu_L(dx_1).$$

Then we have

$$\begin{aligned}
&\int_{\mathbb{R}^n} f(x)\frac{\partial\zeta}{\partial x_1}(x)\,\mu_L^n(dx)\\
&= \int_{\mathbb{R}_2\times\cdots\times\mathbb{R}_n}\left[\int_{\mathbb{R}_1} f(x_1,x_2,\dots,x_n)\frac{\partial\zeta}{\partial x_1}(x_1,x_2,\dots,x_n)\,\mu_L(dx_1)\right]\,\mu_L^{n-1}(d(x_2,\dots,x_n))\\
&= \int_{\mathbb{R}_2\times\cdots\times\mathbb{R}_n}\left[-\int_{\mathbb{R}_1} \frac{\partial f}{\partial x_1}(x_1,x_2,\dots,x_n)\zeta(x_1,x_2,\dots,x_n)\,\mu_L(dx_1)\right]\,\mu_L^{n-1}(d(x_2,\dots,x_n))\\
&= -\int_{\mathbb{R}} \frac{\partial f}{\partial x_1}(x)\zeta(x)\,\mu_L^n(dx)
\end{aligned}$$

where the first and third equalities are by Fubini's Theorem and the second equality is by (6). This proves (2). For any $i\in\{1,\dots,n\}$, (1) is proved exactly the same way as (2). ■

Lemma 13.28. *Let f be a Lipschitz mapping of an open set $\Omega\subset\mathbb{R}^n$ into $\mathbb{R}$. Then for every $v\in S(\mathbf{0},1)$ we have*

$$\text{(1)}\qquad D_v f(x) = \langle \operatorname{grad} f(x), v\rangle \quad \textit{for } \mu_L^n\textit{-a.e. } x\in\Omega.$$

Proof. According to Theorem 13.12, every Lipschitz mapping f of Ω into $\mathbb{R}$ can be extended to a Lipschitz mapping of $\mathbb{R}^n$ into $\mathbb{R}$. Thus no generality is lost if we assume that f is a Lipschitz mapping of $\mathbb{R}^n$ into $\mathbb{R}$.

By Lemma 13.25, for every $v\in S(\mathbf{0},1)$ the directional derivative $D_v f(x)$ exists for μ_L^n-a.e. $x\in\mathbb{R}^n$. By Lemma 13.26, grad $f(x)$ exists for μ_L^n-a.e. $x\in\mathbb{R}^n$. To prove (1), it suffices to show that for every $\zeta\in C_c^1(\mathbb{R}^n)$ we have

$$\text{(2)}\qquad \int_{\mathbb{R}^n} D_v f(x)\,\zeta(x)\,\mu_L^n(dx) = \int_{\mathbb{R}^n} \langle \operatorname{grad} f(x), v\rangle\,\zeta(x)\,\mu_L^n(dx).$$

Recall that by Definition 13.19 we have

$$\text{(3)}\qquad D_v f(x) = \lim_{t\to 0}\frac{f(x+tv)-f(x)}{t} = \lim_{k\to\infty}\frac{f(x+\frac{1}{k}v)-f(x)}{\frac{1}{k}}.$$

Then

$$(4) \qquad \int_{\mathbb{R}^n} D_v f(x)\,\zeta(x)\,\mu_L^n(dx) = \int_{\mathbb{R}^n} \lim_{k\to\infty} \frac{f(x+\frac{1}{k}v)-f(x)}{\frac{1}{k}}\zeta(x)\,\mu_L^n(dx).$$

By Definition 13.6 and Theorem 13.7 we have

$$(5) \qquad \left|\frac{f(x+\frac{1}{k}v)-f(x)}{\frac{1}{k}}\right| \le \Lambda(f) = \mathrm{Lip}(f).$$

Note also that

$$(6) \qquad \frac{f(x+\frac{1}{k}v)-f(x)}{\frac{1}{k}}\zeta(x) = 0 \quad \text{for } x \in (\operatorname{supp}\zeta)^c.$$

Then we have

$$\int_{\mathbb{R}^n} \lim_{k\to\infty} \frac{f(x+\frac{1}{k}v)-f(x)}{\frac{1}{k}}\,\zeta(x)\,\mu_L^n(dx) = \lim_{k\to\infty}\int_{\mathbb{R}^n} \frac{f(x+\frac{1}{k}v)-f(x)}{\frac{1}{k}}\,\zeta(x)\,\mu_L^n(dx)$$

$$= \lim_{k\to\infty}\left\{\int_{\mathbb{R}^n} \frac{f(x+\frac{1}{k}v)}{\frac{1}{k}}\,\zeta(x)\,\mu_L^n(dx) - \int_{\mathbb{R}^n}\frac{f(x)}{\frac{1}{k}}\,\zeta(x)\,\mu_L^n(dx)\right\}$$

$$= \lim_{k\to\infty}\left\{\int_{\mathbb{R}^n} \frac{f(x)}{\frac{1}{k}}\,\zeta(x-\frac{1}{k}v)\,\mu_L^n(dx) - \int_{\mathbb{R}^n}\frac{f(x)}{\frac{1}{k}}\,\zeta(x)\,\mu_L^n(dx)\right\}$$

$$= \lim_{k\to\infty}\int_{\mathbb{R}^n} f(x)\,\frac{\zeta(x-\frac{1}{k}v)-\zeta(x)}{\frac{1}{k}}\,\mu_L^n(dx) = \int_{\mathbb{R}^n}\lim_{k\to\infty} f(x)\,\frac{\zeta(x-\frac{1}{k}v)-\zeta(x)}{\frac{1}{k}}\,\mu_L^n(dx)$$

$$= \int_{\mathbb{R}^n} f(x) D_{-v}\zeta(x)\,\mu_L^n(dx) = -\int_{\mathbb{R}^n} f(x) D_v\zeta(x)\,\mu_L^n(dx)$$

where the first equality is the by Dominated Convergence Theorem which is applicable by (5) and (6) and the third equality is by the translation invariance of the Lebesgue measure space $(\mathbb{R}^n, \mathfrak{M}_L^n, \mu_L^n)$. Thus we have

$$(7) \qquad \int_{\mathbb{R}^n} D_v f(x)\,\zeta(x)\,\mu_L^n(dx) = -\int_{\mathbb{R}^n} f(x) D_v\zeta(x)\,\mu_L^n(dx).$$

For $\zeta \in C_c^1(\mathbb{R}^n)$, Theorem 13.24 and Proposition 13.23 imply that ζ is differentiable at every $x \in \mathbb{R}^n$ and the differential is given by $d\zeta(x,\cdot) = \langle \operatorname{grad}\zeta(x), \cdot\rangle$ and for any $v \in S(\mathbf{0},1)$ the directional derivative $D_v\zeta(x)$ exists and is given by

$$(8) \qquad D_v\zeta(x) = \langle \operatorname{grad}\zeta(x), v\rangle = \frac{\partial\zeta}{\partial x_1}(x)u_1 + \cdots + \frac{\partial\zeta}{\partial x_n}(x)u_n$$

where $v = (u_1, \dots, u_n)$. Substituting (8) in (7) we have

(9) $$\int_{\mathbb{R}^n} D_v f(x)\zeta(x)\, \mu_L^n(dx) = -\sum_{i=1}^n u_i \int_{\mathbb{R}^n} f(x)\frac{\partial \zeta}{\partial x_i}(x)\, \mu_L^n(dx).$$

By Lemma 13.27 (Integration by Parts), we have for each $i \in \{1, \dots, n\}$

(10) $$\int_{\mathbb{R}^n} f(x)\frac{\partial \zeta}{\partial x_i}(x)\, \mu_L^n(dx) = -\int_{\mathbb{R}^n} \frac{\partial f}{\partial x_i}(x)\zeta(x)\, \mu_L^n(dx).$$

Substituting (10) in (9) we have

$$\int_{\mathbb{R}^n} D_v f(x)\zeta(x)\, \mu_L^n(dx) = \sum_{i=1}^n u_i \int_{\mathbb{R}^n} \frac{\partial f}{\partial x_i}(x)\zeta(x)\, \mu_L^n(dx)$$

$$= \int_{\mathbb{R}^n} \langle v, \operatorname{grad} f(x)\rangle\, \zeta(x)\, \mu_L^n(dx).$$

This proves (2). ■

Theorem 13.29. (Rademacher's Theorem) *Let f be a Lipschitz mapping of an open set $\Omega \subset \mathbb{R}^n$ into $\mathbb{R}$. Then f is differentiable at μ_L^n-a.e. $x \in \Omega$ and the differential $df(x, \cdot)$ at x is given by*

$$df(x, \cdot) = \langle \operatorname{grad} f(x), \cdot \rangle\,.$$

Proof. According to Theorem 13.12, every Lipschitz mapping f of Ω into $\mathbb{R}$ can be extended to a Lipschitz mapping of $\mathbb{R}^n$ into $\mathbb{R}$. Thus no generality is lost if we assume that f is a Lipschitz mapping of $\mathbb{R}^n$ into $\mathbb{R}$.

According to Lemma 13.25, for every $v \in S(\mathbf{0}, 1)$ the directional derivative $D_v f(x)$ exists for μ_L^n-a.e. $x \in \mathbb{R}^n$. According to Lemma 13.26, grad $f(x)$ exists for μ_L^n-a.e. $x \in \mathbb{R}^n$. According to Lemma 13.28, for every $v \in S(\mathbf{0}, 1)$ we have

(1) $$D_v f(x) = \langle \operatorname{grad} f(x), v\rangle \quad \text{for } \mu_L^n\text{-a.e. } x \in \mathbb{R}^n\,.$$

Let $\{v_k : k \in \mathbb{N}\}$ be a countable dense subset of $S(\mathbf{0}, 1)$. For each $k \in \mathbb{N}$, let

(2) $$A_k = \{x \in \mathbb{R}^n : \exists D_{v_k} f(x), \exists \operatorname{grad} f(x), D_{v_k} f(x) = \langle \operatorname{grad} f(x), v_k\rangle\,\}.$$

By (1) and (2) we have

(3) $$\mu_L^n(A_k^c) = 0 \quad \text{for every } k \in \mathbb{N}.$$

Let us define

(4) $$A = \bigcap_{k \in \mathbb{N}} A_k.$$

Then we have

$$\mu_L^n(A^c) = \mu_L^n\left(\left(\bigcap_{k\in\mathbb{N}} A_k\right)^c\right) = \mu_L^n\left(\bigcup_{k\in\mathbb{N}} A_k^c\right) \le \sum_{k\in\mathbb{N}} \mu_L^n(A_k^c) = 0. \tag{5}$$

To prove the Theorem we show that f is differentiable at every $x \in A$ and the differential $df(x,\cdot)$ is given by $df(x,\cdot) = \langle \text{grad}\, f(x), \cdot\rangle$. Thus we are to show that

$$\lim_{z\to\mathbf{0}} \frac{f(x+z) - f(x) - \langle \text{grad}\, f(x), z\rangle}{|z|} = 0. \tag{6}$$

Let $z \in \mathbb{R}^n$ be such that $z \neq \mathbf{0}$. Let $v = \frac{1}{|z|}z \in S(\mathbf{0}, 1)$ and let $t = |z| > 0$. Then $z = tv$. Then we have

$$\begin{aligned}
&\frac{f(x+z) - f(x) - \langle \text{grad}\, f(x), z\rangle}{|z|} \\
&= \frac{f(x+tv) - f(x) - \langle \text{grad}\, f(x), tv\rangle}{t} \\
&= \frac{f(x+tv) - f(x)}{t} - \langle \text{grad}\, f(x), v\rangle \quad \text{where } t > 0.
\end{aligned}$$

For $x \in A$ and $v \in S(\mathbf{0}, 1)$, let us define a function

$$Q(x, v, t) = \frac{f(x+tv) - f(x)}{t} - \langle \text{grad}\, f(x), v\rangle \quad \text{for } t \in \mathbb{R} \setminus \{0\}. \tag{7}$$

Then (6) is equivalent to $\lim_{t\to 0, t>0} Q(x, v, t) = 0$. To prove (6) we show that we actually have the stronger condition

$$\lim_{t\to 0} Q(x, v, t) = 0. \tag{8}$$

To prove (8), note first that since $x \in A$ we have $x \in A_k$ for every $k \in \mathbb{N}$ by the definition of A by (4). Then by the definition of A_k by (2), $D_{v_k} f(x)$ and grad $f(x)$ exist and moreover $D_{v_k} f(x) = \langle \text{grad}\, f(x), v_k\rangle$. Then by the definition of the directional derivative $D_{v_k} f(x)$ by Definition 13.19 we have

$$\lim_{t\to 0} \frac{f(x+tv_k) - f(x)}{t} = D_{v_k} f(x) = \langle \text{grad}\, f(x), v_k\rangle$$

so that

$$\lim_{t\to 0} \frac{f(x+tv_k) - f(x)}{t} - \langle \text{grad}\, f(x), v_k\rangle = 0.$$

Recalling the definition of $Q(x, v, t)$ by (7), we have for every $k \in \mathbb{N}$

$$\lim_{t\to 0} Q(x, v_k, t) = 0. \tag{9}$$

We show next that (9) implies (8). Now for every $k \in \mathbb{N}$ we have from the definition of $Q(x, v, t)$ by (7)

$$(10) \quad \begin{aligned} |Q(x,v,t) - Q(x,v_k,t)| &= \left| \frac{f(x+tv) - f(x+tv_k)}{t} \right| + |\langle \operatorname{grad} f(x), v - v_k \rangle| \\ &\le \frac{1}{|t|} \operatorname{Lip}(f) |t(v - v_k)| + |\operatorname{grad} f(x)||v - v_k| \\ &= \{ \operatorname{Lip}(f) + |\operatorname{grad} f(x)| \} |v - v_k|. \end{aligned}$$

Now by Definition 13.6 and Theorem 13.7 we have $\left|\frac{\partial f}{\partial x_i}(x)\right| \le \Lambda(f) = \operatorname{Lip}(f)$. Thus

$$|\operatorname{grad} f(x)| = \sqrt{\left|\frac{\partial f}{\partial x_1}(x)\right|^2 + \cdots + \left|\frac{\partial f}{\partial x_n}(x)\right|^2} \le \sqrt{n \operatorname{Lip}(f)^2} = \sqrt{n} \operatorname{Lip}(f).$$

Substituting this in (10) we have

$$(11) \quad |Q(x,v,t) - Q(x,v_k,t)| \le \{\sqrt{n} + 1\} \operatorname{Lip}(f) |v - v_k|.$$

Now let $\varepsilon > 0$ be arbitrarily given. Since $\{v_k : k \in \mathbb{N}\}$ is a countable dense subset of $S(\mathbf{0}, 1)$ there exists $k_0 \in \mathbb{N}$ such that

$$(12) \quad |v - v_{k_0}| \le \frac{\varepsilon}{2\{\sqrt{n} + 1\} \operatorname{Lip}(f)}.$$

Since $\lim_{t \to 0} Q(x, v_{k_0}, t) = 0$ by (9), there exists $\delta > 0$ such that

$$(13) \quad |Q(x, v_{k_0}, t)| < \frac{\varepsilon}{2} \quad \text{for } 0 < |t| < \delta.$$

Then for $0 < |t| < \delta$, we have

$$\begin{aligned} |Q(x,v,t)| &\le |Q(x,v_{k_0},t)| + |Q(x,v,t) - Q(x,v_{k_0},t)| \\ &< \frac{\varepsilon}{2} + \{\sqrt{n} + 1\} \operatorname{Lip}(f) |v - v_{k_0}| \\ &< \varepsilon, \end{aligned}$$

where the second inequality is by (13) and (11) and the last inequality is by (12). This shows that $\lim_{t \to 0} Q(x, v, t) = 0$ and proves (8). ∎

§14 Integration with Respect to a Measure on a Metric Space

[I] Continuous Extensions of Continuous Mappings of a Metric Space

Theorem 14.1. (Urysohn) *Let (X, ρ) be a metric space and let A and B be disjoint non-empty closed sets in X.*
(a) *There exists a real-valued continuous function g on X such that $g(x) = 0$ for $x \in A$, $g(x) = 1$ for $x \in B$ and $g(x) \in (0, 1)$ for $x \in (A \cup B)^c$.*
(b) *For $a, b \in \mathbb{R}$ such that $a < b$, there exists a real-valued continuous function h on X such that $h(x) = a$ for $x \in A$, $h(x) = b$ for $x \in B$ and $h(x) \in (a, b)$ for $x \in (A \cup B)^c$.*

Proof. 1. Let us define a function g on X by letting

$$(1) \qquad g(x) = \frac{\rho(x, A)}{\rho(x, A) + \rho(x, B)} \quad \text{for } x \in X.$$

Then g has the properties stated in (a).

2. Let $a, b \in \mathbb{R}$ and $a < b$. Let us define a function φ on $[0, 1]$ by letting

$$(2) \qquad \varphi(\xi) = (b - a)\xi + b \quad \text{for } \xi \in [0, 1].$$

Then φ is real-valued, strictly increasing and continuous on $[0, 1]$ with $\varphi(0) = a$ and $\varphi(1) = b$. With the function g as defined by (1), let us define a function h on X by letting

$$(3) \qquad h(x) = \varphi \circ g(x) \quad \text{for } x \in X.$$

Since g is a continuous mapping of X into $[0, 1]$ and φ is a continuous mapping of $[0, 1]$ into $[a, b]$, $h = \varphi \circ g$ is a continuous mapping of X into $[a, b]$. Moreover for $x \in A$ we have $h(x) = \varphi \circ g(x) = \varphi(0) = a$; for $x \in B$ we have $h(x) = \varphi \circ g(x) = \varphi(1) = b$; and for $x \in (A \cup B)^c$ we have $h(x) = \varphi \circ g(x) \in (a, b)$ since $g(x) \in (0, 1)$. ■

Theorem 14.2. (Tietze Extension Theorem) *Let (X, ρ) be a metric space and let F be a closed subset of X. Let f be a real-valued continuous function defined on F with respect to the relative topology on F. Then there exists a real-valued continuous function $\widetilde{f}$ on X such that $\widetilde{f}(x) = f(x)$ for $x \in F$ and $\inf_{x \in F} f(x) \leq \inf_{x \in X} \widetilde{f}(x)$ and $\sup_{x \in X} \widetilde{f} \leq \sup_{x \in F} f(x)$.*

Proof. 1. Consider first the case that f is a real-valued continuous function on F with $\inf_{x \in F} f(x) = -1$ and $\sup_{x \in F} f(x) = 1$. Let $f_0 = f$ on F and let

$$A_0 = \left\{x \in F : f_0(x) \leq -\tfrac{1}{3}\right\}, \quad B_0 = \left\{x \in F : f_0(x) \geq \tfrac{1}{3}\right\}.$$

Then A_0 and B_0 are disjoint non-empty closed sets in F and then since F is a closed set in X, A_0 and B_0 are closed sets in X. Thus by Theorem 14.1 (Urysohn) there exists a continuous function g_0 on X such that $g_0(X) \subset \left[-\tfrac{1}{3}, \tfrac{1}{3}\right]$ and $g(A_0) = -\tfrac{1}{3}$ and $g(B_0) = \tfrac{1}{3}$.

Let $f_1 = f_0 - g_0$ on F. For $x \in A_0$ we have $f_0(x) \in \left[-1, -\frac{1}{3}\right]$ and $g_0(x) = -\frac{1}{3}$ so that $f_1(x) = f_0(x) - g_0(x) \in \left[-\frac{2}{3}, 0\right]$ and $\inf_{x \in A_0} f_1(x) = -\frac{2}{3}$. For $x \in B_0$ we have $f_0(x) \in \left[\frac{1}{3}, 1\right]$ and $g_0(x) = \frac{1}{3}$ so that $f_1(x) = f_0(x) - g_0(x) \in \left[0, \frac{2}{3}\right]$ and $\sup_{x \in B_0} f_1(x) = \frac{2}{3}$. For $x \in F \setminus (A_0 \cup B_0)$ we have $f_0(x) \in \left[-\frac{1}{3}, \frac{1}{3}\right]$ and $g_0(x) \in \left[-\frac{1}{3}, \frac{1}{3}\right]$ also so that $f_1(x) = f_0(x) - g_0(x) \in \left[-\frac{2}{3}, \frac{2}{3}\right]$. Thus $\inf_{x \in F} f_1(x) = -\frac{2}{3}$ and $\sup_{x \in F} f_1(x) = \frac{2}{3}$. Let

$$A_1 = \left\{x \in F : f_1(x) \leq -\tfrac{1}{3}\left(\tfrac{2}{3}\right)\right\}, \quad B_1 = \left\{x \in F : f_1(x) \geq \tfrac{1}{3}\left(\tfrac{2}{3}\right)\right\}.$$

Then A_1 and B_1 are disjoint non-empty closed sets in X so that by Theorem 14.1 there exists a real-valued continuous function g_1 on X such that $g_1(X) \in \left[-\frac{1}{3}\left(\frac{2}{3}\right), \frac{1}{3}\left(\frac{2}{3}\right)\right]$ with $g_1(A_1) = -\frac{1}{3}\left(\frac{2}{3}\right)$ and $g_1(B_1) = \frac{1}{3}\left(\frac{2}{3}\right)$.

Let $f_2 = f_1 - g_1$ on F. By the same argument as above we have $f_2(x) = \left[-\left(\frac{2}{3}\right)^2, \left(\frac{2}{3}\right)^2\right]$ and $\inf_{x \in F} f_2(x) = -\left(\frac{2}{3}\right)^2$ and $\sup_{x \in F} f_2(x) = \left(\frac{2}{3}\right)^2$.

Continuing in this manner we obtain a sequence $(f_0, f_1, f_2, \ldots)$ of real-valued continuous functions on F and a sequence $(g_0, g_1, g_2, \ldots)$ of real-valued continuous functions on X such that

$$\inf_{x \in F} f_n(x) = -\left(\tfrac{2}{3}\right)^n, \quad \sup_{x \in F} f_n(x) = \left(\tfrac{2}{3}\right)^n,$$
$$\inf_{x \in X} g_n(x) = -\tfrac{1}{3}\left(\tfrac{2}{3}\right)^n, \quad \sup_{x \in X} g_n(x) = \tfrac{1}{3}\left(\tfrac{2}{3}\right)^n,$$

and

$$f_n = f_0 - (g_0 + g_1 + \cdots + g_{n-1}) \quad \text{on } F.$$

Let

$$s_n = g_0 + g_1 + \cdots + g_{n-1} \quad \text{on } X.$$

Since $|g_n(x)| \leq \frac{1}{3}\left(\frac{2}{3}\right)^n$ for all $x \in X$ and since $\sum_{n \in \mathbb{Z}_+} \frac{1}{3}\left(\frac{2}{3}\right)^n = 1$, s_n converges uniformly on X to a function $\widetilde{f}$ such that $\widetilde{f}(x) \in [-1, 1]$ for $x \in X$. Continuity of s_n and uniformity of convergence on X imply that $\widetilde{f}$ is a continuous function on X. Now we have

$$s_n(x) \geq \inf_{x \in X} g_0(x) + \inf_{x \in X} g_1(x) + \cdots + \inf_{x \in X} g_{n_1}(x) = -\tfrac{1}{3} - \tfrac{1}{3}\left(\tfrac{2}{3}\right) - \cdots - \tfrac{1}{3}\left(\tfrac{2}{3}\right)^{n-1}$$

and then

$$\widetilde{f}(x) = \lim_{n \to \infty} s_n(x) \geq -1 = \inf_{x \in F} f(x)$$

so that

$$\inf_{x \in X} \widetilde{f}(x) \geq \inf_{x \in F} f(x).$$

Similarly we have

$$s_n(x) \leq \sup_{x \in X} g_0(x) + \sup_{x \in X} g_1(x) + \cdots + \sup_{x \in X} g_{n_1}(x) = \tfrac{1}{3}\tfrac{1}{3}\left(\tfrac{2}{3}\right) \cdots \tfrac{1}{3}\left(\tfrac{2}{3}\right)^{n-1}$$

and then

$$\widetilde{f}(x) = \lim_{n\to\infty} s_n(x) \le 1 = \inf_{x\in F} f(x)$$

so that

$$\sup_{x\in X} \widetilde{f}(x) \le \sup_{x\in F} f(x).$$

We have $f_0(x) - s_0(x) = f_n(x)$ for $x \in F$. Now since $|f_n(x)| \le \left(\frac{2}{3}\right)^n$, we have $\lim_{n\to\infty} |f_n(x)| = 0$ for $x \in F$. Thus $\lim_{n\to\infty} \left|f_0(x) - s_0(x)\right| = 0$ and hence we have $f_0(x) - \lim_{n\to\infty} s_n(x) = 0$, that is, $f_0(x) = \lim_{n\to\infty} s_n(x) = \widetilde{f}(x)$ for $x \in F$. Since $f_0 = f$ on F, this shows that $\widetilde{f}(x) = f(x)$ for $x \in F$. This completes the proof of the Theorem for the particular case that $\inf_{x\in F} f(x) = -1$ and $\sup_{x\in F} f(x) = 1$.

2. Let $a, b \in \mathbb{R}$ and $a < b$. Let f be a real-valued continuous function on F with $\inf_{x\in F} f(x) = a$ and $\sup_{x\in F} f(x) = b$. Consider a real-valued function φ on $[a, b]$ given by

$$\varphi(\xi) = \frac{2}{b-a}\xi - \frac{2a}{b-a} - 1 \quad \text{for } \xi \in [a, b].$$

Thus φ is a strictly increasing continuous function mapping $[a, b]$ one-to-one onto $[-1, 1]$ with $\varphi(a) = -1$ and $\varphi(b) = 1$. Consider the function on F defined by $h = \varphi \circ f$. Then h is a continuous function on F with $\inf_{x\in F} h(x) = -1$ and $\sup_{x\in F} h(x) = 1$. Thus by our result in **1** there exists a real-valued continuous function $\widetilde{h}$ on X such that

$$\widetilde{h} = h \text{ on } F, \quad \inf_{x\in F} h(x) \le \inf_{x\in X} \widetilde{h}(x), \quad \sup_{x\in X} \widetilde{h}(x) \le \sup_{x\in F} h(x).$$

The inverse mapping φ^{-1} of φ is a strictly increasing continuous function on $[-1, 1]$ mapping $[-1, 1]$ one-to-one onto $[a, b]$ with $\varphi^{-1}(-1) = a$ and $\varphi^{-1}(1) = b$. Consider $\varphi^{-1} \circ \widetilde{h}$, a continuous mapping of X into $[a, b]$. For $x \in F$ we have $\widetilde{h}(x) = h(x) = \varphi \circ f(x)$. Thus for $x \in F$ we have $\varphi^{-1} \circ \widetilde{h}(x) = \varphi^{-1} \circ \varphi \circ f(x) = f(x)$. This shows that $\varphi^{-1} \circ \widetilde{h}$ is a continuous extended of f to X. We have also

$$\begin{aligned}\inf_{x\in X} \varphi^{-1} \circ \widetilde{h}(x) &= \varphi^{-1}\left(\inf_{x\in X} \widetilde{h}(x)\right) \ge \varphi^{-1}\left(\inf_{x\in F} h(x)\right) \\ &= \varphi^{-1}(-1) = a = \inf_{x\in F} f(x).\end{aligned}$$

Similarly we have

$$\begin{aligned}\sup_{x\in X} \varphi^{-1} \circ \widetilde{h}(x) &= \varphi^{-1}\left(\sup_{x\in X} \widetilde{h}(x)\right) \le \varphi^{-1}\left(\sup_{x\in F} h(x)\right) \\ &= \varphi^{-1}(1) = b = \sup_{x\in F} f(x).\end{aligned}$$

This shows that $\varphi^{-1}\circ\widetilde{h}$ is a real-valued continuous function on X such that $\varphi^{-1}\circ\widetilde{h}(x) = f(x)$ for $x \in F$ and $\inf_{x\in F} f(x) \le \inf_{x\in X} \varphi^{-1}\circ\widetilde{h}(x)$ and $\sup_{x\in X}\varphi^{-1}\circ\widetilde{h}(x) \le \sup_{x\in F} f(x)$.

3. Let f be a real-valued continuous function on F. Let φ be a strictly increasing continuous mapping of $\mathbb{R}$ onto $(-1,1)$. Let $h = \varphi\circ f$ on F. Then h is a real-valued continuous function on F with $a = \inf_{x\in F} h(x) \ge -1$ and $b = \sup_{x\in F} h(x) \le 1$. Then by our result in **2** there exists a real-valued continuous function $\widetilde{h}$ on X such that

$$\widetilde{h} = h \text{ on } F,\quad \inf_{x\in F} h(x) \le \inf_{x\in X}\widetilde{h}(x),\quad \sup_{x\in X}\widetilde{h}(x) \le \sup_{x\in F} h(x).$$

Now φ^{-1} is a strictly increasing continuous function mapping $(-1,1)$ one-to-one onto $\mathbb{R}$. Then $\varphi^{-1}\circ\widetilde{h}$ is a continuous mapping of X into $\mathbb{R}$. Now $\widetilde{h}(x) = h(x) = \varphi\circ f(x)$ for $x \in F$. Thus $\varphi^{-1}\circ\widetilde{h}(x) = \varphi^{-1}\circ\varphi\circ f(x) = f(x)$ for $x\in F$. This shows that $\varphi^{-1}\circ\widetilde{h}$ is a continuous extension of f to X. We have also

$$\begin{aligned}\inf_{x\in X}\varphi^{-1}\circ\widetilde{h}(x) &= \varphi^{-1}\Big(\inf_{x\in X}\widetilde{h}(x)\Big) \ge \varphi^{-1}\Big(\inf_{x\in F} h(x)\Big)\\ &= \varphi^{-1}\Big(\inf_{x\in F}\varphi\circ f(x)\Big) = \inf_{x\in F} f(x).\end{aligned}$$

Similarly we have

$$\begin{aligned}\sup_{x\in X}\varphi^{-1}\circ\widetilde{h}(x) &= \varphi^{-1}\Big(\sup_{x\in X}\widetilde{h}(x)\Big) \le \varphi^{-1}\Big(\sup_{x\in F} h(x)\Big)\\ &= \varphi^{-1}\Big(\sup_{x\in F}\varphi\circ f(x)\Big) = \sup_{x\in F} f(x).\end{aligned}$$

This shows that $\varphi^{-1}\circ\widetilde{h}$ is a real-valued continuous function on X such that $\varphi^{-1}\circ\widetilde{h}(x) = f(x)$ for $x \in F$ and $\inf_{x\in F} f(x) \le \inf_{x\in X}\varphi^{-1}\circ\widetilde{h}(x)$ and $\sup_{x\in X}\varphi^{-1}\circ\widetilde{h}(x) \le \sup_{x\in F} f(x)$. ∎

The following continuous extension theorem is a particular case of Theorem 14.2 (Tietze). Its construction however is different from that of Theorem 14.2.

Theorem 14.3. *Let (X,ρ) be a metric space and let $K \subset X$ be a compact set which has a countable dense subset in the sense that there exists $\{s_j : j \in \mathbb{N}\} \subset K$ such that every open ball with center $s \in K$ contains some s_j. Let f be a real-valued continuous function on K with respect to the relative topology on K. Then there exists a real-valued continuous function $\widetilde{f}$ on X such that $\widetilde{f} = f$ on K.*

Proof. Let $U = X \setminus K$, an open set in X. For $x \in X$, let us write $\rho(x,K)$ for the distance between x and K, that is,

$$\rho(x,K) = \inf_{s\in K}\rho(x,s).$$

Corresponding to each $s \in K$ let us define a real-valued function φ_s on U by setting

(1) $$\varphi_s(x) = \max\left\{2 - \frac{rho(x,s)}{\rho(x,K)}, 0\right\} \quad \text{for } x \in U.$$

If $x \in U$ and $s \in K$ then $x \neq s$ and then $rho(x,s) > 0$.
If $x \in U$ then $x \notin K$ and then since K is a closed set we have $\rho(x,K) > 0$.
If $x \in U$ and $s \in K$ then $\rho(x,K) = \inf_{s \in K} \rho(x,s) \leq \rho(x,s)$ so that $\frac{\rho(x,s)}{\rho(x,K)} \in [1,\infty)$ and then

$$2 - \frac{\rho(x,s)}{\rho(x,K)} \in (-\infty,-1].$$

Thus we have

(2) $$\varphi_s(x) \in [0,1] \quad \text{for } x \in U.$$

Now $\rho(x,s)$ and $\rho(x,K)$ are continuous functions of $x \in U$. Since $\rho(x,K) > 0$, $\frac{\rho(x,s)}{\rho(x,K)}$ is a continuous function of $x \in U$. Then from (1) we have

(3) $$\varphi_s \text{ is a continuous function on } U.$$

For $x \in U$ and $s \in K$ we have from (1)

(4) $$\varphi_s(x) = 0 \Leftrightarrow 2 - \frac{\rho(x,s)}{\rho(x,K)} \leq 0 \Leftrightarrow 2\rho(x,K) \leq \rho(x,s)$$

and hence

(5) $$\varphi_s(x) > 0 \Leftrightarrow 2\rho(x,K) > \rho(x,s).$$

Let $\{s_j : j \in \mathbb{N}\}$ be a countable dense subset of K and define

(6) $$\tau(x) = \sum_{j \in \mathbb{N}} \frac{1}{2^j} \varphi_{s_j}(x) \quad \text{for } x \in U.$$

According to Weierstrass M-test, if $(g_j : j \in \mathbb{N})$ is a sequence of real-valued functions on a set E such that $|g_j(x)| \leq M_j$ for $x \in E$ and $j \in \mathbb{N}$ and if $\sum_{j \in \mathbb{N}} M_j < \infty$, then $\sum_{j \in \mathbb{N}} g_j(x)$ converges uniformly for $x \in E$.
Since $|\frac{1}{2^j}\varphi_{s_j}(x)| \leq \frac{1}{2^j}$ for $x \in U$ and $j \in \mathbb{N}$ and $\sum_{j \in \mathbb{N}} \frac{1}{2^j} = 1$, $\tau(x) = \sum_{j \in \mathbb{N}} \frac{1}{2^j}\varphi_{s_j}(x)$ converges uniformly on U by Weierstrass M-test. Then the continuity of $\frac{1}{2^j}\varphi_{s_j}$ on U implies

(7) $$\tau \text{ is a continuous function on } U.$$

Let us show that

(8) $$\tau(x) \in (0,1] \quad \text{for } x \in U.$$

By (2) we have $\varphi_{s_j}(x) \in [0,1]$ for $x \in U$ and $j \in \mathbb{N}$. Thus we have $\tau(x) \in [0,1]$ for $x \in U$. It remains to show that $\tau(x) > 0$ for every $x \in U$. For this it suffices to show that for $x \in U$ there exists $j \in \mathbb{N}$, depending on x, such that $\varphi_{s_j}(x) > 0$. Let $x \in U$. As we noted above, $\rho(x, K) > 0$ so that $2\rho(x, K) > \rho(x, K)$. Since $\rho(x, K) = \inf_{s \in K} \rho(x, s)$, there exists $s_0 \in K$ such that $\rho(x, s_0) < 2\rho(x, K)$ and hence $\varphi_{s_0}(x) > 0$ by (5). Now there exists $\delta > 0$ such that $\rho(x, s_0) + \delta < 2\rho(x, K)$. Let $s \in B_\delta(s_0) \cap K$. Then

$$\rho(x, s) \leq \rho(x, s_0) + \rho(s_0, s) < \rho(x, s_0) + \delta < 2\rho(x, K)$$

and then $\varphi_s(x) > 0$ by (5). Since $\{s_j : j \in \mathbb{N}\}$ is a dense subset of K there exists $j \in \mathbb{N}$ such that $s_j \in B_\delta(s_0)$. Then $\varphi_{s_j}(x) > 0$. This proves that $\tau(x) > 0$.

Let us define a sequence of functions on U, $\{\psi_j : j \in \mathbb{N}\}$, by setting

$$(9) \qquad \psi_j(x) = \frac{1}{\tau(x)} \frac{1}{2^j} \varphi_{s_j}(x) \quad \text{for } x \in U.$$

Since φ_{s_j} and τ are continuous on U and $\tau > 0$, we have

$$(10) \qquad \psi_j \text{ is nonnegative and continuous on } U.$$

Note also that

$$(11) \qquad \sum_{j \in \mathbb{N}} \psi_j(x) = \frac{1}{\tau(x)} \sum_{j \in \mathbb{N}} \frac{1}{2^j} \varphi_{s_j}(x) = \frac{\tau(x)}{\tau(x)} = 1 \quad \text{for } x \in U.$$

Let us define a real-valued function $\widetilde{f}$ on X by setting

$$(12) \qquad \widetilde{f}(x) = \begin{cases} f(x) & \text{for } x \in K, \\ \sum_{j \in \mathbb{N}} f(s_j)\psi_j(x) & \text{for } x \in U. \end{cases}$$

Since f is continuous on K, $\widetilde{f}$ is continuous on K.

Let us show that $\widetilde{f}$ is continuous on U. Let us observe first that since f is continuous on the compact set K, $f(K)$ is a compact set in $\mathbb{R}$ and therefore it is a bounded closed set in $\mathbb{R}$. Then there exists $M > 0$ such that

$$(13) \qquad |f(s)| \leq M \quad \text{for every } s \in K.$$

Now τ is a continuous function on the open set U with $\tau(x) \in (0,1]$ for $x \in U$. For each $\ell \in \mathbb{N}$, let

$$U_\ell = \left\{x \in U : \tau(x) \in \left(\tfrac{1}{\ell}, \infty\right)\right\} = \tau^{-1}\left(\tfrac{1}{\ell}, \infty\right).$$

As the preimage of the open set $\left(\frac{1}{\ell}, \infty\right)$ in $\mathbb{R}$ by the continuous mapping τ, U_ℓ is an open set in X. For every $x \in U$ we have $\tau(x) \in (0,1]$ so that $\tau(x) \in \left(\frac{1}{\ell}, \infty\right)$ for some $\ell \in \mathbb{N}$ and thus $x \in \tau^{-1}\left(\frac{1}{\ell}, \infty\right) = U_\ell$. Therefore we have $U = \bigcup_{\ell \in \mathbb{N}} U_\ell$. Thus to show that $\widetilde{f}$ is continuous on U it suffices to show that $\widetilde{f}$ is continuous on U_ℓ for every $\ell \in \mathbb{N}$. Let $\ell \in \mathbb{N}$ be fixed. For $x \in U_\ell$ we have $\tau(x) \in \left(\frac{1}{\ell}, \infty\right)$ so that $\frac{1}{\tau(x)} < \ell$. Thus for $x \in U_\ell$ we have

$$|f(s_j)\psi_j(x)| = \left|f(s_j)\frac{1}{\tau(x)}\frac{1}{2^j}\varphi_{s_j}(x)\right| \leq M\ell\frac{1}{2^j}.$$

Now $\sum_{j\in\mathbb{N}} M\ell\frac{1}{2^j} = M\ell < \infty$. Thus by Weierstrass M-test, $\widetilde{f}(x) = \sum_{j\in\mathbb{N}} f(s_j)\psi_j(x)$ converges uniformly on U_ℓ. Then the continuity of $f(s_j)\psi_j$ on U_ℓ for every $j \in \mathbb{N}$ implies that $\widetilde{f}$ is continuous on U_ℓ. Since this true for every $\ell \in \mathbb{N}$ and $U = \bigcup_{\ell\in\mathbb{N}} U_\ell$, $\widetilde{f}$ is continuous on U.

Now $\widetilde{f}$ is continuous on the bounded closed set K and $\widetilde{f}$ is continuous on the open set $U = X \setminus K$. To show that $\widetilde{f}$ is continuous on X and $\widetilde{f} = f$ on K it suffices to show that

(14) $$\lim_{x\to a, x\in U} \widetilde{f}(x) = f(a) \quad \text{for every } a \in K.$$

Let $a \in K$. Then for $x \in U$, applying (11) we have

$$\widetilde{f}(x) - f(a) = \sum_{j\in\mathbb{N}} f(s_j)\psi_j(x) - f(a) = \sum_{j\in\mathbb{N}} f(s_j)\psi_j(x) - \left\{\sum_{j\in\mathbb{N}} \psi_j(x)\right\} f(a).$$

Thus recalling that ψ_j is nonnegative, we have

(15) $$|\widetilde{f}(x) - f(a)| \leq \sum_{j\in\mathbb{N}} \psi_j(x)|f(s_j) - f(a)|.$$

Since f is continuous on K and $a \in K$, for every $\varepsilon > 0$ there exists $\delta > 0$ such that

(16) $$|f(s_j) - f(a)| < \varepsilon \quad \text{for all } s_j \in B_\delta(a).$$

Let us estimate $|\widetilde{f}(x) - f(a)|$ in (15) for $x \in U \cap B_{\delta/4}(a)$. Let $x \in U \cap B_{\delta/4}(a)$. For s_j such that $s_j \notin B_\delta(a)$, that is, $ho(s_j, a) \geq \delta$, we have

$$\delta \leq \rho(s_j, a) \leq \rho(s_j, x) + \rho(x, a) < \rho(s_j, x) + \frac{\delta}{4}$$

so that

$$\rho(s_j, x) > \frac{3}{4}\delta \geq 2\rho(x, a) \geq 2\rho(x, K) \quad \text{since } a \in K$$

and thus $\varphi_{s_j}(x) = 0$ by (5) and then $\psi_j(x) = 0$. Thus for $x \in U \cap B_{\delta/4}(a)$, on the right side of (15) we have $\psi_j(x) = 0$ for $j \in \mathbb{N}$ such that $s_j \notin B_\delta(a)$ and hence we sum over $j \in \mathbb{N}$ such that $s_j \in B_\delta(a)$ for which we have $|f(s_j) - f(a)| < \varepsilon$ by (16). Therefore for $x \in U \cap B_{\delta/4}(a)$ we have from (15) and (11)

$$|\widetilde{f}(x) - f(a)| \leq \sum_{j\in\mathbb{N}} \psi_j(x)|f(s_j) - f(a)| \leq \sum_{j\in\mathbb{N}} \psi_j(x)\varepsilon = \varepsilon.$$

This proves (14). ∎

Theorem 14.4. *Consider a metric space (X,d) and the metric space $(\mathbb{R}^m,\rho)$ where ρ is the Euclidean metrics on $\mathbb{R}^m$. Let F be a closed subset of X. Let f be a continuous mapping of F into $\mathbb{R}^m$ with respect to the relative topology on K. Then there exists a continuous mapping $\widetilde{f}$ of X into $\mathbb{R}^m$ such that $\widetilde{f}=f$ on F.*

Proof. Note that the metric topology on $\mathbb{R}^m$ by the Euclidean metric ρ on $\mathbb{R}^m$ is equivalent to the product topology on $\mathbb{R}^m=\mathbb{R}_1\times\cdots\times\mathbb{R}_m$ where the topology on $\mathbb{R}_i$ is the metric topology on $\mathbb{R}_i$ by the Euclidean metric on $\mathbb{R}_i$ for $i=1,\ldots,m$.

Let π_i be the projection of $\mathbb{R}^m=\mathbb{R}_1\times\cdots\times\mathbb{R}_m$ onto $\mathbb{R}_i$ for $i=1,\ldots,m$. Consider the mapping $f_i:=\pi_i\circ f$ of F into $\mathbb{R}_i$ for $i=1,\ldots,m$. Since f is a continuous mapping of F into $\mathbb{R}^m$ and π_i is a continuous mapping of $\mathbb{R}^m$ into $\mathbb{R}_i$, $f_i=\pi_i\circ f$ is a continuous mapping of F into $\mathbb{R}_i$. Then by Theorem 14.2 (Tietze), there exists a continuous mapping $\widetilde{f_i}$ of X into $\mathbb{R}_i$ such that $\widetilde{f_i}=f_i$ on F. If we let $\widetilde{f}=\widetilde{(f_1,\ldots,f_m)}$ then $\widetilde{f}$ is a continuous mapping of X into $\mathbb{R}^m$. Since $\widetilde{f_i}=f_i$ on F for $i=1,\ldots,m$ we have $\widetilde{f}=f$ on F. ∎

[II] Approximation of Measurable Functions by Continuous Functions on a Metric Space

Lemma 14.5. *Let $(X,\mathfrak{A},\mu)$ be a measure space with $\mu(X)<\infty$. Let f be an extended real-valued $\mathfrak{A}$-measurable function on X which is finite a.e. on X. Then for every $\varepsilon>0$ there exists a bounded $\mathfrak{A}$-measurable function h on X such that $\mu\{x\in X: h(x)\neq f(x)\}<\varepsilon$.*

Proof. Let

$$A_k=\{x\in X:|f(x)|>k\}\quad\text{for } k\in\mathbb{N},$$

$$A_\infty=\{x\in X:|f(x)|=\infty\}.$$

Since f is finite a.e. on X we have $\mu(A_\infty)=0$. Now $(A_k:k\in\mathbb{N})$ is a decreasing sequence in $\mathfrak{A}$ and $A_\infty=\bigcap_{k\in\mathbb{N}}A_k=\lim_{k\to\infty}A_k$. Then since $\mu(X)<\infty$ we have

$$\lim_{n\to\infty}\mu(A_k)=\mu\left(\lim_{k\to\infty}A_k\right)=\mu(A_\infty)=0.$$

Thus there exists $k_0\in\mathbb{N}$ such that $\mu(A_{k_0})<\varepsilon$. Let us define a function h on X by setting

$$h(x)=\begin{cases} f(x) & \text{for } x\in A_{k_0}^c\\ 0 & \text{for } x\in A_{k_0}.\end{cases}$$

Then h is a real-valued $\mathfrak{A}$-measurable function on X and $|h(x)|\leq k_0$ for all $x\in X$. Also we have $\{x\in X:h(x)\neq f(x)\}\subset A_{k_0}$ so that $\mu\{x\in X:h(x)\neq f(x)\}\leq\mu(A_{k_0})<\varepsilon$.
∎

Lemma 14.6. *Let $(X, \mathfrak{A}, \mu)$ be a measure space. Let $(f_n : n \in \mathbb{N})$ be a sequence of extended real-valued $\mathfrak{A}$-measurable functions on a set $D \in \mathfrak{A}$ and let f be an extended real-valued $\mathfrak{A}$-measurable function on D which is finite a.e. on D. Then $f_n \overset{\mu}{\to} f$ on D if and only if for every $\delta > 0$ there exists $N_\delta \in \mathbb{N}$ such that*

$$\mu\{D : |f_n - f| \geq \delta\} < \delta \quad \text{for } n \geq N_\delta.$$

Proof. See Lemma 6.17, [LRA]. ■

Theorem 14.7. *Let μ be a Borel outer measure on a metric space (X, d). Consider the measure space $(X, \mathfrak{M}(\mu), \mu)$. Assume that μ satisfies the following conditions:*
1° $\mu(X) < \infty$.
2° for every $E \in \mathfrak{M}(\mu)$ and $\varepsilon > 0$ there exists a closed set $F \subset E$ such that $\mu(E \setminus F) < \varepsilon$.
Let f be an extended real-valued $\mathfrak{M}(\mu)$-measurable function which is finite a.e. on X.
(a) *For every $\varepsilon > 0$ and $\eta > 0$ there exists a bounded continuous function g on X such that*

$$\mu\{x \in X : |f(x) - g(x)| \geq \eta\} < \varepsilon.$$

Moreover g can be so chosen that $|g| \leq M \leq \sup_{x \in X} |f(x)|$ on X.
(b) *There exists a sequence $(g_k : k \in \mathbb{N})$ of bounded continuous functions on X such that*

$$g_k \overset{\mu}{\to} f \quad \text{on } X.$$

Moreover g_k can be so chosen that $|g_k| \leq M \leq \sup_{x \in X} |f(x)|$ on X.
(c) *There exists a sequence $(g_k : k \in \mathbb{N})$ of bounded continuous functions on X such that*

$$\lim_{k \to \infty} g_k = f \quad \text{a.e. on } X.$$

Moreover g_k can be so chosen that $|g_k| \leq M \leq \sup_{x \in X} |f(x)|$ on X.

Proof. 1. Let us prove (a). Let $\varepsilon > 0$ and $\eta > 0$ be arbitrarily given. Consider first the case that $|f| \leq M$ on X. Then there exists a sequence $(\varphi_n : n \in \mathbb{N})$ of simple functions on $(X, \mathfrak{M}(\mu))$, bounded by M, such that $\lim_{n \to \infty} \varphi_n(x) = f(x)$ uniformly on X. (See 2° of Lemma 8.6, [LRA].) Thus for $\eta > 0$ there exists a simple function φ, bounded by M, such that

$$(1) \qquad |\varphi(x) - f(x)| < \eta \quad \text{for all } x \in X.$$

Let $c_1, \ldots, c_m$ be the distinct real values that φ assumes and let $E_1, \ldots, E_m$ be the subsets of X on which these values are assumed respectively. Then $\{E_1, \ldots, E_m\}$ is a disjoint collection in $\mathfrak{M}(\mu)$ and $X = \bigcup_{i=1}^m E_i$. By 2° there exist closed sets $F_1, \ldots, F_m$ such that $F_i \subset E_i$ and

$$\mu(E_i \setminus F_i) < \frac{\varepsilon}{m} \quad \text{for } i = 1, \ldots, m.$$

Let $F = \bigcup_{i=1}^m F_i$. Then F is a closed set in X. Now for the fact that $F_i \subset E_i$ we have

$$X \setminus F = \bigcup_{i=1}^m E_i \setminus \bigcup_{i=1}^m F_i = \bigcup_{i=1}^m (E_i \setminus F_i).$$

Thus we have

$$\mu(X \setminus F) \leq \sum_{i=1}^m \mu(E_i \setminus F_i) < \sum_{i=1}^m \frac{\varepsilon}{m} = \varepsilon. \tag{2}$$

Let us show that φ is a continuous function on F with respect to the relative topology of F. To show this we show that for every $x_0 \in F$ and every sequence $(x_n : n \in \mathbb{N})$ in F such that $\lim_{n\to\infty} x_n = x_0$ we have $\lim_{n\to\infty} \varphi(x_n) = \varphi(x_0)$. Let $x_0 \in F$. Since F is the union of disjoint sets $F_1, \ldots, F_m$, we have $x_0 \in F_i$ for a unique $i \in \{1, \ldots, m\}$. For simplicity in notation let us assume that $x_0 \in F_1$. Then $x_0 \notin \bigcup_{i=2}^m F_i$. Since $\bigcup_{i=2}^m F_i$ is a closed set we have $B_r(x_0) \cap \bigcup_{i=2}^m F_i = \emptyset$ for sufficiently small $r > 0$. Then since $\lim_{n\to\infty} x_n = x_0$ there exists $N \in \mathbb{N}$ such that $x_n \in B_r(x_0)$ for $n \geq N$. Since $B_r(x_0) \cap \bigcup_{i=2}^m F_i = \emptyset$ we have $x_n \notin \bigcup_{i=2}^m F_i$ for $n \geq N$ and hence $x_n \in F_1$ for $n \geq N$. Then since φ assumes the constant value c_1 on F_1, we have $\varphi(x_n) = c_1 = \varphi(x_0)$ for $n \geq N$. Thus $\lim_{n\to\infty} \varphi(x_n) = \varphi(x_0)$. This verifies that φ is a continuous function on F with respect to the relative topology of F. As defined above we have $|\varphi(x)| \leq M$ for $x \in F$. Thus by Theorem 14.2 (Tietze), there exists a real-valued continuous function g on X such that $g(x) = \varphi(x)$ for $x \in F$ and $|g(x)| \leq M$for $x \in X$. Then by (1) we have

$$|g(x) - \varphi(x)| < \eta \quad \text{for } x \in F. \tag{3}$$

This implies that $\{x \in X : |g(x) - f(x)| \geq \eta\} \subset X \setminus F$ and then we have

$$\mu\{x \in X : |g(x) - f(x)| \geq \eta\} \leq \mu(X \setminus F) < \varepsilon. \tag{4}$$

This proves (a) for the particular case that f is bounded.

Let us consider the general case, that is, f is not assumed to be bounded. According to Lemma 14.5 there exist a positive real number $M \leq \sup_{x\in X} |f(x)|$ and a bounded $\mathfrak{M}(\mu)$-measurable function h on X such that $|h(x)| \leq M$ for $x \in X$ and

$$\mu\{x \in X : f(x) \neq h(x)\} < \tfrac{\varepsilon}{2}.$$

By our result above for bounded function f, there exists a bounded continuous function g on X such that $|g(x)| \leq M$ for $x \in X$ and

$$\mu\{x \in X : |h(x) - g(x)| \geq \tfrac{\eta}{2}\} < \tfrac{\varepsilon}{2}.$$

Now we have

$$\begin{aligned}
&\{x \in X : |f(x) - g(x)| \geq \eta\} \\
&\subset \{x \in X : |f(x) - h(x)| + |h(x) - g(x)| \geq \eta\} \\
&\subset \{x \in X : |f(x) - h(x)| \geq \tfrac{\eta}{2}\} \cup \{x \in X : |h(x) - g(x)| \geq \tfrac{\eta}{2}\} \\
&\subset \{x \in X : f(x) \neq h(x) \geq \tfrac{\eta}{2}\} \cup \{x \in X : |h(x) - g(x)| \geq \tfrac{\eta}{2}\}
\end{aligned}$$

and then

$$\begin{aligned}
&\mu\{x \in X : |f(x) - g(x)| \geq \eta\} \\
&\leq \mu\{x \in X : f(x) \neq h(x) \geq \tfrac{\eta}{2}\} + \mu\{x \in X : |h(x) - g(x)| \geq \tfrac{\eta}{2}\} \\
&< \frac{\varepsilon}{2} + \frac{\varepsilon}{2} = \varepsilon.
\end{aligned}$$

Note that $|g| \leq M \leq \sup_{x \in X} |f(x)|$. This completes the proof of (a).

2. Let us prove (b). According to (a), for every $\varepsilon > 0$ and $\eta > 0$ there exists a bounded continuous function g on X such that

$$\mu\{x \in X : |f(x) - g(x)| \geq \eta\} < \varepsilon,$$

and g can be so chosen that $|g| \leq M \leq \sup_{x \in X} |f(x)|$ on X. Thus for every $k \in \mathbb{N}$ there exists a bounded continuous function g_k on X such that $|g_k| \leq M \leq \sup_{x \in X} |f(x)|$ on X and

$$\mu\{x \in X : |g_k(x) - f(x)| \geq \tfrac{1}{k}\} < \tfrac{1}{k}.$$

Consider the sequence $(g_k : k \in \mathbb{N})$. For $\delta > 0$ let $N \in \mathbb{N}$ be so large that $\frac{1}{N} < \delta$. Then for $k \geq N$ we have

$$\begin{aligned}
&\mu\{x \in X : |g_k(x) - f(x)| \geq \delta\} \leq \mu\{x \in X : |g_k(x) - f(x)| \geq \tfrac{1}{N}\} \\
&\leq \mu\{x \in X : |g_k(x) - f(x)| \geq \tfrac{1}{k}\} < \tfrac{1}{k} < \delta.
\end{aligned}$$

According to Lemma 14.6, this proves $g_k \xrightarrow{\mu} f$ on X.

3. (c) is an immediate consequence of (b) since every sequence that converges in measure has a subsequence that converges a.e. ■

Theorem 14.8. (Lusin's Theorem) *Let μ be a Borel outer measure on a metric space (X, d). Consider the measure space $(X, \mathfrak{M}(\mu), \mu)$. Assume that μ satisfies the following conditions:*

1° $\mu(X) < \infty$.

2° *for every* $E \in \mathfrak{M}(\mu)$ *and* $\varepsilon > 0$ *there exists a closed set* $F \subset E$ *such that* $\mu(E \setminus F) < \varepsilon$. *Let* f *be an extended real-valued* $\mathfrak{M}(\mu)$*-measurable function which is finite a.e. on* X. *Then for every* $\varepsilon > 0$, *there exist a closed set* F *with* $\mu(F^c) < \varepsilon$ *and a continuous function* g *on* X *such that* $f(x) = g(x)$ *for* $x \in F$. *Moreover* g *can be so chosen that*

$$\inf_{x \in F} f(x) \leq \inf_{x \in X} g(x) \leq \sup_{x \in X} g(x) \leq \sup_{x \in F} f(x).$$

Proof. By Theorem 14.7, there exists a sequence $(g_k : k \in \mathbb{N})$ of bounded continuous functions on X which converges to f a.e. on X. Since $\mu(X) < \infty$, Egorov's Theorem (see Theorem 6.12, [LRA]) is applicable and for every $\varepsilon > 0$ there exists $E \in \mathfrak{M}(\mu)$ with $\mu(E^c) < \frac{\varepsilon}{2}$ such that $(g_k : k \in \mathbb{N})$ converges to f uniformly on E. By 2° there exists a closed set $F \subset E$ such that $\mu(E \setminus F) < \frac{\varepsilon}{2}$. Now $F^c = X \setminus F = (X \setminus E) \cup (E \setminus F)$ and $\mu(F^c) < \frac{\varepsilon}{2} + \frac{\varepsilon}{2} = \varepsilon$.

Let us write $f|_F$ for the restriction of f to the closed set F. Since $(g_k : k \in \mathbb{N})$ converges to f uniformly on $E \supset F$, $(g_k : k \in \mathbb{N})$ converges to $f|_F$ uniformly on F. The uniformity of convergence on F and the continuity of g_k on $X \supset F$ imply that the limit function $f|_F$ is continuous on F. By Theorem 14.2 (Tietze), for the continuous function $f|_F$ on the closed set F there exists a continuous function g on X such that $g(x) = f|_F(x) = f(x)$ for $x \in F$, $\inf_{x \in X} g(x) \geq \inf_{x \in F} f|_F(x) = \inf_{x \in X} f(x)$ and $\sup_{x \in X} g(x) \leq \sup_{x \in F} f|_F(x) = \sup_{x \in F} f(x)$. ∎

Theorem 14.9. *Let* μ *be a Borel outer measure on a metric space* (X, ρ). *Consider the measure space* $(X, \mathfrak{M}(\mu), \mu)$. *Assume that* μ *satisfies the following conditions:*
1° $\mu(X) < \infty$.
2° *for every* $E \in \mathfrak{M}(\mu)$ *and* $\varepsilon > 0$ *there exists a closed set* $F \subset E$ *such that* $\mu(E \setminus F) < \varepsilon$. *Consider the measure space* $(X, \mathfrak{M}(\mu), \mu)$. *Let* $f \in L^1(X, \mathfrak{M}(\mu), \mu)$. *Then for every* $\varepsilon > 0$ *there exists a continuous function* g *on* X *such that*

$$\int_X |f - g| \, d\mu < \varepsilon.$$

Proof. The collection of simple functions on $(X, \mathfrak{M}(\mu), \mu)$ is dense in $L^1(X, \mathfrak{M}(\mu), \mu)$. (See Proposition 18.3, [LRA].) Thus for an arbitrary $\varepsilon > 0$ there exists a simple function φ such that

$$\int_X |f - \varphi| \, d\mu < \frac{\varepsilon}{2}.$$

Let $\alpha = \sup_{x \in X} |\varphi(x)|$. If $\alpha = 0$ then $\varphi = 0$ on X so that φ is trivially a continuous function on X and then with the choice $g = \varphi$ we are done. Consider the case $\alpha > 0$. By Theorem 14.8 (Lusin)there exist a closed set F with $\mu(F^c) < \frac{\varepsilon}{4\alpha}$ and a continuous function g on X such that $g = \varphi$ on F with $\inf_{x \in F} \varphi(x) \leq \inf_{x \in X} g(x) \leq \sup_{x \in X} g(x) \leq \sup_{x \in X} \varphi(x)$. Then we have

$$\sup_{x \in X} |g(x)| \leq \sup_{x \in F} |\varphi(x)| \leq \sup_{x \in X} |\varphi(x)| = \alpha.$$

Then we have

$$(1)\qquad \int_X |f-g|\,d\mu \le \int_X |f-\varphi|\,d\mu + \int_X |\varphi-g|\,d\mu < \frac{\varepsilon}{2} + \int_X |\varphi-g|\,d\mu.$$

Now

$$\begin{aligned}(2)\qquad \int_X |\varphi-g|\,d\mu &= \int_{\{X:\varphi\neq g\}} |\varphi-g|\,d\mu \le \int_{F^c} |\varphi-g|\,d\mu \\ &\le \int_{F^c} |\varphi|\,d\mu + \int_{F^c} |g|\,d\mu \\ &\le \Big\{\sup_{x\in X}|\varphi(x)| + \sup_{x\in X}|g(x)|\Big\}\mu(F^c) \\ &\le 2\alpha\frac{\varepsilon}{4\alpha} = \frac{\varepsilon}{2}.\end{aligned}$$

Then substituting (2) into (1) we have

$$\int_X |f-g|\,d\mu < \frac{\varepsilon}{2} + \frac{\varepsilon}{2} = \varepsilon. \ ■$$

[III] Approximation of Measurable Mappings of a Metric Space into a Metric Space

Proposition 14.10. (Lindelöf) *Let (X,ρ) be a separable metric space. Let $E \subset X$. Then every open cover of E has a countable subcover, that is, if $\{G_\alpha : \alpha \in A\}$ is an arbitrary collection of open sets such that $E \subset \bigcup_{\alpha\in A} G_\alpha$ then there exists a countable subcollection $\{G_{\alpha_j} : j \in \mathbb{N}\}$ such that $E \subset \bigcup_{j\in\mathbb{N}} G_{\alpha_j}$.*

Proof. Since (X,ρ) is a separable metric space there exists a countable dense subset of X. Let $\{x_m : m \in \mathbb{N}\}$ be a countable dense subset of X. Let $\{r_n : n \in \mathbb{N}\}$ be the collection of all positive rational numbers. Consider the collection of open balls

$$\mathcal{B} := \big\{B(x_m, r_n) : m \in \mathbb{N}, n \in \mathbb{N}\big\}.$$

This is a countable collection.

Let $E \subset X$ and let $\{G_\alpha : \alpha \in A\}$ is an arbitrary open cover of E. Let us show first that for every $x \in E$ there exist $\alpha \in A$ and an open ball in the collection $\mathcal{B}$, which we denote by B_x, such that

$$(1)\qquad x \in B_x \subset G_\alpha.$$

(Note that while B_x contains x its center need not be at x.) Now since $x \in E \subset \bigcup_{\alpha\in A} G_\alpha$, we have $x \in G_\alpha$ for some $\alpha \in A$. Since G_α is an open set containing x there exists $\varepsilon > 0$ such that $B(x,\varepsilon) \subset G_\alpha$. Now there exists $n \in \mathbb{N}$ such that $0 < r_n < \varepsilon$. Then we have

$$x \in B(x, r_n) \subset B(x,\varepsilon) \subset G_\alpha.$$

Since $\{x_m : m \in \mathbb{N}\}$ is a dense subset of X, there exists $m \in \mathbb{N}$ such that $x_m \in B\big(x, \frac{r_n}{2}\big)$. Consider $B\big(x_m, \frac{r_n}{2}\big)$. For every $y \in B\big(x_m, \frac{r_n}{2}\big)$, we have

$$\rho(y,x) \le \rho(y,x_m) + \rho(x_m,x) < \frac{r_n}{2} + \frac{r_n}{2} = r_n$$

so that $y \in B(x, r_n)$. This shows that $B\big(x_m, \frac{r_n}{2}\big) \subset B(x, r_n)$. Also $x_m \in B\big(x, \frac{r_n}{2}\big)$ implies $\rho(x_m, x) < \frac{r_n}{2}$ and then $x \in B\big(x_m, \frac{r_n}{2}\big)$. Therefore we have

$$x \in B\big(x_m, \tfrac{r_n}{2}\big) \subset B(x, r_n) \subset B(x, \varepsilon) \subset G_\alpha.$$

Now $B\big(x_m, \frac{r_n}{2}\big) \in \mathcal{B}$. This proves (1) with the choice $B_x := B\big(x_m, \frac{r_n}{2}\big)$.

By (1) we have

$$E \subset \bigcup_{x\in E} B_x \subset \bigcup_{\alpha \in A} G_\alpha. \tag{2}$$

Now consider the collection $\{B_x : x \in E\} \subset \mathcal{B}$. Since $\mathcal{B}$ is a countable collection, its subcollection $\{B_x : x \in E\}$ is a countable collection. Let us label $\{B_x : x \in E\}$ as $\{B_j; j \in \mathbb{N}\}$. According to (1), for each $j \in \mathbb{N}$, B_j is contained in G_α for some $\alpha \in A$. Pick an arbitrary G_α containing B_j and label it as G_{α_j}. Then $\{G_{\alpha_j} : j \in \mathbb{N}\}$ is a countable subcollection of $\{G_\alpha : \alpha \in A\}$ and

$$E \subset \bigcup_{x\in E} B_x = \bigcup_{j\in\mathbb{N}} B_j \subset \bigcup_{j\in\mathbb{N}} G_{\alpha_j}.$$

Thus $\{G_{\alpha_j} : j \in \mathbb{N}\}$ is a countable subcover of E. ∎

Lemma 14.11. *Let (X, ρ) be a separable metric space. Then for every $\varepsilon > 0$ there exists a countable disjoint collection $\{E_j : j \in \mathbb{N}\}$ in $\mathfrak{B}_X$ such that $X = \bigcup_{nj\in\mathbb{N}} E_j$ and the diameter $|E_j| < \varepsilon$ for all $j \in \mathbb{N}$.*

Proof. Let $\varepsilon > 0$. Consider the collection of open balls $\big\{B_{\frac{\varepsilon}{2}}(x) : x \in X\big\}$. We have $X = \bigcup_{x\in X} B_{\frac{\varepsilon}{2}}(x)$. Thus the collection is an open cover of X. By Proposition 14.10 there exists a countable subcover. Thus there exists $\{x_j : j \in \mathbb{N}\} \subset X$ such that

$$X = \bigcup_{j\in\mathbb{N}} B_{\frac{\varepsilon}{2}}(x_j).$$

Let

$$E_1 = B_{\frac{\varepsilon}{2}}(x_1) \quad \text{and} \quad E_j = B_{\frac{\varepsilon}{2}}(x_j) \setminus \bigcup_{i=1}^{j-1} B_{\frac{\varepsilon}{2}}(x_i) \text{ for } j \ge 2.$$

Then $\{E_j : j \in \mathbb{N}\}$ is a disjoint collection in $\mathfrak{B}_X$ and

$$\bigcup_{j\in\mathbb{N}} E_j = \bigcup_{j\in\mathbb{N}} B_{\frac{\varepsilon}{2}}(x_j) = X.$$

Moreover since $E_j \subset B_{\frac{\varepsilon}{2}}(x_j)$ we have $|E_j| \le |B_{\frac{\varepsilon}{2}}(x_j)| = \varepsilon$ for every $j \in \mathbb{N}$. ∎

Theorem 14.12. (Lusin's Theorem) *Let (X, ρ) be a metric space and (Y, ρ') be a separable metric space. Let μ be a Borel regular outer measure on X. Let f be a $\mathfrak{M}(\mu)/\mathfrak{B}_Y$-measurable mapping of X into Y. Let $A \in \mathfrak{M}(\mu)$ and $\mu(A) < \infty$. Then for every $\varepsilon > 0$ there exists a bounded closed set $C \subset A$ such that*
1° $\mu(A \setminus C) < \varepsilon$.
2° $f|_C$ *is continuous on* C.

Proof. Let $A \in \mathfrak{M}(\mu)$ and $\mu(A) < \infty$. Let $\nu := \mu|_A$. Then ν is a finite Borel regular outer measure on X by Theorem 3.24 and moreover $\mathfrak{M}(\mu) \subset \mathfrak{M}(\nu)$ by Theorem 2.34.

Now that ν is a finite Borel regular outer measure on the separable metric space X, Theorem 3.34 implies that for every $E \in \mathfrak{M}(\nu)$ we have

$$\nu(E) = \sup\{\nu(C) : C \subset E,\ C \text{ is a bounded closed set}\}.$$

Thus for $E \in \mathfrak{M}(\nu)$ and $\varepsilon > 0$ there exists a bounded closed set $C \subset E$ such that $\nu(E) - \varepsilon < \nu(C)$ and then since $\nu(C) \le \nu(A) < \infty$ we have

$$\nu(E \setminus C) = \nu(E) - \nu(C) < \varepsilon. \tag{1}$$

Now consider Y. By Lemma 14.11, for each $i \in \mathbb{N}$ there exists a countable disjoint collection $\{B_{i,j} : j \in \mathbb{N}\}$ in $\mathfrak{B}_Y$ such that $Y = \bigcup_{j \in \mathbb{N}} B_{i,j}$ and the diameter of $B_{i,j}$, $|B_{i,j}| < \frac{1}{i}$ for all $j \in \mathbb{N}$. Let

$$A_{i,j} = A \cap f^{-1}(B_{i,j}) \quad \text{for } (i,j) \in \mathbb{N} \times \mathbb{N}. \tag{2}$$

Since f is a $\mathfrak{M}(\mu)/\mathfrak{B}_Y$-measurable mapping of X into Y and $B_{i,j} \in \mathfrak{B}_Y$, we have $f^{-1}(B_{i,j}) \in \mathfrak{M}(\mu)$. Then since $A \in \mathfrak{M}(\mu)$ we have $A_{i,j} \in \mathfrak{M}(\mu)$. Since $\{B_{i,j} : j \in \mathbb{N}\}$ is a disjoint collection in $\mathfrak{B}_Y$, $\{f^{-1}(B_{i,j}) : j \in \mathbb{N}\}$ is a disjoint collection in $\mathfrak{M}(\mu)$ and then $\{A_{i,j} : j \in \mathbb{N}\}$ is a disjoint collection in $\mathfrak{M}(\mu)$. We have also

$$\begin{aligned} \bigcup_{j \in \mathbb{N}} A_{i,j} &= \bigcup_{j \in \mathbb{N}} f^{-1}(B_{i,j}) \cap A = f^{-1}\Big(\bigcup_{j \in \mathbb{N}} B_{i,j}\Big) \cap A \\ &= f^{-1}(Y) \cap A = X \cap A = A. \end{aligned}$$

Now by (1) for each $A_{i,j}$ there exists a bounded closed set $C_{i,j} \subset A_{i,j}$ such that

$$\nu(A_{i,j} \setminus C_{i,j}) < \frac{\varepsilon}{2^{i+j}}.$$

Then since $\nu = \mu|_A$ we have

$$\mu\left(A \setminus \bigcup_{j\in\mathbb{N}} C_{i,j}\right) = \nu\left(A \setminus \bigcup_{j\in\mathbb{N}} C_{i,j}\right) = \nu\left(\bigcup_{j\in\mathbb{N}} A_{i,j} \setminus \bigcup_{j\in\mathbb{N}} C_{i,j}\right)$$
$$\leq \nu\left(\bigcup_{j\in\mathbb{N}} (A_{i,j} \setminus C_{i,j})\right) = \sum_{j\in\mathbb{N}} \nu(A_{i,j} \setminus C_{i,j})$$
$$< \sum_{j\in\mathbb{N}} \frac{\varepsilon}{2^{i+j}} = \frac{\varepsilon}{2^i}.$$

Now consider the decreasing sequence in $\mathfrak{M}(\mu)$, $(A \setminus \bigcup_{j=1}^{N} C_{i,j} : N \in \mathbb{N})$, with $\lim_{N\to\infty} (A \setminus \bigcup_{j=1}^{N} C_{i,j}) = A \setminus \bigcup_{j\in\mathbb{N}} C_{i,j}$. Since $\mu(A) < \infty$ we have

$$\lim_{N\to\infty} \mu\left(A \setminus \bigcup_{j=1}^{N} C_{i,j}\right) = \mu\left(\lim_{N\to\infty} A \setminus \bigcup_{j=1}^{N} C_{i,j}\right) = \mu\left(A \setminus \bigcup_{j\in\mathbb{N}} C_{i,j}\right) < \frac{\varepsilon}{2^i}.$$

Thus there exists $N(i) \in \mathbb{N}$ such that

$$\mu\left(A \setminus \bigcup_{j=1}^{N(i)} C_{i,j}\right) < \frac{\varepsilon}{2^i}. \tag{3}$$

Let

$$D_i = \bigcup_{j=1}^{N(i)} C_{i,j}. \tag{4}$$

Then D_i is a bounded closed set. Let us define a mapping g_i of D_i into Y as follows. For each $i \in \mathbb{N}$ and $j \in \mathbb{N}$, select $b_{i,j} \in B_{i,j}$ arbitrarily and then define

$$g_i(x) = b_{i,j} \quad \text{for } x \in C_{i,j}, j = 1, \dots, N(i). \tag{5}$$

We claim that g_i is continuous on D_i with respect to the relative topology on D_i. To show that g_i is continuous at $x_0 \in D_i$ we show that for every $\varepsilon > 0$ there exists $\delta > 0$ such that

$$g_i(x) \in B_\varepsilon\big(g_i(x_0)\big) \quad \text{for all } x \in B_\delta(x_0) \cap D_i. \tag{6}$$

Let $x_0 \in D_i$. Since D_i is the union of the disjoint collection $\{C_{i,1}, \dots, C_{i,N(i)}\}$, x_0 is in just one of these sets. For simplicity in notation let us assume that $x_0 \in C_{1,1}$. Then x_0 is not in the closed set $C_{i,2} \cup \dots \cup C_{i,N(i)}$. Thus there exists $\delta > 0$ such that $B_\delta(x_0)$ and $C_{i,2} \cup \dots \cup C_{i,N(i)}$ are disjoint and then $B_\delta(x_0) \cap D_i$ and $C_{i,2} \cup \dots \cup C_{i,N(i)}$ are disjoint. Thus if $x \in B_\delta(x_0) \cap D_i$ then $x_0 \in C_{i,1}$ and hence $g_i(x) = b_{i,1} \in B_\varepsilon(b_{1,i}) = B_\varepsilon(g_i(x_0))$ since $x_0 \in C_{i,1}$ so that $g_i(x_0) = b_{i,1}$. This verifies (6) and proves the continuity of g_i at every $x_0 \in D_i$.

Now consider the sequence $(D_i : i \in \mathbb{N})$ of bounded closed sets and the sequence $(g_i : i \in \mathbb{N})$ of continuous mappings of D_i respectively into Y. Let

$$C = \bigcap_{i \in \mathbb{N}} D_i. \tag{7}$$

Then C is a bounded closed set. Now

$$A \setminus C = A \cap C^c = A \cap \left(\bigcup_{n \in \mathbb{N}} D_i^c \right) = \bigcup_{i \in \mathbb{N}} (A \cap D_i^c) = \bigcup_{i \in \mathbb{N}} (A \setminus D_i)$$

and then recalling (3) and (4) we have

$$\mu(A \setminus C) \le \sum_{i \in \mathbb{N}} \mu(A \setminus D_i) < \sum_{i \in \mathbb{N}} \frac{\varepsilon}{2^i} = \varepsilon. \tag{8}$$

To show that $f|_C$ is continuous on C we show that the sequence of continuous mappings $(g_i : i \in \mathbb{N})$ converges uniformly on C to f. Let $x \in C$. Then by (7) we have $x \in D_i$ for every $i \in \mathbb{N}$. Then by (4) we have $x \in C_{i,j}$ for some $j = 1, \dots, N(i)$. Now $x \in C_{i,j} \subset A_{i,j} = A \cap f^{-1}(B_{i,j})$ so that $f(x) \in B_{i,j}$. On the other hand since $x \in C_{i,j}$ we have $g_i(x) = b_{i,j} \in B_{i,j}$. Since the diameter $|B_{i,j}| < \frac{1}{i}$, we have $\rho\big(g_i(x), f(x)\big) < \frac{1}{i}$. Thus we have shown that

$$\rho\big(g_i(x), f(x)\big) < \frac{1}{i} \quad \text{for all } i \in \mathbb{N} \text{ for all } x \in C.$$

This shows that $(g_i : i \in \mathbb{N})$ converges uniformly on C to$f$. ∎

Theorem 14.13. *Let (X, d) be a metric space and (Y, ρ) be a separable metric space. Let μ be a σ-finite Borel regular outer measure on X. Let f be a $\mathfrak{M}(\mu)/\mathfrak{B}_Y$-measurable mapping of X into Y.*

(a) *There exists a countable disjoint collection of bounded closed sets in X, $\{K_i : i \in \mathbb{N}\}$, such that*

$$\mu\left(X \setminus \bigcup_{i \in \mathbb{N}} K_i \right) = 0, \tag{1}$$

$$f|_{K_i} \text{ is continuous on } K_i \text{ for every } i \in \mathbb{N}. \tag{2}$$

(b) *There exists a $\mathfrak{B}_X/\mathfrak{B}_Y$-measurable mapping of X into Y such that*

$$g = f \quad \text{a.e. on } (X, \mathfrak{B}_X, \mu).$$

Proof. Since $(X, \mathfrak{M}(\mu), \mu)$ is a σ-finite measure space, there exists a countable collection $\{E_i : i \in \mathbb{N}\}$ in $\mathfrak{M}(\mu)$ such that $X = \bigcup_{i \in \mathbb{N}} E_i$ and $\mu(E_i) < \infty$ for all $i \in \mathbb{N}$. Let

$$A_1 = E_1 \quad \text{and} \quad A_i = E_i \setminus \bigcup_{\ell=1}^{i-1} E_\ell \text{ for } i \ge 2.$$

Then $\{A_i : i \in \mathbb{N}\}$ is a disjoint collection in $\mathfrak{M}(\mu)$, $\bigcup_{i\in\mathbb{N}} A_i = \bigcup_{i\in\mathbb{N}} E_i = X$ and $\mu(A_i) \le \mu(E_i) < \infty$ for all $i \in \mathbb{N}$. If $\mu(A_i) = 0$ for some $i \in \mathbb{N}$ then we adjoin A_i to the first successor A_j in the sequence such that $\mu(A_j) > 0$. Thus we can assume that $\mu(A_i) \in (0, \infty)$ for all $i \in \mathbb{N}$.

Now consider our $A_i \in \mathfrak{M}(\mu)$ with $\mu(A_i) \in (0, \infty)$. Let $\varepsilon_i = \mu(A_i) > 0$. By Theorem 14.12, given $\frac{\varepsilon_i}{2} > 0$ there exists a bounded closed set $C_{i,1} \subset A_i$ such that

$$\mu(A_i \setminus C_{i,1}) < \frac{\varepsilon_i}{2},$$

$$f|_{C_{i,1}} \text{ is continuous on } C_{i,1}.$$

Since $\mu(C_{i,1}) + \mu(A_i \setminus C_{i,1}) = \varepsilon_i$ we have

$$\mu(C_{i,1}) = \varepsilon_i - \mu(A_i \setminus C_{i,1}) \ge \frac{\varepsilon_i}{2}.$$

By Theorem 14.12, given $\frac{\varepsilon_i}{2^2} > 0$ there exists a bounded closed set $C_{i,2} \subset A \setminus C_{i,1}$ such that

$$\mu(A_i \setminus C_{i,1} \setminus C_{i,2}) < \frac{\varepsilon_i}{2^2},$$

$$f|_{C_{i,2}} \text{ is continuous on } C_{i,2}.$$

Now $\mu(C_{i,2}) + \mu(A_i \setminus C_{i,1} \setminus C_{i,2}) = \mu(A_i \setminus C_{i,1})$ so that

$$\begin{aligned}\mu(C_{i,2}) &= \mu(A_i \setminus C_{i,1}) + \mu(A_i \setminus C_{i,1} \setminus C_{i,2}) \\ &\ge \frac{\varepsilon_i}{2} - \frac{\varepsilon_i}{2^2} = \frac{\varepsilon_i}{2^2}.\end{aligned}$$

Continuing in this manner we have a disjoint sequence of bounded closed sets $(C_{i,j} : j \in \mathbb{N})$ contained in A_i with $\mu(C_{i,j}) \ge \frac{\varepsilon_i}{2^j}$. Then we have

$$\mu\left(\bigcup_{j\in\mathbb{N}} C_{i,j}\right) = \sum_{j\in\mathbb{N}} \mu(C_{i,j}) \ge \sum_{j\in\mathbb{N}} \frac{\varepsilon_i}{2^j} = \varepsilon_i = \mu(A_i).$$

On the other hand, since $\bigcup_{j\in\mathbb{N}} C_{i,j} \subset A_i$ we have $\mu\big(\bigcup_{j\in\mathbb{N}} C_{i,j}\big) \le \mu(A_i)$. Therefore we have $\mu\big(\bigcup_{j\in\mathbb{N}} C_{i,j}\big) = \mu(A_i)$. Thus we have

$$\mu\left(A_i \setminus \bigcup_{j\in\mathbb{N}} C_{i,j}\right) = \mu(A_i) - \mu\left(\bigcup_{j\in\mathbb{N}} C_{i,j}\right) = 0.$$

Consider the disjoint collection of bounded closed sets $\{C_{i,j} : i \in \mathbb{N}, j \in \mathbb{N}\}$. We have

$$\begin{aligned}X \setminus \bigcup_{i\in\mathbb{N}}\left(\bigcup_{j\in\mathbb{N}} C_{i,j}\right) &= \bigcup_{i\in\mathbb{N}} A_i \setminus \bigcup_{i\in\mathbb{N}}\left(\bigcup_{j\in\mathbb{N}} C_{i,j}\right) \\ &= \bigcup_{i\in\mathbb{N}}\left(A_i \setminus \bigcup_{j\in\mathbb{N}} C_{i,j}\right),\end{aligned}$$

and then

$$\begin{aligned}\mu\left(X\setminus\bigcup_{i\in\mathbb{N}}\left(\bigcup_{j\in\mathbb{N}}C_{i,j}\right)\right)&=\mu\left(\bigcup_{i\in\mathbb{N}}A_i\setminus\bigcup_{i\in\mathbb{N}}\left(\bigcup_{j\in\mathbb{N}}C_{i,j}\right)\right)\\&=\sum_{i\in\mathbb{N}}\mu\left(A_i\setminus\bigcup_{j\in\mathbb{N}}C_{i,j}\right)\\&=0.\end{aligned}$$

Let us relabel the countable disjoint collection of bounded closed sets constructed above $\{C_{i,j}:i\in\mathbb{N},j\in\mathbb{N}\}$ as $\{K_i:i\in\mathbb{N}\}$. Then we have (1) and (2). This proves (a).

Let us prove (b). Let $N=X\setminus\bigcup_{i\in\mathbb{N}}\left(\bigcup_{j\in\mathbb{N}}C_{i,j}\right)$. Then $N\in\mathfrak{B}_X$ and $\mu(N)=0$, that is, N is a null set in $(X,\mathfrak{B}_X,\mu)$. Let us define a mapping g of X into Y by setting

$$g(x)=\begin{cases}f|_{C_{i,j}}(x) & \text{for } x\in C_{i,j} \text{ for } i\in\mathbb{N},j\in\mathbb{N},\\ y_0 & \text{for } x\in N,\end{cases}$$

where $y_0\in Y$. Since $f|_{C_{i,j}}$ is a continuous mapping of $C_{i,j}$ into Y, it is a $\mathfrak{B}_X/\mathfrak{B}_Y$-measurable mapping of $C_{i,j}$ into Y. Thus g is $\mathfrak{B}_X/\mathfrak{B}_Y$-measurable on $C_{i,j}$ for every $i\in\mathbb{N}$ and $j\in\mathbb{N}$. As a constant, g is $\mathfrak{B}_X/\mathfrak{B}_Y$-measurable on N. Then since we have $X=\bigcup_{i\in\mathbb{N}}\left(\bigcup_{j\in\mathbb{N}}C_{i,j}\right)\cup N$, the mapping g is $\mathfrak{B}_X/\mathfrak{B}_Y$-measurable on X. Moreover $g=f$ on N^c. This shows that $g=f$ a.e. on $(X,\mathfrak{B}_X,\mu)$. ■

Theorem 14.14. *Consider a metric spaces (X,d) and the metric space $(\mathbb{R}^m,\rho)$ where ρ is the Euclidean metrics on $\mathbb{R}^m$. Let μ be a Borel regular outer measure on X. Let f be a $\mathfrak{M}(\mu)/\mathfrak{B}_{\mathbb{R}^m}$-measurable mapping of X into $\mathbb{R}^m$. Let A be an arbitrary subset of X such that $A\in\mathfrak{M}(\mu)$ with $\mu(A)<\infty$. Then for every $\varepsilon>0$ there exists a continuous mapping $\widetilde{f}$ of X into $\mathbb{R}^m$ such that*

$$\mu\{x\in A:\widetilde{f}(x)\neq f(x)\}<\varepsilon.$$

Proof. By Theorem 14.12 (Lusin), there exists a bounded closed set $C\subset A$ such that $\mu(A\setminus C)<\varepsilon$ and $f|_C$ is continuous on C. Then by Theorem 14.4, there exists a continuous mapping $\widetilde{f}$ of X into $\mathbb{R}^m$ such that $\widetilde{f}|_C=f|_C$. Then $\{x\in A:\widetilde{f}(x)\neq f(x)\}\subset A\setminus C$ so that

$$\mu\{x\in A:\widetilde{f}(x)\neq f(x)\}\leq\mu(A\setminus C)<\varepsilon. \ ■$$

[IV] Egorov's Theorem

Theorem 14.15. (Egorov) *Let (X,d) be a metric space and (Y,ρ) be a separable metric space. Let μ be a Borel regular outer measure on X. Consider the measure space*

$(X, \mathfrak{M}(\mu), \mu)$. Let $(f_n : n \in \mathbb{N})$ and g be $\mathfrak{M}(\mu)/\mathfrak{B}_Y$-measurable mappings of X into Y. Let $A \in \mathfrak{M}(\mu)$ and $\mu(A) < \infty$. Suppose

$$\lim_{n\to\infty} f_n(x) = g(x) \quad \text{for } (\mathfrak{M}(\mu), \mu)\text{-a.e. } x \in A. \tag{1}$$

Then for every $\varepsilon > 0$ there exists $B \in \mathfrak{M}(\mu)$ such that $B \subset A$ with $\mu(A \setminus B) < \varepsilon$ and

$$\lim_{n\to\infty} f_n(x) = g(x) \quad \text{uniformly for } x \in B. \tag{2}$$

Proof. Let $x \in X$. Then $\lim_{n\to\infty} f_n(x) = g(x)$ if and only if for every $i \in \mathbb{N}$ there exists $j \in \mathbb{N}$ such that $\rho\big(f_n(x), g(x)\big) < \frac{1}{2^i}$ for all $n \geq j$. Thus we have

$$\{x \in X : \lim_{n\to\infty} f_n(x) = g(x)\} = \bigcap_{i\in\mathbb{N}} \bigcup_{j\in\mathbb{N}} \bigcap_{n\geq j} \{x \in X : \rho\big(f_n(x), g(x)\big) < \tfrac{1}{2^i}\}.$$

Then we have

$$\begin{aligned}
&\big\{x \in X : \lim_{n\to\infty} f_n(x) = g(x) \text{ is not valid}\big\} \\
=&\big\{x \in X : \lim_{n\to\infty} f_n(x) = g(x)\big\}^c \\
=&\Big(\bigcap_{i\in\mathbb{N}} \bigcup_{j\in\mathbb{N}} \bigcap_{n\geq j} \{x \in X : \rho\big(f_n(x), g(x)\big) < \tfrac{1}{2^i}\}\Big)^c \\
=&\bigcup_{i\in\mathbb{N}} \bigcap_{j\in\mathbb{N}} \bigcup_{n\geq j} \{x \in X : \rho\big(f_n(x), g(x)\big) \geq \tfrac{1}{2^i}\}.
\end{aligned}$$

Thus

$$\begin{aligned}
&big\{x \in A : \lim_{n\to\infty} f_n(x) = g(x) \text{ is not valid}\big\} \\
=&\bigcup_{i\in\mathbb{N}} \bigcap_{j\in\mathbb{N}} \bigcup_{n\geq j} \{x \in A : \rho\big(f_n(x), g(x)\big) \geq \tfrac{1}{2^i}\}.
\end{aligned}$$

Then (1) implies

$$\begin{aligned}
&\mu\Big(\bigcup_{i\in\mathbb{N}} \bigcap_{j\in\mathbb{N}} \bigcup_{n\geq j} \{x \in A : \rho\big(f_n(x), g(x)\big) \geq \tfrac{1}{2^i}\}\Big) \\
=&\mu\big(\{x \in A : \lim_{n\to\infty} f_n(x) = g(x) \text{ is not valid}\}\big) = 0.
\end{aligned} \tag{3}$$

This implies

$$\mu\Big(\bigcap_{j\in\mathbb{N}} \bigcup_{n\geq j} \{x \in A : \rho\big(f_n(x), g(x)\big) \geq \tfrac{1}{2^i}\}\Big) = 0 \quad \text{for } i \in \mathbb{N}. \tag{4}$$

For brevity let us write

(5) $$E_{i,n} = \big\{x \in A : \rho\big(f_n(x), g(x)\big) \geq \tfrac{1}{2^i}\big\} \quad \text{for } i \in \mathbb{N}, n \in \mathbb{N},$$

and

(6) $$C_{i,j} = \bigcup_{n \geq j} E_{i,n} \quad \text{for } i \in \mathbb{N}, j \in \mathbb{N}.$$

Then (4) is rewritten as

(7) $$\mu\left(\bigcap_{n \in \mathbb{N}} C_{i,j}\right) = 0 \quad \text{for } i \in \mathbb{N}.$$

For each fixed $i \in \mathbb{N}$, $(C_{i,j}; j \in \mathbb{N})$ is a decreasing sequence and

(8) $$\lim_{j\to\infty} C_{i,j} = \bigcap_{j \in \mathbb{N}} C_{i,j} = \bigcap_{j\in\mathbb{N}} \bigcup_{n \geq j} E_{i,n} = \limsup_{n\to\infty} E_{i,n}.$$

Now $x \in \limsup_{n\to\infty} E_{i,n}$ if and only if $x \in E_{i,n}$ for infinitely many $n \in \mathbb{N}$, that is to say that $\rho\big(f_n(x), g(x)\big) \geq \frac{1}{2^i}$ for infinitely many $n \in \mathbb{N}$. Thus if $x \in \limsup_{n\to\infty} E_{i,n}$ then $\lim_{n\to\infty} f_n(x) = g(x)$ does not hold. Therefore we have

(9) $$\limsup_{n\to\infty} E_{i,n} \subset \left\{x \in A : \lim_{n\to\infty} f_n(x) = g(x) \text{ is not valid}\right\}.$$

By (8) and (9) we have

(10) $$\mu\left(\lim_{j\to\infty} C_{i,j}\right) \leq \mu\left(\left\{x \in A : \lim_{n\to\infty} f_n(x) = g(x) \text{ is not valid}\right\}\right) = 0.$$

Now $(C_{i,j} : j \in \mathbb{N})$ is a decreasing sequence contained in A and $\mu(A) < \infty$. Thus we have

(11) $$\lim_{j\to\infty} \mu(C_{i,j}) = \mu\left(\lim_{j\to\infty} C_{i,j}\right) = 0 \quad \text{for } i \in \mathbb{N}.$$

Let $\varepsilon > 0$. By (11), for each $i \in \mathbb{N}$ there exists $J(i) \in \mathbb{N}$ such that

(12) $$\mu\big(C_{i,J(i)}\big) < \frac{\varepsilon}{2^i}.$$

Let us define

(13) $$B = A \setminus \bigcup_{i \in \mathbb{N}} C_{i,J(i)}.$$

For any $E \subset A$ let us write E^* for the complement of E in A, that is, $E^* = A \setminus E$. Then

$$\begin{aligned} B &= \left(\bigcup_{i\in\mathbb{N}} C_{i,J(i)}\right)^* = \left(\bigcup_{i\in\mathbb{N}} \bigcup_{n\geq J(i)} E_{i,n}\right)^* \\ &= \bigcap_{i\in\mathbb{N}} \bigcap_{n\geq J(i)} E^*_{i,n} \\ &= \bigcap_{i\in\mathbb{N}} \bigcap_{n\geq J(i)} \{x \in A : \rho\big(f_n(x), g(x)\big) < \tfrac{1}{2^i}\}. \end{aligned}$$

Thus we have $\rho\big(f_n(x), g(x)\big) < \frac{1}{2^i}$ for $n \geq J(i)$ for every $i \in \mathbb{N}$ for all $x \in B$. This shows that $\lim_{n\to\infty} f_n(x) = g(x)$ uniformly for $x \in B$.

By (13) and (12), we have

$$\mu(A \setminus B) = \mu\left(\bigcup_{i\in\mathbb{N}} C_{i,J(i)}\right) \leq \sum_{i\in\mathbb{N}} \mu\big(C_{i,J(i)}\big) < \sum_{i\in\mathbb{N}} \frac{\varepsilon}{2^i} = \varepsilon.$$

This completes the proof. ■

[V] Measurable Mapping into a Metric Space

Observation 14.16. Let $(y, \mathfrak{F})$ be a measurable space and let Z be a non-empty subset of Y. Let $\mathfrak{G} := \mathfrak{F} \cap Z$. Then $\mathfrak{G}$ is a σ-algebra of subsets of Z and thus $(Z, \mathfrak{G})$ is a measurable space.

Proof. Let us show that $\mathfrak{G}$ is a σ-algebra of subsets of Z.

1. Let us show that $Z \in \mathfrak{G}$. Now $Z = Y \cap Z$. Since $\mathfrak{F}$ is a σ-algebra of subsets of Y we have $Y \in \mathfrak{F}$. Then $Z = Y \cap Z \in \mathfrak{F} \cap Z = \mathfrak{G}$.

2. Let us show that if $E \in \mathfrak{G}$ then $Z \setminus E \in \mathfrak{G}$. Now if $E \in \mathfrak{G} = \mathfrak{F} \cap Z$ then there exists $F \in \mathfrak{F}$ such that $E = F \cap Z$. Then $Z \setminus E = Z \setminus (F \cap Z)$. Now we have

$$\begin{aligned} Z \setminus E &= Z \setminus (F \cap Z) = Z \cap (F \cap Z)^c \\ &= Z \cap (F^c \cup Z^c) = Z \cap F^c \\ &\in \mathfrak{F} \cap Z \quad \text{since } F^c \in \mathfrak{F} \\ &= \mathfrak{G}. \end{aligned}$$

3. Let us show that if $\{E_n : n \in \mathbb{N}\} \subset \mathfrak{G}$ then $\bigcup_{n \in \mathbb{N}} E_n \in \mathfrak{G}$. Now $E_n \in \mathfrak{G} = \mathfrak{F} \cap Z$ implies that there exists $F_n \in \mathfrak{F}$ such that $E_n = F_n \cap Z$. Then we have

$$\bigcup_{n \in \mathbb{N}} E_n = \bigcup_{n \in \mathbb{N}} (F_n \cap Z) = \Big(\bigcup_{n \in \mathbb{N}} F_n \Big) \cap Z \in \mathfrak{F} \cap Z = \mathfrak{G}$$

since $\{F_n : n \in \mathbb{N}\} \subset \mathfrak{F}$ implies $\bigcup_{n \in \mathbb{N}} F_n \in \mathfrak{F}$. ∎

Theorem 14.17. *Let $(X, \mathfrak{A})$ and $(Y, \mathfrak{F})$ be measurable spaces. Let $D \in \mathfrak{A}$ and let f be a mapping of D into Y. Let $Z = f(D) \subset Y$ and let $\mathfrak{G} = \mathfrak{F} \cap Z$. Then f is $\mathfrak{A}/\mathfrak{F}$-measurable if and only if f is $\mathfrak{A}/\mathfrak{G}$-measurable.*

Proof. Note that by Observation 14.16, $(Z, \mathfrak{G})$ is a measurable space and the mapping f of $D \in \mathfrak{A}$ into Y is also a mapping of D into Z.

1. Let $F \in \mathfrak{F}$. Then $F = (F \cap Z) \cup (F \cap Z^c)$ so that

$$f^{-1}(F) = f^{-1}(F \cap Z) \cup f^{-1}(F \cap Z^c).$$

Since $Z = f(D)$ we have $f(D) \cap Z^c = \emptyset$ and hence $f^{-1}(Z \cap Z^c) = \emptyset$ and then we have $f^{-1}(F \cap Z^c) \subset f^{-1}(Z^c) = \emptyset$. Thus we have

$$f^{-1}(F) = f^{-1}(F \cap Z). \tag{1}$$

2. Suppose f is $\mathfrak{A}/\mathfrak{F}$-measurable, that is, $f^{-1}(\mathfrak{F}) \subset \mathfrak{A}$. Let us show that f is $\mathfrak{A}/\mathfrak{G}$-measurable, that is, $f^{-1}(\mathfrak{G}) \subset \mathfrak{A}$. Let $E \in \mathfrak{G}$. Then $E = F \cap Z$ where $F \in \mathfrak{F}$. Now since f is $\mathfrak{A}/\mathfrak{F}$-measurable so that $f^{-1}(F) \in \mathfrak{A}$. Thus by (1) we have

$$f^{-1}(E) = f^{-1}(F \cap Z) = f^{-1}(F) \in \mathfrak{A}. \tag{2}$$

Since (2) holds for every $E \in \mathfrak{G}$, we have $f^{-1}(\mathfrak{G}) \subset \mathfrak{A}$.

3. Conversely suppose f is $\mathfrak{A}/\mathfrak{G}$-measurable, that is, $f^{-1}(\mathfrak{G}) \subset \mathfrak{A}$. Let us show that f is $\mathfrak{A}/\mathfrak{F}$-measurable, that is, $f^{-1}(\mathfrak{F}) \subset \mathfrak{A}$. Let $F \in \mathfrak{F}$. Then $F \cap Z \in \mathfrak{F} \cap Z = \mathfrak{G}$. Since f is $\mathfrak{A}/\mathfrak{G}$-measurable, we have $f^{-1}(F \cap Z) \in \mathfrak{A}$. Then by (1) we have $f^{-1}(F) \in \mathfrak{A}$. Since this holds for every $F \in \mathfrak{F}$, we have $f^{-1}(\mathfrak{F}) \subset \mathfrak{A}$. ∎

Proposition 14.18. *Let (Y, ρ) be a metric space, let Z be a non-empty subset of Y and consider the metric space (Z, ρ). Let $\mathfrak{O}_Y$ and $\mathfrak{O}_Z$ be the metric topologies of Y and Z respectively. Then we have:*
(a) *$\mathfrak{O}_Z = \mathfrak{O}_Y \cap Z$, that is, the metric topology on Z is identical with the relative topology.*
(b) *$\mathfrak{B}_Z = \mathfrak{B}_Y \cap Z$.*
(c) *If (Y, ρ) is a separable metric space, then (Z, ρ) is a separable metric space.*

Proof. 1. The metric ρ on Y restricted to Z is a metric on Z. Thus (Z, ρ) is a metric space. Let us write $B^Y(y_0, r)$ and $B^Z(z_0, r)$ for open balls in Y and Z respectively. Thus

$$B^Y(y_0, r) = \{y \in Y : \rho(y_0, y) < r\} \quad \text{for } y_0 \in Y \text{ and } r > 0,$$
$$B^Z(z_0, r) = \{z \in Z : \rho(z_0, z) < r\} \quad \text{for } z_0 \in Z \text{ and } r > 0.$$

Then we have

$$(1) \qquad \begin{aligned} B^Z(z_0, r) &= \{z \in Z : \rho(z_0, z) < r\} = \{y \in Y : \rho(z_0, y) < r\} \cap Z \\ &= B^Y(z_0, r) \cap Z, \end{aligned}$$

that is, an open ball in Z is always the intersection of an open ball in Y with Z. On the other hand the intersection of an open ball in Y, $B^Y(y_0, r) = \{y \in Y : \rho(y_0, y) < r\}$, with Z is an open ball in Z if and only if $y_0 \in Z$.

The metric topology $\mathfrak{O}_Y$ on Y is the collection of all arbitrary unions of open balls in Y and the metric topology $\mathfrak{O}_Z$ on Z is the collection of all arbitrary unions of open balls in Z.

2. Let us prove (a). Let $W \in \mathfrak{O}_Z$. Then

$$\begin{aligned} W &= \bigcup_{z_0 \in A} B^Z(z_0, r(z_0)) = \bigcup_{z_0 \in A} \{B^Y(z_0, r(z_0)) \cap Z\} \\ &= \Big(\bigcup_{z_0 \in A} B^Y(z_0, r(z_0)) \Big) \cap Z \\ &\in \mathfrak{O}_Y \cap Z \end{aligned}$$

where A is a subset of Z and the second equality is by (1). This shows that if $W \in \mathfrak{O}_Z$ then $W \in \mathfrak{O}_Y \cap Z$. Therefore we have $\mathfrak{O}_Z \subset \mathfrak{O}_Y \cap Z$.

Conversely let $W \in \mathfrak{O}_Y \cap Z$. Then $W = V \cap Z$ where $V \in \mathfrak{O}_Y$. Now for every $y \in V$ there exists $r(y) > 0$ such that $B^Y(y, r(y)) \subset V$. Then $\bigcup_{y \in V} B^Y(y, r(y)) = V$. Thus

$$\begin{aligned} W = V \cap Z &= \left(\bigcup_{y \in V} B^Y(y, r(y)) \right) \cap Z \\ &= \bigcup_{y \in V} \{B^Y(y, r(y)) \cap Z\} \in \mathfrak{O}_Z. \end{aligned}$$

This shows that if $W \in \mathfrak{O}_Y \cap Z$ then $W \in \mathfrak{O}_Z$. Thus we have $\mathfrak{O}_Y \cap Z \subset \mathfrak{O}_Z$. This completes the proof that $\mathfrak{O}_Z = \mathfrak{O}_Y \cap Z$.

3. To prove (b) we observe that

$$\mathfrak{B}_Z = \sigma(\mathfrak{O}_Z) = \sigma(\mathfrak{O}_Y \cap Z) = \sigma(\mathfrak{O}_Y) \cap Z = \mathfrak{B}_Y \cap Z,$$

where the second equality is by (a) and the third equality is by Theorem 1.15, [LRA].

4. Let us prove (c). Now a metric space is separable (that is, it has a countable dense subset) if and only if it satisfies the Second Axiom of Countability (that is, it has a countable open base). Suppose (Y, ρ) is separable. Then (Y, ρ) has a countable open base. According to (b) we have $\mathfrak{O}_Z = \mathfrak{O}_Y \cap Z$. Thus (Z, ρ) has a countable open base and thus it is a separable metric space. ■

Lemma 14.19. *Let (Y, ρ) be a separable metric space. Then there exists a countable open base consisting of open balls in Y. Indeed if (Y, ρ) is a separable metric space then there exists a countable dense subset $\{y_i : i \in \mathbb{N}\}$ of Y and if we let $\{r_j : j \in \mathbb{N}\}$ be the collection of all positive rational numbers, then $\mathcal{B} := \{B(y_i, r_j) : i \in \mathbb{N}, j \in \mathbb{N}\}$ is a countable open base for Y.*

Proof. As defined above, $\mathcal{B}$ is a countable collection of open balls. It remains to show that $\mathcal{B}$ is an open base for Y, that is, every open set in Y is a union of members of $\mathcal{B}$.

Let O be a non-empty open set in Y. Let $y \in O$. Then there exists $r_j > 0$ such that $B(y, r_j) \subset O$. Since $\{y_i : i \in \mathbb{N}\}$ is a dense subset of Y there exists $y_i \in B(y, r_j/3)$. Then $y \in B(y_i, r_j/3) \subset B(y, r_j) \subset O$. Thus for every $y \in O$ we can select a member of $\mathcal{B}$ which contains y and is contained in O. Call this open ball B_y. Then $O = \bigcup_{y \in O}\{y\} \subset \bigcup_{y \in O} B_y$. On the other hand $B_y \subset O$ so that $\bigcup_{y \in O} B_y \subset O$. Thus we have $O = \bigcup_{y \in O} B_y$. This shows that $\mathcal{B}$ is an open base for Y. ■

Proposition 14.20. *Let $(X, \mathfrak{A})$ be a measurable space and let (Y, ρ) be a separable metric space. Let f be a mapping of $D \in \mathfrak{A}$ into Y. Then f is $\mathfrak{A}/\mathfrak{B}_Y$-measurable if and only if $f^{-1}(B_r(y)) \in \mathfrak{A}$ for every $y \in Y$ and $r > 0$.*

Proof. 1. If f is $\mathfrak{A}/\mathfrak{B}_Y$-measurable, that is, $f^{-1}(\mathfrak{B}_Y) \subset \mathfrak{A}$, then since $B_r(y) \in \mathfrak{B}_Y$ for every $y \in Y$ and $r > 0$, we have $f^{-1}(B_r(y)) \in \mathfrak{A}$.

2. Conversely suppose $f^{-1}(B_r(y)) \in \mathfrak{A}$ for every $y \in Y$ and $r > 0$. According to Lemma 14.19, there exists a countable collection of open balls $\mathcal{B} = \{B_i : i \in \mathbb{N}\}$ which is an open base for Y. Thus if O is a non-empty open set in Y then O is a union of members of $\mathcal{B}$. Let us write $O = \bigcup_{i \in J} B_i$ where J is a subset of $\mathbb{N}$. Since we assume that $f^{-1}(B_r(y)) \in \mathfrak{A}$ for every $y \in Y$ and $r > 0$, we have $f^{-1}(B_i) \in \mathfrak{A}$ for every $i \in \mathbb{N}$. Then we have

$$(1) \qquad f^{-1}(O) = f^{-1}\left(\bigcup_{i \in J} B_i\right) = \bigcup_{i \in J} f^{-1}(B_i) \in \mathfrak{A}$$

since $f^{-1}(B_i) \in \mathfrak{A}$ and $\mathfrak{A}$ is closed under countable unions. Let $\mathfrak{O}_Y$ be the collection of all open sets in Y. Then (1) implies

$$(2) \qquad f^{-1}(\mathfrak{O}_Y) \subset \mathfrak{A}.$$

Now $\mathfrak{B}_Y = \sigma(\mathfrak{O}_Y)$. Thus

$$(3) \qquad f^{-1}(\mathfrak{B}_Y) = f^{-1}\big(\sigma(\mathfrak{O}_Y)\big) = \sigma\big(f^{-1}(\mathfrak{O}_Y)\big) \subset \sigma(\mathfrak{A}) = \mathfrak{A},$$

where the second equality is by Theorem 1.14, [LRA] and the set inclusion is by (2). Thus we have shown that $f^{-1}(\mathfrak{B}_Y) \subset \mathfrak{A}$ and this shows that f is $\mathfrak{A}/\mathfrak{B}_Y$-measurable. ∎

Combining Theorem 14.17 and Proposition 14.20, we have:

Theorem 14.21. *Let $(X, \mathfrak{A})$ be a measurable space and let (Y, ρ) be a separable metric space. Let f be a mapping of $D \in \mathfrak{A}$ into Y. Then f is $\mathfrak{A}/\mathfrak{B}_Y$-measurable if and only if*

$$f^{-1}\big(B_r(f(x))\big) \in \mathfrak{A} \quad \textit{for every } x \in D \textit{ and } r > 0.$$

Proof. Let $Z = f(D) \subset Y$. Consider the metric space (Z, ρ). Let $\mathfrak{O}_Y$ and $\mathfrak{O}_Z$ be the metric topologies of Y and Z respectively. By Proposition 14.18, we have $\mathfrak{B}_Z = \mathfrak{B}_Y \cap Z$ and the fact that (Y, ρ) be a separable metric space implies that (Z, ρ) is a separable metric space. Now since $Z = f(D) \subset Y$, the mapping f of $D\mathfrak{A}$ into Y is actually a mapping of D into Z. According to Theorem 14.17 we have

$$\begin{aligned}(1) \qquad f \text{ is } \mathfrak{A}/\mathfrak{B}_Y\text{-measurable} &\Leftrightarrow f \text{ is } \mathfrak{A}/\mathfrak{B}_Y \cap Z\text{-measurable} \\ &\Leftrightarrow f \text{ is } \mathfrak{A}/\mathfrak{B}_Z\text{-measurable}\end{aligned}$$

since $\mathfrak{B}_Z = \mathfrak{B}_Y \cap Z$. Since (Z, ρ) is a separable metric space, Proposition 14.20 is applicable and thus we have

$$\begin{aligned}(2) \qquad f \text{ is } \mathfrak{A}/\mathfrak{B}_Z\text{-measurable} &\Leftrightarrow f^{-1}\big(B_r(z)\big) \in \mathfrak{A} \quad \text{for every } z \in Z \text{ and } r > 0 \\ &\Leftrightarrow f^{-1}\big(B_r(f(x))\big) \in \mathfrak{A} \quad \text{for every } x \in D \text{ and } r > 0\end{aligned}$$

since $Z = f(D)$ and $z \in Z$ if and only if there exists $x \in D$ such that $f(x) = z$. By (1) and (2), we have

$$f \text{ is } \mathfrak{A}/\mathfrak{B}_Y\text{-measurable} \Leftrightarrow f^{-1}\big(B_r\big(f(x)\big)\big) \in \mathfrak{A} \quad \text{for every } x \in D \text{ and } r > 0.$$

This completes the proof. ∎

Printed in the United States
By Bookmasters